CODES FROM
DIFFERENCE SETS

CODES FROM DIFFERENCE SETS

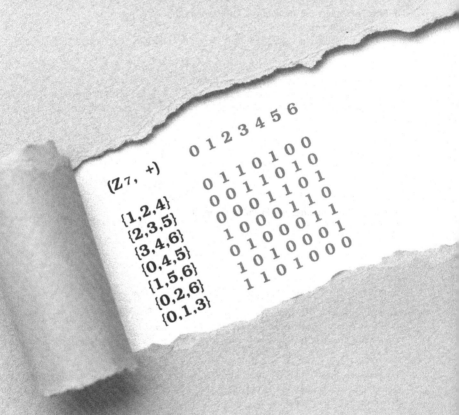

Cunsheng Ding

The Hong Kong University of Science and Technology, Hong Kong

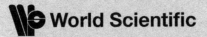 **World Scientific**

NEW JERSEY · LONDON · SINGAPORE · BEIJING · SHANGHAI · HONG KONG · TAIPEI · CHENNAI

Published by

World Scientific Publishing Co. Pte. Ltd.

5 Toh Tuck Link, Singapore 596224

USA office: 27 Warren Street, Suite 401-402, Hackensack, NJ 07601

UK office: 57 Shelton Street, Covent Garden, London WC2H 9HE

Library of Congress Cataloging-in-Publication Data
Ding, C. (Cunsheng), 1962–
    Codes from difference sets / Cunsheng Ding, The Hong Kong University of Science and
Technology, Hong Kong.
    pages cm
    Includes bibliographical references and index.
    ISBN 978-9814619356 (hardbound : alk. paper) -- ISBN 9814619353 (hardbound : alk. paper) --
1. Combinatorial designs and configurations. 2. Coding theory. 3. Difference sets. I. Title.
    QA166.25.D56 2014
    511'.6--dc23

                                                    2014021577

British Library Cataloguing-in-Publication Data
A catalogue record for this book is available from the British Library.

In-house Editors: Sutha Surenddar/Chandra Nugraha

Typeset by Stallion Press
Email: enquiries@stallionpress.com

Printed in Singapore

Preface

There are a number of books and monographs on coding theory. Quite a few books on design theory are also available. Connections and interactions between combinatorial designs and error correcting codes are dealt with to some extent in books on combinatorial designs and coding theory. The most comprehensive reference on interactions between the two areas is the monograph entitled "Designs and Their Codes" by Assmus and Key. However, in all these books and monographs, only a very small amount of work on interactions between difference sets and their codes is covered. In addition, little about almost difference sets and their codes is dealt within these references. The primary motivation of this monograph is to provide the reader with a comprehensive reference on codes from almost difference sets and difference sets.

Chapter 1 is a collection of mathematical foundations for subsequent chapters. It covers the basics of group actions, the ring of integers modulo n, finite fields, cyclotomy, generalized cyclotomy, finite geometries, planar functions, and sequences. The purpose is to introduce the basic concepts and results without providing a proof. However, a reference is given, so that the reader is able to find a proof of each key result.

Since there are many books on coding theory, Chapter 2 summarizes the fundamentals of linear codes without providing proofs. For basic results on coding theory, a reference is provided. This chapter presents only the topics of linear codes that are needed in later chapters. At the end of this chapter, a combinatorial approach to cyclic codes is introduced.

Chapter 3 is a brief introduction to designs and their codes, and covers incidence structures, incidence matrices, t-designs and their codes. In addition, the concept of t-adesigns is coined, and the codes of t-adesigns are introduced.

In view that there are comprehensive references on difference sets, Chapter 4 outlines the theory of difference sets and summarizes important constructions of difference sets without providing proofs in most cases. As usual, a reference is provided for each fundamental result. It is necessary to introduce these difference sets as their codes will be investigated later.

Chapter 5 aims at surveying the current constructions of almost difference sets. Sometimes a proof of the almost difference set property is provided. Otherwise a reference for a proof is given.

Chapter 6 deals with the codes of many classes of difference sets and a few classes of almost difference sets. This is one of the major chapters of this monograph. Proofs of the major results in this chapter are usually provided.

Chapter 7 studies the cyclic codes of a few families of cyclic almost difference sets and provides a number of open problems.

Codebooks have important applications in communication systems. Chapter 8 investigates complex codebooks from both difference sets and almost difference sets.

There is a database containing tables of best known linear codes currently maintained by Markus Grassl at http://www.codetables.de/. In this monograph, we call this the *Database*. However, there were no such tables of best cyclic codes. In the Appendix of this monograph, tables of best BCH codes and best cyclic codes over GF(2) and GF(3) of length up to 125 and 79 are collected, respectively. This is meant as a repository for benchmarking newly constructed binary and ternary cyclic codes.

In summary, this is neither a comprehensive book on coding theory nor a book on difference sets, but a monograph on codes from (almost) difference sets. The presentations of the materials in this monograph are elementary. This monograph is intended as a reference for postgraduates and other researchers who work on coding theory and/or combinatorial designs.

Finally, I would like to thank Dr. Qi Wang for reading the initial draft of this monograph and making valuable comments, Dr. Yin Tan for working out the proof of Theorem 5.12, Dr. Fei Gao for helping me with the computation of two tables of best BCH and cyclic codes in the Appendix, and The Hong Kong Research Grants Council for supporting this project (Project No. 601311). Thanks also go to the staffs, especially Dr. Chandra Nugraha, Ms. Sutha Surenddar and Ms. Mary Lau, at the World Scientific for helping me with this project.

Cunsheng Ding
March 2014

Contents

Chapter 1

Mathematical Foundations

It would be impossible to deal with linear codes without basic tools in certain mathematical areas. In this chapter, we give a brief introduction to group actions, finite fields, cyclotomy, finite geometry and sequences, which are the foundations of subsequent chapters.

1.1 Group Actions

In this monograph, basics of group actions are required. The purpose of this section is to introduce concepts and fundamental results of group actions. Proofs of the results presented in this section can be found in Aschbacher (2000).

In group theory, an *elementary abelian group* is a finite abelian group, where every nonidentity element has order p, where p is a prime. By the classification of finitely generated abelian groups, every elementary abelian group must be of the form $(\mathbb{Z}_p^n, +)$ for n a nonnegative integer. Here and hereafter $\mathbb{Z}_m$ denotes the *ring of integers modulo m*.

Example 1.1. The elementary abelian group $(\mathbb{Z}_2^2, +)$ has four elements $\{(0, 0), (0, 1), (1, 0), (1, 1)\}$. Addition is performed componentwise, taking the result modulo 2. For instance, $(1, 0) + (1, 1) = (0, 1)$.

The *symmetric group* on a finite set S, denoted $\mathrm{Sym}(S)$, is the group whose elements are all permutations on S and whose group operation is that of function composition. When $S = \{1, 2, \ldots, n\}$, the symmetric group on S is denoted by Sym_n.

An *action* of a group A on a set S is a permutation $\pi_a : S \to S$, for each $a \in A$, such that the following two conditions are met:

(a) (Identity) π_e is the identity, i.e., $\pi_e(s) = s$ for all $s \in S$, where e is the identity of A.

(b) (Compatibility) For every a_1 and a_2 in A, $\pi_{a_1} \circ \pi_{a_2} = \pi_{a_1 a_2}$.

1

Example 1.2. Let Sym_n act on $S = \{1, 2, \ldots, n\}$ in the natural way, i.e., $\pi_\sigma(i) = \sigma(i)$ for all $i \in S$, where $\sigma \in \mathrm{Sym}_n$.

Example 1.3. Any group A acts on itself by left multiplication functions.

A group A may act on a set S in many ways. For simplicity, we get rid of the notation π_a and write $\pi_a(s)$ as $a \cdot s$ or as. With this notation, the two conditions above become the following:

(a) (Identity) For each $s \in S$, $e \cdot s = s$, where e is the identity of A.
(b) (Compatibility) For every a_1 and a_2 in A and $s \in S$, $a_1 \cdot (a_2 \cdot s) = (a_1 a_2) \cdot s$.

Group actions can be classified into different types. A group action of A on S is called

- *faithful* if different elements of A act on S in different ways, i.e., when $a_1 \neq a_2$, there is an $s \in S$ such that $a_1 \cdot s \neq a_2 \cdot s$;
- *transitive* if for any s_1, s_2 in S there exists an $a \in A$ such that $a \cdot s_1 = s_2$;
- *free* (or *semiregular*) if, given a_1, a_2 in A, the existence of an $s \in S$ with $a_1 \cdot s = a_2 \cdot s$ implies that $a_1 = a_2$;
- *regular* (or *simply transitive* or *sharply transitive*) if it is both transitive and free.

Example 1.4. The action of a group A on itself is faithful, as different elements send the identity element e to different elements.

Example 1.5. Let $\mathbb{R}$ denotes the set of all real numbers, and let $\mathrm{GL}_n(\mathbb{R})$ be the set of all $n \times n$ invertible matrices over $\mathbb{R}$. Then $\mathrm{GL}_n(\mathbb{R})$ is a group under the matrix multiplication, and is called the *general linear group* of degree n.

This group acts on the n-dimensional vector space $\mathbb{R}^n$ as follows:

$$A \cdot \mathbf{x} = A\mathbf{x},$$

where $A\mathbf{x}$ means the multiplication of the matrix A with the $n \times 1$ vector $\mathbf{x} \in \mathbb{R}^n$, and $A \in \mathrm{GL}_n(\mathbb{R})$. The axioms of a group action are properties of matrix-vector multiplication.

It is easily seen that this group action is not transitive, not free, not regular, but faithful.

Let A act on S. The *orbit* of an element $s \in S$ is defined by

$$\mathrm{Orb}_s = \{a \cdot s : a \in A\} \subseteq S$$

and the *stabilizer* of $s \in S$ is

$$\mathrm{Stab}_s = \{a \in A : a \cdot s = s\} \subseteq A.$$

Example 1.6. Let A be a group, and let A act on itself as $a \cdot s = asa^{-1}$ for all a and s in A. Then $\mathrm{Orb}_s = \{asa^{-1} : a \in A\}$, which is called the *conjugate class* of s. The stabilizer $\mathrm{Stab}_s = \{a \in A : as = sa\}$, which is referred to as the *centralizer* of s.

The following theorem is classical and can be easily proved.

Theorem 1.1. *Let A act on S. Then the following statements hold*:

(a) *Different orbits are disjoint.*
(b) *For each $s \in S$, Stab_s is a subgroup of A and $\mathrm{Stab}_{a \cdot s} = a\mathrm{Stab}_s a^{-1}$.*
(c) *$a \cdot s = a' \cdot s$ if and only if a and a' lie in the same left coset of Stab_s. In particular, the length of the orbit of s is given by*

$$|\mathrm{Orb}_s| = [A : \mathrm{Stab}_s] = |A|/|\mathrm{Stab}_s|.$$

As corollaries of Theorem 1.1, we have the following.

Corollary 1.1. *Let A act on S. Then the length of every orbit divides the size of A. In addition, points in a common orbit have conjugate stabilizers, and in particular the size of the stabilizer is the same for all points in an orbit.*

Corollary 1.2. *Let A act on S, where A and S are finite. Let the different orbits of S be represented by $s_1, s_2, \ldots, s_t$. Then*

$$|S| = \sum_{i=1}^{t} |\mathrm{Orb}_{s_i}| = \sum_{i=1}^{t} [A : \mathrm{Stab}_{s_i}].$$

In a group action of A on S, the length of an orbit divides $|A|$, but the number of orbits usually does not divide $|A|$. There is an interesting relation between the number of orbits and the group action as follows.

Lemma 1.1 (Burnside's lemma). *Let a finite group A act on a finite set S with r orbits. Then r is the average number of fixed points of the elements of the group:*

$$r = \frac{1}{|A|} \sum_{a \in A} |\mathrm{Fix}_a(S)|,$$

where $\mathrm{Fix}_a(S) = \{s \in S : a \cdot s = s\}$ is the set of elements of S fixed by a.

1.2 The Rings $\mathbb{Z}_n$

For every positive integer $n \geq 1$, Euler's function $\phi(n)$ is defined to be the number of integers a such that $\gcd(a, n) = 1$, where $1 \leq a < n$. This function has the following properties:

(1) If p is a prime, then $\phi(p) = p - 1$.
(2) For any prime p, $\phi(p^k) = p^{k-1}(p-1)$.
(3) If $m, n \geq 1$ and $\gcd(m, n) = 1$, then $\phi(mn) = \phi(m)\phi(n)$, that is, ϕ is a *multiplicative function*.
(4) For any integer with the canonical factorization $n = \prod_p p^k$, $\phi(n) = \prod_p p^{k-1}(p-1)$.

Proofs of these properties are easy and can be found in most books about number theory.

An integer g is said to be a *primitive root* of (or modulo) n if $\mathrm{ord}_n(g) = \phi(n)$, where $\mathrm{ord}_n(g)$ denotes the multiplicative order of g modulo n. If $g \equiv g' \pmod{n}$, then g is a primitive root of N if and only if g' is a primitive root of N.

Throughout this monograph, $\mathbb{Z}_n = \{0, 1, 2, \ldots, n - 1\}$ denotes the ring of integers modulo n, and $\mathbb{Z}_n^*$ consists of all the units of the ring $\mathbb{Z}_n$. Hence, $\mathbb{Z}_n^*$ is a multiplicative group of order $\phi(n)$.

For any subset D of $\mathbb{Z}_n$, the *Hall polynomial* of D is defined by

$$D(x) = \sum_{i \in D} x^i,$$

which can be viewed as a polynomial over any ring or field, depending on the specific application under consideration.

1.3   Finite Fields

Our objective of this section is to introduce necessary notations and results of finite fields. We will recall and summarize them without providing technical proofs. The reader may work out the proofs as an exercise or find out the proofs in Lidl and Niederreiter (1997*a*).

1.3.1   *Introduction to finite fields*

A *field* is a set $\mathbb{F}$ associated with two operations: $+$, called addition, and $\cdot$, called multiplication, which satisfy the following axioms. The set $\mathbb{F}$ is an abelian group under $+$ with additive identity called zero and denoted 0; the set $\mathbb{F}^*$ of all nonzero elements of $\mathbb{F}$ is also an abelian group under multiplication with multiplicative identity called one and denoted 1; and multiplication distributes over addition. For convenience, we will usually omit the symbol for multiplication and write ab for the product $a \cdot b$. The field is *finite* if $\mathbb{F}$ has a finite number of elements; the number of elements in F is called the order of $\mathbb{F}$. We will denote a field with q elements by GF(q). If p is a prime, the integers modulo p form a field, which is then denoted GF(p). These are the simplest examples of finite fields.

It is known that all finite fields with the same number of elements are isomorphic [Lidl and Niederreiter (1997a)][Theorem 2.5]. Let GF(q) be a finite field with q elements. The following is a list of basic properties of the finite field GF(q) [Lidl and Niederreiter (1997a)][Chapters 1 and 2].

- $q = p^m$ for some prime p and some positive integer m, and p is called the *characteristic* of GF(q).
- GF(q) contains GF(p) as a subfield.
- GF(q) is a vector space over GF(p) of dimension m.
- Every subfield of GF(q) has order p^t for some integer t dividing m.
- The multiplicative group GF(q)* of nonzero elements of GF(q) is cyclic. A generator of the multiplicative group GF(q)* is called a *primitive element* of GF(q).

Let GF(q^m) be an extension of GF(q) and let $a \in$ GF(q^m). The monic polynomial $f(x) \in$ GF(q)[x] such that $f(a) = 0$ with the least degree is called the *minimal polynomial* of a over GF(q) and is denoted by $\mathbb{M}_a(x)$. It is known that the minimal polynomial of a over GF(q) is unique and irreducible over GF(q) [Lidl and Niederreiter (1997a)][Theorem 1.82].

1.3.2 Traces, norms, and bases

Let GF(q^m) be an extension of GF(q) and let $a \in$ GF(q^m). The elements $a, a^q, a^{q^2}, \ldots, a^{q^{m-1}}$ are called the *conjugates* of a with respect to GF(q).

The *trace function* from GF(q^m) to GF(q) is defined by

$$\mathrm{Tr}_{q^m/q}(x) = x + x^q + x^{q^2} + \cdots + x^{q^{m-1}}.$$

When q is prime, $\mathrm{Tr}_{q^m/q}(x)$ is called the *absolute trace* of x.

The following theorem summarizes basic properties of the trace function $\mathrm{Tr}_{q^m/q}(x)$ whose proofs can be found in Lidl and Niederreiter (1997a)[Theorems 2.23 and 2.24].

Theorem 1.2. *The trace function* $\mathrm{Tr}_{q^m/q}(x)$ *has the following properties*:

(a) $\mathrm{Tr}_{q^m/q}(x + y) = \mathrm{Tr}_{q^m/q}(x) + \mathrm{Tr}_{q^m/q}(y)$ *for all* $x, y \in$ GF(q^m).
(b) $\mathrm{Tr}_{q^m/q}(cx) = c\mathrm{Tr}_{q^m/q}(x)$ *for all* $x \in$ GF(q^m) *and all* $c \in$ GF(q).
(c) $\mathrm{Tr}_{q^m/q}$ *is a linear transformation from* GF(q^m) *to* GF(q), *when both* GF(q^m) *and* GF(q) *are viewed as vector spaces over* GF(q).

(d) *Every linear transformation from* $\mathrm{GF}(q^m)$ *to* $\mathrm{GF}(q)$ *can be expressed as* $\mathrm{Tr}(ax)$ *for some* $a \in \mathrm{GF}(q^m)$.

(e) $\mathrm{Tr}_{q^m/q}(a) = am$ *for all* $a \in \mathrm{GF}(q)$.

(f) $\mathrm{Tr}_{q^m/q}(x^q) = \mathrm{Tr}_{q^m/q}(x)$ *for all* $x \in \mathrm{GF}(q^m)$.

The *norm* $\mathrm{N}_{q^m/q}(a)$ of an element $a \in \mathrm{GF}(q^m)$ is defined by

$$\mathrm{N}_{q^m/q}(a) = a^{(q^m-1)/(q-1)}.$$

It follows that $\mathrm{N}_{q^m/q}(a)$ is always an element of $\mathrm{GF}(q)$.

The following theorem summarizes basic properties of the norm function $\mathrm{N}_{q^m/q}(x)$ whose proofs can be found in Lidl and Niederreiter (1997a)[Theorem 2.28].

Theorem 1.3. *The norm function* $\mathrm{N}_{q^m/q}(x)$ *has the following properties*:

(i) $\mathrm{N}_{q^m/q}(xy) = \mathrm{N}_{q^m/q}(x)\mathrm{N}_{q^m/q}(y)$ *for all* $x, y \in \mathrm{GF}(q^m)$.

(ii) $\mathrm{N}_{q^m/q}$ *maps* $\mathrm{GF}(q^m)$ *onto* $\mathrm{GF}(q)$ *and* $\mathrm{GF}(q^m)^*$ *onto* $\mathrm{GF}(q)^*$.

(iii) $\mathrm{N}_{q^m/q}(a) = a^m$ *for all* $a \in \mathrm{GF}(q)$.

(iv) $\mathrm{N}_{q^m/q}(x^q) = \mathrm{N}_{q^m/q}(x)$ *for all* $x \in \mathrm{GF}(q^m)$.

Two bases $\{\alpha_1, \ldots, \alpha_m\}$ and $\{\beta_1, \ldots, \beta_m\}$ of $\mathrm{GF}(q^m)$ over $\mathrm{GF}(q)$ are said to be *dual basis* if for $1 \leq i, j \leq m$ we have

$$\mathrm{Tr}_{q^m/q}(\alpha_i \beta_j) = \begin{cases} 0 & \text{if } i \neq j, \\ 1 & \text{if } i = j. \end{cases}$$

For any basis $\{\alpha_1, \ldots, \alpha_m\}$ of $\mathrm{GF}(q^m)$ over $\mathrm{GF}(q)$, it can be easily proved that there exists a dual basis $\{\beta_1, \ldots, \beta_m\}$.

Let α be a primitive element of $\mathrm{GF}(q^m)$. Then $\{1, \alpha, \alpha^2, \ldots, \alpha^{m-1}\}$ is called a *polynomial basis* of $\mathrm{GF}(q^m)$ over $\mathrm{GF}(q)$.

A basis of $\mathrm{GF}(q^m)$ over $\mathrm{GF}(q)$ of the form $\{\beta, \beta^q, \beta^{q^2}, \ldots, \beta^{q^{m-1}}\}$ is called a *normal basis*. It is known that there exists a normal basis of $\mathrm{GF}(q^m)$ over $\mathrm{GF}(q)$ [Lidl and Niederreiter (1997a)][Theorem 2.35].

1.3.3 *Polynomials over finite fields*

1.3.3.1 *Permutation Polynomials*

A polynomial $f \in \mathrm{GF}(r)[x]$ is called a *permutation polynomial* if the associated polynomial function $f : a \mapsto f(a)$ from $\mathrm{GF}(r)$ to $\mathrm{GF}(r)$ is a permutation of $\mathrm{GF}(r)$. Obviously, f is a permutation polynomial of $\mathrm{GF}(r)$ if and only if the equation $f(x) = a$ has exactly one solution $x \in \mathrm{GF}(r)$ for each $a \in \mathrm{GF}(r)$.

Example 1.7. Every linear polynomial ax is a permutation polynomial of $GF(r)$, where $a \in GF(r)^*$.

The following is a general criterion for the permutation property of polynomials over $GF(r)$, but not a very useful one [Lidl and Niederreiter (1997a)][Theorem 7.4].

Theorem 1.4 (Hermite's Criteria). *Let* $GF(r)$ *be of characteristic* p. *Then* $f \in GF(r)[x]$ *is a permutation polynomial of* $GF(r)$ *if and only if the following two conditions hold:*

(i) $f(x) = 0$ *has exactly one solution* $x \in GF(r)$;
(ii) *for each integer* t *with* $1 \le t \le r - 2$ *and* $t \not\equiv 0$ (mod p), *the reduction of* $f(x)^t$ mod $(x^r - x)$ *has degree at most* $r - 2$.

The Hermite Criterion above can be modified into the following [Lidl and Niederreiter (1997a)][Theorem 7.6].

Theorem 1.5. *Let* $GF(r)$ *be of characteristic* p. *Then* $f \in GF(r)[x]$ *is a permutation polynomial of* $GF(r)$ *if and only if the following two conditions hold:*

(i) *the reduction of* $f(x)^{r-1}$ mod $(x^r - x)$ *has degree* $r - 1$;
(ii) *for each integer* t *with* $1 \le t \le r - 2$ *and* $t \not\equiv 0$ (mod p), *the reduction of* $f(x)^t$ mod $(x^r - x)$ *has degree at most* $r - 2$.

The two criteria above are not very useful as the two conditions in each theorem are not easy to check. For special types of polynomials over $GF(r)$ there are simple conditions for checking the permutation property.

It is easily seen that the monomial x^n is a permutation polynomial of $GF(r)$ if and only if $\gcd(n, r - 1) = 1$. For p-polynomials we have the following.

Theorem 1.6. *Let* $GF(r)$ *be of characteristic* p. *Then the* p-*polynomial*

$$L(x) = \sum_{i=0}^{m} a_i x^{p^i} \in GF(r)[x]$$

is a permutation polynomial of $GF(r)$ *if and only if* $L(x)$ *only has the root* 0 *in* $GF(r)$.

For more information on permutation polynomials, the reader is referred to Lidl and Niederreiter (1997a)[Chapter 7].

1.3.3.2 Dickson Polynomials

One hundred and sixteen years ago, Dickson introduced the following family of polynomials over the finite field GF(r) [Dickson (1896)]:

$$D_h(x, a) = \sum_{i=0}^{\lfloor \frac{h}{2} \rfloor} \frac{h}{h-i} \binom{h-i}{i} (-a)^i x^{h-2i}, \qquad (1.1)$$

where $a \in \text{GF}(r)$ and $h \geq 0$ is called the *order* of the polynomial. This family is referred to as the *Dickson polynomials of the first kind*.

It is known that Dickson polynomials of the first kind satisfy the following recurrence relation:

$$D_{h+2}(x, a) = x D_{h+1}(x, a) - a D_h(x, a) \qquad (1.2)$$

with the initial state $D_0(x, a) = 2$ and $D_1(x, a) = x$.

A proof of the following theorem can be found in Lidl, Mullen and Turnwald (1993)[Theorem 3.2].

Theorem 1.7. $D_h(x, a)$ *is a permutation polynomial over* GF(r) *if and only if* $\gcd(h, r^2 - 1) = 1$.

Dickson polynomials of the second kind over the finite field GF(r) are defined by

$$E_h(x, a) = \sum_{i=0}^{\lfloor \frac{h}{2} \rfloor} \binom{h-i}{i} (-a)^i x^{h-2i}, \qquad (1.3)$$

where $a \in \text{GF}(r)$ and $h \geq 0$ is called the *order* of the polynomial. This family is referred to as the *Dickson polynomials of the second kind*.

It is known that Dickson polynomials of the second kind satisfy the following recurrence:

$$E_{h+2}(x, a) = x E_{h+1}(x, a) - a E_h(x, a) \qquad (1.4)$$

with the initial state $E_0(x, a) = 1$ and $E_1(x, a) = x$.

Dickson polynomials are an interesting topic of mathematics, and have many applications. For example, the Dickson polynomials $D_5(x, u) = x^5 - ux - u^2x$ over GF(3^m) are employed to construct a family of planar functions [Ding and Yuan (2006)], and those planar functions give two families of commutative presemifields, planes, several classes of linear codes [Carlet, Ding and Yuan (2005)], and two families of skew Hadamard difference sets [Ding and Yuan (2006)]. The reader is referred to Lidl, Mullen and Turnwald (1993) for detailed information about Dickson polynomials.

1.3.4 *Additive and multiplicative characters*

Let A be a finite abelian group (written multiplicatively) of order $|A|$ with identity 1_A. A *character* χ of A is a homomorphism from A into the multiplicative group U of complex numbers of absolute value 1, i.e.,

$$\chi(a_1 a_2) = \chi(a_1)\chi(a_2)$$

for all $a_1, a_2 \in A$.

For any finite abelian group A, we have the *trivial* (also called *principal*) character χ_0 defined by $\chi_0(a) = 1$ for all $a \in A$. All other characters of A are called *nontrivial* or *nonprincipal*. For each character χ of A, its *conjugate* $\overline{\chi}$ is defined by $\overline{\chi}(a) = \overline{\chi(a)}$. Given any finitely many characters $\chi_1, \ldots, \chi_t$ of A, we define their product character $\chi_1 \ldots \chi_t$ by

$$(\chi_1 \ldots \chi_t)(a) = \chi_1(a) \ldots \chi_t(a)$$

for all $a \in A$. If $\chi_1 = \cdots = \chi_t = \chi$, we write χ^t for $\chi_1 \ldots \chi_t$. It is easily seen that the set $A^\wedge$ of all characters of A form an abelian group under this multiplication of characters. The *order* of a character χ is the least positive integer ℓ such that $\chi^\ell = \chi_0$.

Example 1.8. Let A be a finite cyclic group of order n. Let a be a generator of A. For any fixed integer j with $0 \le j \le n - 1$, define

$$\chi_j(a^k) = e^{2\pi \sqrt{-1} jk/n}, \quad k = 0, 1, \ldots, n - 1.$$

Then χ_j is a character of A. On the other hand, the set $\{\chi_0, \chi_1, \ldots, \chi_{n-1}\}$ contains all characters of A.

Let p be the characteristic of $\mathrm{GF}(q)$. Then the prime field of $\mathrm{GF}(q)$ is $\mathrm{GF}(p)$, which is identified with $\mathbb{Z}_p$. The function χ_1 defined by

$$\chi_1(x) = e^{2\pi \sqrt{-1} \mathrm{Tr}_{q/p}(x)/p} \quad \text{for all } x \in \mathrm{GF}(q)$$

is a character of the additive group of $\mathrm{GF}(q)$, and is called the *canonical character* of $\mathrm{GF}(q)$. For any $b \in \mathrm{GF}(q)$, the function defined by $\chi_b(x) = \chi_1(bx)$ is a character of $(\mathrm{GF}(q), +)$. On the other hand, every character of $(\mathrm{GF}(q), +)$ can be expressed as $\chi_b(x)$ for some $b \in \mathrm{GF}(q)$. These $\chi_b(x)$ are called *additive characters* of $\mathrm{GF}(q)$.

Since the multiplicative group $GF(q)^*$ is cyclic, by Example 1.8, all the characters of the multiplicative group $GF(q)^*$ are given by

$$\psi_j(a^k) = e^{2\pi\sqrt{-1}jk/(q-1)}, \quad k = 0, 1, \ldots, q - 2,$$

where $0 \le j \le q - 2$. These ψ_j are called *multiplicative characters* of $GF(q)$, and form a group of order $q - 1$ with identity element ψ_0. When q is odd, the character $\psi_{(q-1)/2}$ is called the *quadratic character* of $GF(q)$, and is usually denoted by η. In other words, the quadratic character is defined by

$$\eta(x) = \left(\frac{x}{q}\right),$$

the Legendre symbol from elementary number theory when q is a prime.

1.3.5 Character sums

Let ψ be a multiplicative and χ an additive character of $GF(q)$. Then the Gaussian sum is defined by

$$G(\psi, \chi) = \sum_{x \in GF(q)^*} \psi(x)\chi(x).$$

The following theorem will be useful in later chapters and its proof can be found in Lidl and Niederreiter (1997a)[Theorem 5.11].

Theorem 1.8. *Let ψ be a multiplicative and χ an additive character of $GF(q)$. The Gaussian sum satisfies*

$$G(\psi, \chi) = \begin{cases} q - 1 & \text{if } \psi = \psi_0, \ \chi = \chi_0, \\ -1 & \text{if } \psi = \psi_0, \ \chi \ne \chi_0, \\ 0 & \text{if } \psi \ne \psi_0, \ \chi = \chi_0, \end{cases}$$

where ψ_0 and χ_0 are the trivial multiplicative and additive character of $GF(q)$, respectively.

If $\psi \ne \psi_0$ and $\chi \ne \chi_0$, then

$$|G(\psi, \chi)| = \sqrt{q}.$$

For certain special characters, the associated Gaussian sums can be evaluated explicitly. The following theorem will be needed in subsequent chapters and its proof can be found in Lidl and Niederreiter (1997a)[Theorem 5.15].

Theorem 1.9. *Let $q = p^s$, where p is an odd prime and s a positive integer. Let η be the quadratic character of $GF(q)$, and let χ_1 be the canonical additive* *character of* GF(q). *Then*

$$G(\eta, \chi_1) = \begin{cases} (-1)^{s-1}\sqrt{q} & \text{if } p \equiv 1 \pmod{4}, \\ (-1)^{s-1}(\sqrt{-1})^s \sqrt{q} & \text{if } p \equiv 3 \pmod{4}. \end{cases}$$

Since $G(\psi, \chi_b) = \bar{\psi}(b)G(\psi, \chi_1)$, we just consider $G(\psi, \chi_1)$, briefly denoted as $G(\psi)$, in the sequel. If $\psi \neq \psi_0$, then

$$|G(\psi)| = q^{1/2}. \tag{1.5}$$

Certain types of character sums can be evaluated exactly. The following two theorems describe such cases [Lidl and Niederreiter (1997a)][Theorems 5.33 and 5.35].

Theorem 1.10. *Let χ be a nontrivial additive character of* GF(q) *with q odd, and let $f(x) = a_2 x^2 + a_1 x + a_0 \in$ GF(q)[x] with $a_2 \neq 0$. Then*

$$\sum_{c \in \mathrm{GF}(q)} \chi(f(c)) = \chi\left(a_0 - a_1^2 (4a_2)^{-1}\right) \eta(a_2) G(\eta, \chi),$$

where η is the quadratic character of GF(q).

Theorem 1.11. *Let $\chi_b(x) = \chi_1(bx)$, where χ_1 is the canonical additive character of* GF(q) *with q even and $b \in$ GF(q)*. Let $f(x) = a_2 x^2 + a_1 x + a_0 \in$ GF(q)[x]. Then*

$$\sum_{c \in \mathrm{GF}(q)} \chi_b(f(c)) = \begin{cases} \chi_b(a_0) & \text{if } a_2 = ba_1^2, \\ 0 & \text{otherwise.} \end{cases}$$

In many cases it is difficult to evaluate character sums, and thus necessary to develop tight bounds on the absolute value of the character sums. An example of such bounds is the Weil bound given in the following theorem [Lidl and Niederreiter (1997a)][Theorem 5.37].

Theorem 1.12 (Weil Bound). *Let $f \in$ GF(q)[x] be of degree $e \geq 1$ with* $\gcd(e, q) = 1$, *and let χ be a nontrivial additive character of* GF(q). *Then*

$$\left| \sum_{x \in \mathrm{GF}(q)} \chi(f(x)) \right| \leq (e - 1)\sqrt{q}.$$

With respect to multiplicative characters, we have the following bound [Lidl and Niederreiter (1997a)][Theorem 5.41].

Theorem 1.13. *Let ψ be a multiplicative character of* GF(q) *with order $t > 1$, and let $f \in$ GF(q)[x] be of positive degree that is not a tth power of a polynomial.*

Let e be the number of distinct roots of f in its splitting field over $\mathrm{GF}(q)$. *Then for each* $a \in \mathrm{GF}(q)$ *we have*

$$\left| \sum_{x \in \mathrm{GF}(q)} \psi(af(x)) \right| \leq (e-1)\sqrt{q}.$$

Another kind of useful character sums is the *Kloosterman sums*, which are defined by

$$K(\chi; a, b) = \sum_{x \in \mathrm{GF}(q)^*} \chi(ax + bx^{-1}), \tag{1.6}$$

where χ is a nontrivial additive character of $\mathrm{GF}(q)$, and $a, b \in \mathrm{GF}(q)$.

Kloosterman sums are related to a lot of mathematical and engineering problems, and have been extensively studied in the literature. Unfortunately, it is very hard to evaluate Kloosterman sums. However, we do have a tight bound on the Kloosterman sums as follows [Lidl and Niederreiter (1997a)][Theorem 5.45].

Theorem 1.14. *Let* χ *be a nontrivial additive character of* $\mathrm{GF}(q)$, *and let* $a, b \in \mathrm{GF}(q)$ *with* $(a, b) \neq (0, 0)$. *Then*

$$|K(\chi; a, b)| \leq 2\sqrt{q}.$$

1.4   Cyclotomy in GF(r)

1.4.1   *Cyclotomy*

Cyclotomy is to divide the circumference of a given unit circle with its center into n equal parts using only a straightedge (i.e., idealized ruler) and a compass, where the straightedge is only for drawing straight lines and the compass is only for drawing circles. It is equivalent to the problem of constructing the regular n-gon using only a straightedge and a compass.

Greek geometers played this puzzle 2000 years ago. About 300 BC, people in Euclid's School found that the regular n-gon is constructible for any $n \geq 3$ of the form

$$n = 2^a 3^b 5^c, \quad a \geq 0, b \in \{0, 1\}, c \in \{0, 1\}.$$

For more than 2000 years mathematicians had been unanimous in their view that for no prime p bigger than 5 can the p-gon be constructed by ruler and compass. The 18-year old Carl Friedrich Gauss proved that the regular 17-gon is constructible [Gauss (1801)]. This achievement of Gauss is one of the most surprising discoveries in mathematics. He asked to have his 17-gon carved on his tombstone! This discovery led him to choose mathematics (rather than philosophy) as his life-time research topic.

Gauss proved that the regular n-gon is constructible when $n = p2^s$, where $p = 2^{2^k} + 1$ is a Fermat prime. In general, we have the following conclusion.

Theorem 1.15. *A regular n-gon in the plane is constructible iff $n = 2^e p_1 p_2 \cdots p_k$ for $e \geq 0$ and distinct Fermat primes $p_1, \ldots, p_k, k \geq 0$.*

The necessity and sufficiency were proved by Gauss in 1796 and Wanzel in 1836, respectively. A detailed proof of Theorem 1.15 can be found in Pollack (2009)[Chapter 2].

The algebraic criterion for the constructibility of regular n-gon is given in the following theorem [Pollack (2009)][Section 2.2].

Theorem 1.16. *A regular n-gon in the plane is constructible iff all the complex roots of $z^n = 1$ can be found out by solving a chain of linear and quadratic equations.*

To have a better understanding of the algebraic aspect of cyclotomy (a beautiful problem in geometry), we look into the case that $n = 5$.

Let γ_i be the complex roots of $x^4 + x^3 + x^2 + x + 1 = 0$, where $1 \leq i \leq 4$. Note that 2 is a primitive root modulo 5. Let

$$C_i = \{2^{2s+i} \bmod 5 : s = 0, 1\}, \quad i = 0, 1,$$

which are the *cyclotomic classes* of order 2 modulo 5. It is obvious that $C_0 \cap C_1 = \emptyset$ and $C_0 \cup C_1 = \mathbb{Z}_5 \backslash \{0\}$.

Define $\eta_i = \sum_{j \in C_i} x^j$, where $i = 0, 1$. These η_i are called *Gaussian periods* of order 2. Then we have

$$\eta_0 + \eta_1 + 1 = x^4 + x^3 + x^2 + x + 1 = 0. \tag{1.7}$$

It is easily verified that

$$\eta_0 \eta_1 = \eta_0 + \eta_1. \tag{1.8}$$

Combining (1.7) and (1.8) proves that η_0 and η_1 are solutions of $\eta^2 + \eta - 1 = 0$. Hence

$$\eta_0 = \frac{-1 \pm \sqrt{5}}{2} \quad \text{and} \quad \eta_1 = \frac{-1 \mp \sqrt{5}}{2}.$$

It is then easy to see that the four roots γ_i are found by solving in chain

$$\eta^2 + \eta - 1 = 0, \quad \gamma^2 - \eta_i \gamma + 1 = 0,$$

where η_0 and η_1 are solutions of $\eta^2 + \eta - 1 = 0$. Hence, the case $n = 5$ is constructible by Theorem 1.16.

It follows from the discussions above that the algebraic aspect of cyclotomy is related to cyclotomic classes and Gaussian periods, which will be the subjects of the next subsection.

1.4.2 *Cyclotomy in* GF(*r*)

Let r be a power of a prime p. Let $r - 1 = nN$ for two positive integers $n > 1$ and $N > 1$, and let α be a fixed primitive element of GF(r). Define $C_i^{(N,r)} = \alpha^i \langle \alpha^N \rangle$ for $i = 0, 1, \ldots, N - 1$, where $\langle \alpha^N \rangle$ denotes the subgroup of GF(r)* generated by α^N. The cosets $C_i^{(N,r)}$ are called the *cyclotomic classes* of order N in GF(r). The *cyclotomic numbers* of order N are defined by

$$(i, j)^{(N,r)} = \left| \left(C_i^{(N,r)} + 1 \right) \cap C_j^{(N,r)} \right|$$

for all $0 \leq i \leq N - 1$ and $0 \leq j \leq N - 1$.

The following theorem describes elementary facts about cyclotomic numbers, which are not hard to prove [Storer (1967)][Lemma 3].

Theorem 1.17. *Let symbols and notations be the same as before. Then the following equations hold.*

(A) $(l, m)^{(N,r)} = (l', m')^{(N,r)}$ *when* $l \equiv l'$ (mod N) *and* $m \equiv m'$ (mod N).

(B) $(l, m)^{(N,r)} = (N - l, m - l)^{(N,r)} = \begin{cases} (m, l)^{(N,r)} \text{ for even } n, \\ (m + N/2, l + N/2)^{(N,r)} \text{ for odd } n. \end{cases}$

(C) $\sum_{m=0}^{N-1}(l, m)^{(N,r)} = n - n_l$, *where*

$$n_l = \begin{cases} 1 & \text{if } l \equiv 0 \pmod{N}, \ n \text{ even}, \\ 1 & \text{if } l \equiv N/2 \pmod{N}, \ n \text{ odd}, \\ 0 & \text{otherwise}. \end{cases}$$

(D) $\sum_{l=0}^{N-1}(l, m)^{(N,r)} = n - k_m$, *where*

$$k_m = \begin{cases} 1 & \text{if } m \equiv 0 \pmod{N}, \\ 0 & \text{otherwise}. \end{cases}$$

(E) $\sum_{l=0}^{N-1} \sum_{m=0}^{N-1}(l, m)^{(N,r)} = Nn - 1 = r - 2.$

(F) $(l, m)^{(N',r)} = (sl, sm)^{(N,r)}$, where $(l, m)^{(N',r)}$ is based on the primitive element $\alpha' \equiv \alpha^s \pmod{N}$; necessarily then s is prime to $r - 1$.

In the sequel we will need the following lemma [Ding and Yin (2008)] whose initial version was proved in Tze, Chanson, Ding, Helleseth and Parker (2003).

Lemma 1.2. *Let* $r - 1 = nN$ *and let* r *be a prime power. Then*

$$\sum_{u=0}^{N-1} (u, u + k)^{(N,r)} = \begin{cases} n - 1 & \text{if } k = 0, \\ n & \text{if } k \neq 0. \end{cases}$$

In general, it is very hard to determine the cyclotomic numbers $(i, j)^{(N,r)}$. But they are known when N is small or under certain conditions [Storer (1967)]. We will introduce cyclotomic numbers of certain orders in the sequel when we really need them.

The *Gaussian periods* are defined by

$$\eta_i^{(N,r)} = \sum_{x \in C_i^{(N,r)}} \chi(x), \quad i = 0, 1, \ldots, N - 1,$$

where χ is the canonical additive character of GF(r).

The following lemma presents some basic properties of Gaussian periods, and will be employed later.

Lemma 1.3 ([Storer (1967)]). *Let symbols be the same as before. Then we have*

(1) $\sum_{i=0}^{N-1} \eta_i = -1$.

(2) $\sum_{i=0}^{N-1} \eta_i \eta_{i+k} = r\theta_k - n$ *for all* $k \in \{0, 1, \ldots, N - 1\}$, *where*

$$\theta_k = \begin{cases} 1 & \text{if } n \text{ is even and } k = 0, \\ 1 & \text{if } n \text{ is odd and } k = N/2, \\ 0 & \text{otherwise}, \end{cases}$$

and equivalently $\theta_k = 1$ *if and only if* $-1 \in C_k^{(N,r)}$.

Gaussian periods are closely related to Gaussian sums. By finite Fourier transform, it is known that

$$\eta_i^{(N,r)} = \frac{1}{N} \sum_{j=0}^{N-1} \epsilon_N^{-ij} G(\psi^j, \chi_1) = \frac{1}{N} \left[-1 + \sum_{j=1}^{N-1} \epsilon_N^{-ij} G(\psi^j, \chi_1) \right], \quad (1.9)$$

where $\epsilon_N = e^{2\pi\sqrt{-1}/N}$ and ψ is a primitive multiplicative character of order N over GF$(r)^*$.

From (1.9), one knows that the values of the Gaussian periods in general are also very hard to compute. However, they can be computed in a few cases. The following lemma follows from Theorems 1.9 and 1.10.

Lemma 1.4. *Let* $r = p^m$. *When* $N = 2$, *the Gaussian periods are given by the following*:

$$\eta_0^{(2,r)} = \begin{cases} \dfrac{-1 + (-1)^{m-1} r^{1/2}}{2} & \text{if } p \equiv 1 \pmod 4, \\[2ex] \dfrac{-1 + (-1)^{m-1}(\sqrt{-1})^m r^{1/2}}{2} & \text{if } p \equiv 3 \pmod 4 \end{cases}$$

and

$$\eta_1^{(2,r)} = -1 - \eta_0^{(2,r)}.$$

The following result is proved in Myerson (1981).

Lemma 1.5. *If* $r \equiv 1$ (mod 4), *we have*

$$(0,0)^{(2,r)} = \frac{r-5}{4}, \quad (0,1)^{(2,r)} = (1,0)^{(2,r)} = (1,1)^{(2,r)} = \frac{r-1}{4}.$$

If $r \equiv 3$ (mod 4), *we have*

$$(0,1)^{(2,r)} = \frac{r+1}{4}, \quad (0,0)^{(2,r)} = (1,0)^{(2,r)} = (1,1)^{(2,r)} = \frac{r-3}{4}.$$

To present further known results on Gaussian periods, we need to introduce period polynomials.

The *period polynomials* $\psi_{(N,r)}(X)$ are defined by

$$\psi_{(N,r)}(X) = \prod_{i=0}^{N-1} \left(X - \eta_i^{(N,r)} \right).$$

It is known that $\psi_{(N,r)}(X)$ is a polynomial with integer coefficients [Myerson (1981)]. Proofs of the following four lemmas can be found in Myerson (1981).

Lemma 1.6. *Let* $N = 3$ *and let* $r = p^m$. *Let* c *and* d *be defined by* $4r = c^2 + 27d^2, c \equiv 1$ (mod 3), *and, if* $p \equiv 1$ (mod 3), *then* $\gcd(c, p) = 1$. *These restrictions determine* c *uniquely, and* d *up to sign. Then we have*

$$\psi_{(3,r)}(X) = X^3 + X^2 - \frac{r-1}{3} X - \frac{(c+3)r - 1}{27}.$$

Lemma 1.7. *Let* $N = 3$ *and* $r = p^m$. *We have the following results on the factorization of* $\psi_{(3,r)}(X)$.

(a) *If $p \equiv 2$ (mod 3), then m is even, and*

$$\psi_{(3,r)}(X) = \begin{cases} 3^{-3}(3X + 1 + 2\sqrt{r})(3X + 1 - \sqrt{r})^2 & \text{if } m/2 \text{ even,} \\ 3^{-3}(3X + 1 - 2\sqrt{r})(3X + 1 + \sqrt{r})^2 & \text{if } m/2 \text{ odd.} \end{cases}$$

(b) *If $p \equiv 1$ (mod 3), and $m \not\equiv 0$ (mod 3), then $\psi_{(3,r)}(X)$ is irreducible over the rationals.*

(c) *If $p \equiv 1$ (mod 3), and $m \equiv 0$ (mod 3), then*

$$\psi_{(3,r)}(X) = \frac{1}{27}(3X + 1 - c_1 r^{\frac{1}{3}})\left(3X + 1 + \frac{1}{2}(c_1 + 9d_1)r^{\frac{1}{3}}\right)$$
$$\times \left(3X + 1 + \frac{1}{2}(c_1 - 9d_1)r^{\frac{1}{3}}\right),$$

where c_1 and d_1 are given by $4p^{m/3} = c_1^2 + 27d_1^2, c_1 \equiv 1$ (mod 3) and $\gcd(c_1, p) = 1$.

The cyclotomic numbers of order 3 are known and are given in the next lemma.

Lemma 1.8. *Let $N = 3$ and let $r = p^m$. Let u and v be defined by $4r = u^2 + 27v^2, u \equiv 1$ (mod 3), and if $p \equiv 1$ (mod 3), then $\gcd(u, p) = 1$; these restrictions determine u uniquely, and v up to sign. Then the relations among the cyclotomic numbers of order 3 are given in Table 1.1, where*

$$9A = r - 8 + u,$$
$$18B = 2r - 4 - u - 9v,$$
$$18C = 2r - 4 - u + 9v,$$
$$9D = r + 1 + u.$$

Lemma 1.9. *Let $N = 4$ and let $r = p^m$. Let u and v be defined by $r = u^2 + 4v^2, u \equiv 1$ (mod 4), and, if $p \equiv 1$ (mod 4), then $\gcd(u, p) = 1$. These restrictions determine u uniquely, and v up to sign.*

Table 1.1 The relations of cyclotomic numbers of order 3.

$(h, l)^{(3,r)}$	0	1	2
0	A	B	C
1	B	C	D
2	C	D	B

If n is even, then

$$\psi_{(4,r)}(X) = X^4 + X^3 - \frac{3r-3}{8}X^2 + \frac{(2u-3)r+1}{16}X$$
$$+ \frac{r^2 - (4u^2 - 8u + 6)r + 1}{256}.$$

If n is odd, then

$$\psi_{(4,r)}(X) = X^4 + X^3 + \frac{r+3}{8}X^2 + \frac{(2u+1)r+1}{16}X$$
$$+ \frac{9r^2 - (4u^2 - 8u - 2)r + 1}{256}.$$

Lemma 1.10. *Let $N = 4$ and let $r = p^m$. We have the following results on the factorization of $\psi_{(4,r)}(X)$.*

(a) *If $p \equiv 3 \pmod 4$, then m is even, and*

$$\psi_{(4,r)}(X) = \begin{cases} 4^{-4}(4X + 1 + 3\sqrt{r})(4X + 1 - \sqrt{r})^3 & \text{if } m/2 \text{ even,} \\ 4^{-4}(4X + 1 - 3\sqrt{r})(4X + 1 + \sqrt{r})^3 & \text{if } m/2 \text{ odd.} \end{cases}$$

(b) *If $p \equiv 1 \pmod 4$, and m is odd, then $\psi_{(4,r)}(X)$ is irreducible over the rationals.*

(c) *If $p \equiv 1 \pmod 4$, and $m \equiv 2 \pmod 4$, then*

$$\psi_{(4,r)}(X) = 4^{-4}\left((4X + 1)^2 + 2\sqrt{r}(4X + 1) - r - 2\sqrt{r}u\right)$$
$$\times \left((4X + 1)^2 - 2\sqrt{r}(4X + 1) - r + 2\sqrt{r}u\right),$$

the quadratics being irreducible, the u is defined in Lemma 1.9.

(d) *If $p \equiv 1 \pmod 4$, and $m \equiv 0 \pmod 4$, then*

$$\psi_{(4,r)}(X) = 4^{-4}\left((4X + 1) + \sqrt{r} + 2r^{1/4}u_1\right)\left((4X + 1) + \sqrt{r} - 2r^{1/4}u_1\right)$$
$$\times \left((4X + 1) - \sqrt{r} + 4r^{1/4}v_1\right)\left((4X + 1) - \sqrt{r} - 4r^{1/4}v_1\right),$$

where u_1 and v_1 are given by $p^{m/2} = u_1^2 + 4v_1^2, u_1 \equiv 1 \pmod 4$ and $\gcd(u_1, p) = 1$.

The cyclotomic numbers of order 4 are known and are given in the next two lemmas [Storer (1967)][pp. 48–51].

Table 1.2 The relations of cyclotomic
numbers of order 4, when n is odd.

$(h, l)^{(4,r)}$	0	1	2	3
0	A	B	C	D
1	E	E	D	B
2	A	E	A	E
3	E	D	B	E

Table 1.3 The relations of cyclotomic
numbers of order 4, when n is even.

$(h, l)^{(4,r)}$	0	1	2	3
0	A	B	C	D
1	B	D	E	E
2	C	E	C	E
3	D	E	E	B

Lemma 1.11. *When $N = 4$ and n is odd, the cyclotomic numbers are determined by the pertinent cyclotomic matrix in Table 1.2, together with the relations*

$$A = \frac{r - 7 + 2u}{16},$$

$$B = \frac{r + 1 + 2u - 8v}{16},$$

$$C = \frac{r + 1 - 6u}{16},$$

$$D = \frac{r + 1 + 2u + 8v}{16},$$

$$E = \frac{r - 3 - 2u}{16},$$

where $r = u^2 + 4v^2$, $u \equiv 1 \pmod 4$, and also $\gcd(u, p) = 1$ if $p \equiv 1 \pmod 4$, where the sign of v is ambiguously determined.

Lemma 1.12. *When $N = 4$ and n is even, the cyclotomic numbers are determined by the pertinent cyclotomic matrix in Table 1.3, together with the relations*

$$A = \frac{r - 11 - 6u}{16},$$

$$B = \frac{r - 3 + 2u + 8v}{16},$$

$$C = \frac{r - 3 + 2u}{16},$$

$$D = \frac{r - 3 + 2u - 8v}{16},$$

$$E = \frac{r + 1 - 2u}{16},$$

where $r = u^2 + 4v^2$, $u \equiv 1$ (mod 4), *and also* $\gcd(u, p) = 1$ *if* $p \equiv 1$ (mod 4), *where the sign of* v *is ambiguously determined.*

The period polynomial $\psi_{(N,r)}(X)$ and its factorization were determined for $N = 5$ in Hoshi (2006), and for $N \in \{6, 8, 12\}$ in Gurak (2004). The expression of $\psi_{(N,r)}(X)$ and its factorization are quite complex. Cyclotomic numbers of orders 6 and 8 for the case that n being odd were given in Storer (1967)[p. 72; p. 79].

The Gaussian periods are also determined in the semiprimitive case and are described in the next theorem.

Theorem 1.18 ([Baumert, Mills and Ward (1982)]). *Assume that* p *is a prime,* $N \geq 2$ *is a positive integer,* $r = p^{2j\gamma}$, *where* $N|(p^j + 1)$ *and* j *is the smallest such positive integer. Then the Gaussian periods of order* N *are given below:*

(a) *If* γ, p, $\frac{p^j+1}{N}$ *are all odd, then*

$$\eta_{N/2}^{(N,r)} = \sqrt{r} - \frac{\sqrt{r}+1}{N}, \quad \eta_i^{(N,r)} = -\frac{1+\sqrt{r}}{N} \quad \text{for all } i \neq \frac{N}{2}.$$

(b) *In all the other cases,*

$$\eta_0^{(N,r)} = -(-1)^\gamma \sqrt{r} + \frac{(-1)^\gamma \sqrt{r} - 1}{N},$$

$$\eta_i^{(N,r)} = \frac{(-1)^\gamma \sqrt{r} - 1}{N} \quad \text{for all } i \neq 0.$$

1.5   Generalized Cyclotomy in $\mathbb{Z}_{n_1 n_2}$

Let n_1 and n_2 be two distinct primes. Define $N = \gcd(n_1 - 1, n_2 - 1)$ and $e = (n_1-1)(n_2-1)/N$. It is well-known that any prime n_1 has $\phi(n_1-1)$ primitive roots. The Chinese Remainder Theorem guarantees that there are common primitive roots of both n_1 and n_2. Let ς be a fixed common primitive root of both n_1 and n_2, and

ϱ be an integer satisfying

$$\varrho \equiv \varsigma \pmod{n_1}, \quad \varrho \equiv 1 \pmod{n_2}.$$

Whiteman proved that

$$\mathbb{Z}_n^* = \{\varsigma^s \varrho^i : s = 0, 1, \ldots, e-1; \; i = 0, 1, \ldots, N-1\},$$

where $\mathbb{Z}_n^*$ denotes the set of all invertible elements of the residue class ring $\mathbb{Z}_n$. The generalized cyclotomic classes W_i of order N with respect to n_1 and n_2 are defined by

$$W_i^{(N)} = \{\varsigma^s \varrho^i : s = 0, 1, \ldots, e-1\}, \quad i = 0, 1, \ldots, N-1.$$

It was proved in Whiteman (1962) that

$$\mathbb{Z}_n^* = \cup_{i=0}^{N-1} W_i^{(N)}, \quad W_i^{(N)} \cap W_j^{(N)} = \emptyset \;\; \text{for } i \neq j.$$

This generalized cyclotomy was introduced by Whiteman (1962). The motivation behind the investigation of the generalized cyclotomy with respect to two primes is the search for residue difference sets. The famous twin-prime difference sets are among such a class of difference sets.

The following is proved in Whiteman (1962).

Lemma 1.13. *Define* $v = (n_1 - 1)(n_2 - 1)/N^2$. *Let symbols be the same as before. Then*

$$-1 = \begin{cases} \varsigma^\mu \varrho^{N/2} & \text{if } v \text{ is even,} \\ \varsigma^{e/2} & \text{if } v \text{ is odd,} \end{cases}$$

where μ *is some fixed integer such that* $0 \le \mu \le e - 1$.

Recall Whiteman's cyclotomic classes $W_i^{(N)}$ of order N defined before. The cyclotomic numbers corresponding to these cyclotomic classes are defined by

$$(i, j)_N = \left| \left(W_i^{(N)} + 1 \right) \cap W_j^{(N)} \right|$$

for any pair of i and j with $0 \le i \le N - 1$ and $0 \le j \le N - 1$.

The following lemma summarizes a number of properties of the cyclotomic numbers of order N [Whiteman (1962)].

Lemma 1.14. *Define* $v = (n_1 - 1)(n_2 - 1)/N^2$. *Then the following properties hold:*

(W0) $(i, j)_N = (i', j')_N$ *when* $i \equiv i' \pmod{N}$ *and* $j \equiv j' \pmod{N}$.
(W1) $(i, j)_N = (N - i, j - i)_N$.

(W2) $(i, j)_N = \begin{cases} (j, i)_N & \text{if } v \text{ is odd,} \\ (j + N/2, i + N/2)_N & \text{if } v \text{ is even.} \end{cases}$

(W3) $\sum_{j=0}^{N-1} (i, j)_N = \frac{(n_1-2)(n_2-2)-1}{N} + \delta_i$, where

$$\delta_i = \begin{cases} 1 & \text{if } i = 0 \text{ and } v \text{ is odd,} \\ 1 & \text{if } i = N/2 \text{ and } v \text{ is even,} \\ 0 & \text{otherwise.} \end{cases}$$

(W4) $\sum_{i=0}^{N-1} (i, j)_N = \frac{(n_1-2)(n_2-2)-1}{N} + \epsilon_j$, where

$$\epsilon_j = \begin{cases} 1 & \text{if } j = 0, \\ 0 & \text{otherwise.} \end{cases}$$

The proof of Lemma 2 in Whiteman (1962) can be slightly modified into a proof of the following lemma.

Lemma 1.15. *For any* $y \in N_i$ *and* $u, v \in \{0, 1, \ldots, (N - 2)/2\}$,

$$\left| \left(W_{2u}^{(N)} + y \right) \cap W_{2v+1}^{(N)} \right| = \frac{(n_1 - 1)(n_2 - 1)}{N^2}.$$

1.6 Finite Geometries

In this section, we present the basics of finite geometries without providing proofs of the statements. The reader may find out the proofs of these results in Lidl and Niederreiter (1997*a*)[Section 9.3].

1.6.1 *Projective planes*

A *projective plane* is a triple $\Pi = (\mathcal{P}, \mathcal{L}, \mathcal{I})$, where $\mathcal{P}$ is a set of *points*, $\mathcal{L}$ consists of *lines* (i.e., sets of points), and $\mathcal{I}$ is a *relation* (also called *incidence relation*) between the points and the lines, subject to the following three conditions:

(a) Every pair of distinct lines incident with a unique point (i.e., to every pair of distinct lines there is one point contained in both lines, called their *intersection*).

(b) Every pair of distinct points is incident with a unique line (i.e., to every pair of distinct points there is exactly one line which contains both points).

(c) There exist four points such that no three of them are incident with a single line (i.e., there exist four points such that no three of them are on the same line).

By the definition of projective planes above, each line contains at least three points and through each point there are at least three lines. When the set $\mathcal{P}$ of points is finite, the projective plane is called *finite*. One can prove the following result [Lidl and Niederreiter (1997a)][Theorem 9.54].

Theorem 1.19. *Let* Π *be a finite projective plane. Then*

- *there is an integer* $m \geq 2$ *such that every point (line) of* Π *is incident with exactly* $m + 1$ *lines (points) of* Π; *and*
- Π *contains exactly* $m^2 + m + 1$ *points (lines)*.

The integer m above is called the *order* of the finite projective plane.

Example 1.9. It follows from Theorem 1.19 that the smallest finite plane has order $m = 2$, which has 7 points and 7 lines exactly. Let the set of points be $\mathcal{P} = \{1, 2, 3, 4, 5, 6, 7\}$, the 7 lines be

$$\{1, 2, 3\}, \{1, 4, 5\}, \{1, 6, 7\}, \{2, 4, 7\}, \{2, 5, 6\}, \{3, 4, 6\}, \{3, 7, 5\}$$

and let the incidence relation be the membership of sets. Then we have the Fano plane depicted in Fig. 1.1, where no three points in the set $\{1, 3, 5, 6\}$ are on the same line.

For every prime power $q = p^m$, there exists a finite projective plane of order q, i.e., $PG(2, GF(q))$ defined later.

1.6.2 *Affine planes*

An *affine plane* is a triple $\Pi = (\mathcal{P}, \mathcal{L}, \mathcal{I})$, where $\mathcal{P}$ is a set of *points*, $\mathcal{L}$ consists of *lines* (i.e., sets of points), and $\mathcal{I}$ is an *incidence relation* between the points and the lines such that

(a) every pair of distinct points is incident with a unique line;
(b) every point $p \in \mathcal{P}$ not on a line $L \in \mathcal{L}$ lies on a unique line $M \in \mathcal{L}$ which does not intersect L; and

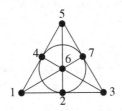

Fig. 1.1 Fano plane.

(c) there exist four points such that no three of them are incident with a single line.

Example 1.10. Let $\mathcal{P} = \mathrm{GF}(q)^2$ be the set of points. For each $(a, b, c) \in \mathrm{GF}(q)^3$ with $(a, b) \neq (0, 0)$, we define a line

$$L_{(a,b,c)} = \{(x, y) : ax + by + c = 0\}.$$

Note that two different triples (a, b, c) may define the same line. Let $\mathcal{L} = \{L_{(a,b,c)} : (a, b, c) \in \mathrm{GF}(q)^3\}$, which does not contain repeated lines. A point $p \in \mathcal{P}$ is incident with a line $L \in \mathcal{L}$ if and only if $p \in L$. Then it is easy to prove that $(\mathcal{P}, \mathcal{L}, \mathcal{I})$ is an affine plane, denoted by $\mathrm{AG}(2, \mathrm{GF}(q))$, and each line of $\mathrm{AG}(2, \mathrm{GF}(q))$ contains exactly q points.

We can construct a projective plane from $\mathrm{AG}(2, \mathrm{GF}(q))$ by adding a line to it. Conversely, we can obtain an affine plane from any projective plane by deleting one line and all the points on it. Points on a line are said to be *collinear*.

1.6.3 *Projective spaces* PG(*m*, GF(*q*))

The points of the *projective space* (also called *projective geometry*) $\mathrm{PG}(m, \mathrm{GF}(q))$ are all the 1-dimensional subspaces of the vector space $\mathrm{GF}(q)^{m+1}$; the lines are the 2-dimensional subspaces of $\mathrm{GF}(q)^{m+1}$, the planes are the 3-dimensional subspaces of $\mathrm{GF}(q)^{m+1}$, and the hyperplanes are the m-dimensional subspaces of $\mathrm{GF}(q)^{m+1}$; and incidence is set-theoretic inclusion.

It is known that the number of subspaces of $\mathrm{GF}(q)^n$ of dimension k, where $0 \leq k \leq n$, is equal to

$$\frac{(q^n - 1)(q^n - q) \cdots (q^n - q^{k-1})}{(q^k - 1)(q^k - q) \cdots (q^k - q^{k-1})}.$$

These numbers are called *Gaussian coefficients*, and are denoted by $\begin{bmatrix} n \\ k \end{bmatrix}_q$.

By definition, the $(m + 1)$-tuples $(ax_0, ax_1, \ldots, ax_m)$ with $a \in \mathrm{GF}(q)^*$ define the same point in $\mathrm{PG}(m, \mathrm{GF}(q))$. A *k-flat* of the projective space $\mathrm{PG}(m-1, \mathrm{GF}(q))$ is the set of all those points whose coordinates satisfy $m - k$ linearly independent homogeneous linear equations

$$
\begin{aligned}
a_{1,0}\, x_0 + \cdots + a_{1,m}\, x_m &= 0 \\
a_{2,0}\, x_0 + \cdots + a_{2,m}\, x_m &= 0 \\
\vdots \qquad\quad \vdots \qquad\quad \vdots \\
a_{m-k,0}\, x_0 + \cdots + a_{m-k,m}\, x_m &= 0
\end{aligned}
$$

whose coefficients $a_{i,j} \in GF(q)$. Hence the number of points in k-flat is

$$\begin{bmatrix} k+1 \\ 1 \end{bmatrix}_q = \frac{q^{k+1}}{q-1}.$$

Our objective of this subsection is to introduce the basic concepts and notation of the projective spaces $PG(m, GF(q))$. We refer the reader to Lidl and Niederreiter (1997a)[Section 9.3] or Assmus and Key (1992a)[Section 3.2] for geometric properties of $PG(m, GF(q))$.

1.6.4 *Affine spaces* AG(m, GF(q))

The *affine space* $AG(m, GF(q))$ consists of all cosets $\mathbf{x} + U$, of all subspaces U of $GF(q)^m$ with incidence defined through the natural containment relation. In this case, the dimension is the same as that of vector space, and if the latter has dimension k, we will call a coset of U a *k-flat*. Thus, the points of $AG(m, GF(q))$ are all the vectors in $GF(q)^m$; the lines are all the 1-dimensional cosets (also called 1-flats); the planes are the 2-dimensional cosets (also called 2-flats); and the hyperplanes are the $(m-1)$-dimensional cosets. Geometric properties of $AG(m, GF(q))$ can be found in Assmus and Key (1992a)[Section 3.2].

1.7 Planar Functions

1.7.1 *Definitions and properties*

A function f from an abelian group $(A, +)$ to an abelian group $(B, +)$ is called *linear* if $f(x + y) = f(x) + f(y)$ for all $x, y \in A$. Hence, linear functions are group homomorphisms. A function $g : A \to B$ is *affine* if $g = f + b$ for a linear function $f : A \to B$ and a constant $b \in B$.

The *Hamming distance* between two functions f and g from an abelian group A to an abelian group B, denoted by $\mathrm{dist}(f, g)$, is defined to be

$$\mathrm{dist}(f, g) = |\{x \in A \mid f(x) - g(x) \neq 0\}|.$$

There are different measures of nonlinearity of functions. The first measure of nonlinearity of a function f from $(A, +)$ to $(B, +)$ is defined by

$$N_f = \min_{l \in L} d(f, l), \tag{1.10}$$

where L denotes the set of all affine functions from $(A, +)$ to $(B, +)$.

The second measure of nonlinearity of a function f from $(A, +)$ to $(B, +)$ is given by

$$P_f = \max_{0 \neq a \in A} \max_{b \in B} \frac{|\{x \in A : f(x + a) - f(x) = b\}|}{|A|}.$$

It is easily seen that $P_f \geq \frac{1}{|B|}$. If the equality is achieved, f is called a *perfect nonlinear function* (PN function for short).

Example 1.11. The function $\mathrm{Tr}_{q^m/q}(x^2)$ from $(\mathrm{GF}(q^m), +)$ to $(\mathrm{GF}(q), +)$ is perfect nonlinear for any odd prime q.

The following result follows from the definition of perfect nonlinear functions.

Theorem 1.20. *Let f be a function from a finite abelian group $(A, +)$ to a finite abelian group $(B, +)$. Then f is perfect if and only if*

$$|\{x \in A : f(x + a) - f(x) = b\}| = \frac{|A|}{|B|} \tag{1.11}$$

for each nonzero element a of A and every $b \in B$.

A perfect nonlinear function from an abelian group $(A, +)$ to an abelian group $(B, +)$ of the same order is called *planar*, i.e., $g_a(x) = f(x + a) - f(x)$ is a one-to-one function from A to B for every nonzero $a \in A$.

Example 1.12. The function x^2 from $(\mathrm{GF}(q), +)$ to itself is planar for any odd q.

Planar functions were introduced by Dembowski and Ostrom for constructing affine and projective planes [Dembowski and Ostrom (1968)]. They have applications in cryptography, coding theory [Carlet, Ding and Yuan (2005)], combinatorics, and some engineering areas. We will look into some of their applications in this monograph.

1.7.2 Some known planar functions

Polynomials over $\mathrm{GF}(q)$ of the form $\sum_{i,j=0}^{m-1} a_{i,j} x^{p^i + p^j}$ are called *Dembowski–Ostrom polynomials*, where $a_{i,j} \in \mathrm{GF}(q)$ for all i and j.

The following theorem characterizes planar Dembowski–Ostrom polynomials [Chen and Polhill (2011)].

Theorem 1.21. *Let q be odd, and let $f(x)$ be a Dembowski–Ostrom polynomial over $\mathrm{GF}(q)$. Then the following are equivalent:*

- *$f(x)$ is planar;*
- *$f(x)$ is a two-to-one map, $f(0) = 0$ and $f(x) \neq 0$ for $x \neq 0$;*

- *there is a permutation polynomial $g(x)$ over* GF(q) *such that $f(x) = g(x^2)$ for all $x \in$ GF(q). (When $q \equiv 3$ (mod 4), $g(-x) = -g(x)$.)*

A list of planar monomials over finite fields is documented in the following theorem.

Theorem 1.22. *The function $f(x) = x^s$ from* GF(p^m) *to* GF(p^m) *is planar when*

- $s = 2$; *or*
- $s = p^k + 1$, *where $m/\gcd(m, k)$ is odd [Dembowski and Ostrom (1968)]; or*
- $s = (3^k + 1)/2$, *where $p = 3, k$ is odd, and $\gcd(m, k) = 1$ [Coulter and Matthews (1997)].*

There are planar binomials over finite fields. Below is a class of such planar binomials [Zha, Kyureghyan and Wang (2009)].

Theorem 1.23. *Let $f(x) = x^{1+p^{k+3\ell}} - u^{p^k-1}x^{p^k+p^{3\ell}}$, where*

- $m = 3k$ *and $k \not\equiv 0$ (mod 3),*
- ℓ *is a positive integer,*
- $k/\gcd(k, \ell)$ *is odd, and*
- u *is a generator of* GF(p^m)*.*

Then $f(x)$ is planar.

The following class of planar functions is based on Dickson polynomials of the first kind.

Theorem 1.24. *The polynomials $f_u(x) = x^{10} - ux^6 - u^2x^2$ over* GF(3^m) *are planar for all $u \in$ GF(3^m), where m is odd.*

The planar function $x^{10} + x^6 - x^2$ over GF(3^m) was presented in Coulter and Matthews (1997). It was extended into the whole class in Ding and Yuan (2006).

More planar monomials can be found in Bierbrauer (2010), Budaghyan and Helleseth (2008) and Zha and Wang (2009).

1.7.3 Planar functions from semifields

A *semifield* consists of a set K and two binary operations $(+, \times) : K \times K \to K$ such that the following axioms hold.

- $(K, +)$ is a group.
- There exists an element 1 of K distinct from zero with $1x = x1 = x$ for each $x \in K$.

- For all $0 \neq a, b \in K$, there is $x \in K$ such that $xa = b$.
- For all $0 \neq a, b \in K$, there is $x \in K$ such that $ax = b$.
- For all $a, b, c \in K$, $(a + b)c = ac + bc$ and $c(a + b) = ca + cb$.

A semifield is a field except that associativity for multiplication may not hold. A semifield is *commutative* if its multiplication is commutative. A *presemifield* is a semifield except that the multiplicative identity may not exist.

Example 1.13. Positive real numbers with the usual addition and multiplication form a commutative semifield.

Example 1.14. Rational functions of the form f/g, where f and g are polynomials in one variable with positive coefficients, form a commutative semifield.

An *isotopism* between two semifields $(F, +, *)$ and $(F', +, \circ)$ is a triple (α, β, γ) of additive bijections $F \to F'$ such that

$$(x*y)\gamma = x\alpha \circ y\beta, \quad \forall x, y, z \in F.$$

If there is an isotopism between two semifields, the two semifields are said *isotopic*. If $\alpha = \beta$ and the above equation holds, the two presemifields are called *strongly isotopic*.

Commutative presemifields can be utilized to construct planar functions, as demonstrated by the next theorem [Kantor (2003)].

Theorem 1.25. *Let $(K, +, \times)$ be a finite presemifield with commutative multiplication. Then the function $f(x) = x \times x$ is a planar function from $(K, +)$ to itself.*

To introduce planar monomials on $\mathrm{GF}(q^2)$ from commutative semifields in the sequel, we do the following preparations.

Let $\{1, \beta\}$ be a basis of $\mathrm{GF}(q^2)$ over $\mathrm{GF}(q)$. Let $x = x_1 + x_2\beta$, where $x_i \in \mathrm{GF}(q)$. It is easily seen that

$$x_1 = \frac{\beta^q x - \beta x^q}{\beta^q - \beta}, \quad x_2 = \frac{x^q - x}{\beta^q - \beta},$$

where q is a prime power.

Planar monomials on $\mathrm{GF}(q^2)$ from the Dickson commutative semifields are given in the following theorem, which follows from Theorem 1.25.

Theorem 1.26. *Assume that q is odd. Let k be a nonsquare in $K = \mathrm{GF}(q)$, and let $1 \neq \sigma \in \mathrm{Aut}(K)$. The* Dickson semifield $(K^2, +, *)$ *has*

$$(a, b)*(c, d) = (ac + jb^{\sigma}d^{\sigma}, ad + bc),$$

where j is a nonsquare in K. Different choices j produce isotopic semifields. The corresponding planar function from $(K^2, +)$ to $(K^2, +)$ is

$$f(a, b) = (a^2 + j(\sigma(b))^2, 2ab),$$

where $(a, b) \in K^2$.
They can be expressed as

$$x^2 + j\left(\sigma\left(\frac{x^q - x}{\beta^q - \beta}\right)\right)^2 - \beta^2\left(\frac{x^q - x}{\beta^q - \beta}\right),$$

where j is a nonsquare in $\mathrm{GF}(q)$, and $1 \neq \sigma \in \mathrm{Aut}(K)$ and $\mathrm{Aut}(K)$ denotes the automorphism group of K.

Example 1.15. If we choose $\sigma(x) = x^q$ in Theorem 1.26, then we have the planar functions

$$(\beta^q - \beta)^2 x^2 + (j - \beta^2)(x^q - x)^2.$$

Planar monomials from the Ganley commutative semifields are described in the next theorem, which is deduced from Theorem 1.25.

Theorem 1.27. *Assume that $q = 3^r, r \geq 3$ odd. Let $K = \mathrm{GF}(q)$. The* Ganley *semifield $(K^2, +, *)$ has*

$$(a, b)*(c, d) = (ac - b^9d - bd^9, ad + bc + b^3d^3).$$

The planar functions on $\mathrm{GF}(q^2)$ defined by the Ganley commutative semifields are

$$(\beta^q - \beta)^{10} x^2 - \beta^2(\beta^q - \beta)^8(x^q - x)^2 + \beta(\beta^q - \beta)^4(x^q - x)^6 + (x^q - x)^{10}.$$

Planar monomials from the Ganley–Cohen commutative semifields are presented in the following theorem, which follows from Theorem 1.25.

Theorem 1.28. *Assume that $q = 3^r, r \geq 2$. Let $K = \mathrm{GF}(q)$, and let $j \in K$ be a nonsquare. The* Ganley–Cohen semifield $(K^2, +, *)$ *has*

$$(a, b)*(c, d) = (ac + jbd + j^3(bd)^9, ad + bc + j(bd)^3).$$

The planar functions on $\mathrm{GF}(q^2)$ defined by the Ganley–Cohen commutative semifields are

$$(\beta^q - \beta)^{18} x^2 - (j - \beta^2)(\beta^q - \beta)^{16}(x^q - x)^2 + \beta j(\beta^q - \beta)^{12}(x^q - x)^6 + j^3(x^q - x)^{18}.$$

1.7.4 Affine planes from planar functions

Planar functions were employed to construct affine planes as follows in the seminal work [Dembowski and Ostrom (1968)].

Theorem 1.29. *Let* $f : (A, +) \to (B, +)$ *be a function. Define* $\mathcal{P} = A \times B$. *The lines are the symbols* $L(a, b)$ *with* $(a, b) \in A \times B$, *together with the symbols* $L(c)$ *with* $c \in A$. *Incidence* $\mathcal{I}$ *is defined by*

- $(x, y) \, \mathcal{I} \, L(a, b)$ *if and only if* $y = f(x - a) + b$; *and*
- $(x, y) \, \mathcal{I} \, L(c)$ *if and only if* $x = c$.

Then f *is planar if and only if* $(\mathcal{P}, \mathcal{L}, \mathcal{I})$ *is an affine plane.*

1.8 Periodic Sequences

Both finite and periodic sequences will play a very important role in dealing with codes derived from difference sets and other combinatorial designs. In this section, we briefly introduce periodic sequences.

1.8.1 *The linear span*

Let $\lambda^\infty = (\lambda_i)_{i=0}^\infty$ be a sequence of period n over $\mathrm{GF}(q)$. The *linear span* (also called *linear complexity*) of λ^∞ is defined to be the smallest positive integer ℓ such that there are constants $c_0 \neq 0, c_1, \ldots, c_\ell \in \mathrm{GF}(q)$ satisfying

$$-c_0 \lambda_i = c_1 \lambda_{i-1} + c_2 \lambda_{i-2} + \cdots + c_l \lambda_{i-\ell} \quad \text{for all } i \geq \ell.$$

The polynomial $c(x) = c_0 + c_1 x + \cdots + c_\ell x^\ell$ is called a *characteristic polynomial* of λ^∞. A characteristic polynomial with the smallest degree is called a *minimal polynomial* of the periodic sequence λ^∞. The degree of a minimal polynomial of λ^∞ is referred to as the *linear span* or *linear complexity* of this sequence. We inform the reader that the minimal polynomials of λ^∞ defined here may be the reciprocals of the minimal polynomials defined in other references.

For periodic sequences, there are a few ways to determine their linear span and minimal polynomials. One of them is given in the following lemma [Ding, Xiao and Shan (1991)][Theorem 2.2].

Lemma 1.16. *Let* λ^∞ *be a sequence of period n over* GF(q). *The* generating polynomial *of* λ^∞ *is defined by* $\Lambda^n(x) = \sum_{i=0}^{n-1} \lambda_i x^i \in$ GF(q)[x]. *Then a minimal polynomial* $m_\lambda(x)$ *of* λ^∞ *is given by*

$$m_\lambda(x) = \frac{x^n - 1}{\gcd(x^n - 1, \Lambda^n(x))}; \tag{1.12}$$

and the linear span $\mathbb{L}_\lambda$ *of* λ^∞ *is given by* $n - \deg(\gcd(x^n - 1, \Lambda^n(x)))$.

The other one is given in the following lemma [Antweiler and Bomer (1992)].

Lemma 1.17. *Any sequence* λ^∞ *over* GF(q) *of period* $q^m - 1$ *has a unique expansion of the form*

$$\lambda_t = \sum_{i=0}^{q^m-2} c_i \alpha^{it} \quad \text{for all } t \geq 0,$$

where $c_i \in$ GF(q^m). *Let the index set be* $I = \{i : c_i \neq 0\}$, *then the minimal polynomial* $\mathbb{M}_\lambda(x)$ *of* λ^∞ *is*

$$\mathbb{M}_\lambda(x) = \prod_{i \in I}(1 - \alpha^i x),$$

and the linear span of λ^∞ *is* $|I|$.

It should be noticed that in most references the reciprocal of $\mathbb{M}_\lambda(x)$ is called the minimal polynomial of the sequence λ^∞. So Lemma 1.17 is a modified version of the original one in Antweiler and Bomer (1992). We are interested in the linear span of periodic sequences, as they are useful in coding theory.

1.8.2 *Correlation functions*

Let χ be an additive character of GF(q), and λ^∞ and γ^∞ be two sequences of period respectively N and M and $P = \text{lcm}\{M, N\}$. Then the *periodic crosscorrelation function* of the two sequences is defined by

$$\text{CC}_{\lambda,\gamma}(l) = \sum_{i=0}^{P-1} \chi(\lambda_i - \gamma_{i+l}) = \sum_{i=0}^{P-1} \chi(\lambda_i)\overline{\chi(\gamma_{i+l})}. \tag{1.13}$$

If the two sequences are identical, then $P = M = N$ and the crosscorrelation function is the so-called *periodic autocorrelation function* of λ^∞ described by

$$\text{AC}_\lambda(l) = \sum_{i=0}^{N-1} \chi(\lambda_i - \lambda_{i+l}) = \sum_{i=0}^{N-1} \chi(\lambda_i)\overline{\chi(\lambda_{i+l})}. \tag{1.14}$$

If $q = 2$, then $\chi(a) = (-1)^a$ is an additive character of GF(2), here we identify GF(2) with $\mathbb{Z}_2$. Then (1.13) and (1.14) are the usual crosscorrelation and autocorrelation functions of binary sequences.

Let λ^∞ be a binary sequence of period N. It is easy to prove the following.

(1) Let $N \equiv 3 \pmod 4$. Then $\max_{1 \leq w \leq N-1} |AC_\lambda(w)| \geq 1$. On the other hand, $\max_{1 \leq w \leq N-1} |AC_\lambda(w)| = 1$ iff $AC_\lambda(w) = -1$ for all $w \not\equiv 0 \pmod N$. In this case, the sequence λ^∞ is said to have *ideal autocorrelation* and *optimal autocorrelation*.

(2) Let $N \equiv 1 \pmod 4$. There is some evidence that there is no binary sequence of period $N > 13$ with $\max_{1 \leq w \leq N-1} |AC_\lambda(w)| = 1$ [Jungnickel and Pott (1999)]. It is then natural to consider the case $\max_{1 \leq w \leq N-1} |AC_\lambda(w)| = 3$. In this case, $AC_\lambda(w) \in \{1, -3\}$ for all $w \not\equiv 0 \pmod N$.

(3) Let $N \equiv 2 \pmod 4$. Then $\max_{1 \leq w \leq N-1} |AC_\lambda(w)| \geq 2$. On the other hand, $\max_{1 \leq w \leq N-1} |AC_\lambda(w)| = 2$ iff $AC_\lambda(w) \in \{2, -2\}$ for all $w \not\equiv 0 \pmod N$. In this case, the sequence λ^∞ is said to have *optimal autocorrelation*.

(4) Let $N \equiv 0 \pmod 4$. We have clearly that $\max_{1 \leq w \leq N-1} |AC_\lambda(w)| \geq 0$. If $\max_{1 \leq w \leq N-1} |AC_\lambda(w)| = 0$, the sequence λ^∞ is called *perfect*. The only known perfect binary sequence up to equivalence is the $(0, 0, 0, 1)$. It is conjectured that there is no perfect binary sequence of period $N \equiv 0 \pmod 4$ greater than 4 [Jungnickel and Pott (1999)]. This conjecture is true for all $N < 108900$ [Jungnickel and Pott (1999)]. Hence, it is natural to construct binary sequences of period $N \equiv 0 \pmod 4$ with $\max_{1 \leq w \leq N-1} |AC_\lambda(w)| = 4$.

Binary sequences with optimal autocorrelation have close connections with certain combinatorial designs. We will get back to their connections in subsequent chapters.

Chapter 2

Linear Codes over Finite Fields

This chapter introduces the basics of linear codes over finite fields. Fundamental results on linear codes and cyclic codes will be summarized. The reader may find out proofs of the theorems and lemmas in this chapter from Huffman and Pless (2003)[Chapters 1, 2 and 4] or work out a proof for the results as an exercise.

2.1 Linear Codes

As usual, let $\mathrm{GF}(q)^n$ denote the vector space of all n-tuples over $\mathrm{GF}(q)$. The vectors in $\mathrm{GF}(q)^n$ are usually denoted by $(a_1, a_2, \ldots, a_n)$ or $(a_1 a_2 \ldots a_n)$, where $a_i \in \mathrm{GF}(q)$. The *Hamming weight* of a vector $\mathbf{a} \in \mathrm{GF}(q)^n$ is the number of nonzero coordinates of $\mathbf{a}$, and is denoted by $\mathrm{wt}(\mathbf{a})$. The *Hamming distance* of two vectors $\mathbf{a}$ and $\mathbf{b}$ in $\mathrm{GF}(q)^n$, denoted by $\mathrm{dist}(\mathbf{a}, \mathbf{b})$ is the Hamming weight of the difference vector $\mathbf{a} - \mathbf{b}$. The *inner product* of two vectors $\mathbf{a}$ and $\mathbf{b}$ in $\mathrm{GF}(q)^n$, denoted by $\mathbf{a} \cdot \mathbf{b}$, is defined by

$$\mathbf{a} \cdot \mathbf{b} = \mathbf{a}\mathbf{b}^T = \sum_{i=1}^{n} a_i b_i,$$

where $\mathbf{b}^T$ denotes the transpose of the vector $\mathbf{b}$. The two vectors $\mathbf{a}$ and $\mathbf{b}$ are *orthogonal* if $\mathbf{a} \cdot \mathbf{b} = 0$.

2.1.1 *Linear codes over* $\mathrm{GF}(q)$

An (n, M, d) *code* $\mathcal{C}$ over $\mathrm{GF}(q)$ is a subset of $\mathrm{GF}(q)^n$ of cardinality M and minimum Hamming distance d. The vectors in a code $\mathcal{C}$ are called *codewords* in $\mathcal{C}$. An $[n, \kappa]$ *linear code* over $\mathrm{GF}(q)$ is a linear subspace of $\mathrm{GF}(q)^n$ with dimension κ. If the minimum Hamming distance of $\mathcal{C}$ is d, we say that $\mathcal{C}$ has parameters $[n, \kappa, d]$. It is easily seen that the minimum Hamming distance of a linear code $\mathcal{C}$ is equal to the minimum nonzero Hamming weight of all codewords in $\mathcal{C}$. By definition, an $[n, \kappa]$ linear code over $\mathrm{GF}(q)$ has q^κ codewords. For simplicity, we sometimes call an $[n, \kappa]$ linear code $\mathcal{C}$ an $[n, \kappa]$ code.

Let A_i denote the number of codewords of weight i in an $[n, \kappa]$ linear code for all i with $0 \leq i \leq n$. The sequence $(A_0, A_1, \ldots, A_n)$ is called the *weight distribution* of $\mathcal{C}$ and the polynomial

$$A_0 + A_1 z + A_2 z^2 + \cdots + A_n z^n$$

is called the *weight enumerator* of $\mathcal{C}$.

The *dual code*, denoted $\mathcal{C}^{\perp}$, of an $[n, \kappa]$ linear code $\mathcal{C}$ over $\mathrm{GF}(q)$ is a linear subspace of $\mathrm{GF}(q)^n$ with dimension $n - \kappa$ and is defined by

$$\mathcal{C}^{\perp} = \{\mathbf{u} \in \mathrm{GF}(q)^n : \mathbf{u} \cdot \mathbf{c} = \mathbf{u}\mathbf{c}^T = 0 \text{ for all } \mathbf{c} \in \mathcal{C}\}.$$

The code $\mathcal{C}$ is said to be *self-orthogonal* if $\mathcal{C} \subseteq \mathcal{C}^{\perp}$, and *self-dual* if $\mathcal{C} = \mathcal{C}^{\perp}$. If $\mathcal{C}$ is self-dual, then the dimension of $\mathcal{C}$ is $n/2$.

A *generator matrix* of an $[n, \kappa]$ linear code over $\mathrm{GF}(q)$ is a $\kappa \times n$ matrix G whose rows form a basis for $\mathcal{C}$ over $\mathrm{GF}(q)$. The generator matrix of the dual code $\mathcal{C}^{\perp}$ is called a *parity check matrix* of $\mathcal{C}$ and is denoted by H. Hence, a linear code $\mathcal{C}$ may be described by a generator matrix or a parity check matrix as follows

$$\mathcal{C} = \{\mathbf{x} \in \mathrm{GF}(q)^n : H\mathbf{x}^T = \bar{\mathbf{0}}\},$$

where $\bar{\mathbf{0}}$ denotes the zero vector. Note that the generator matrix and the parity check matrix of a linear code are not unique.

Example 2.1. Take $q = 2$ and $n = 4$. Then the set

$$\mathcal{C} = \{(0000), (1100), (0011), (1111)\}$$

is a binary $[4, 2, 2]$ linear code with the weight enumerator $1 + 2z^2 + z^4$ and weight distribution

$$A_0 = 1, \quad A_1 = 0, \quad A_2 = 2, \quad A_3 = 0, \quad A_4 = 1.$$

A generator matrix of $\mathcal{C}$ is

$$G = \begin{bmatrix} 1 & 1 & 0 & 0 \\ 0 & 0 & 1 & 1 \end{bmatrix}.$$

This matrix G is also a parity check matrix of $\mathcal{C}$. Hence, the code $\mathcal{C}$ is self-dual.

A vector $\mathbf{x} = (x_1 x_2 \ldots x_n)$ in $\mathrm{GF}(q)^n$ is *even-like* provided that $\sum_{i=1}^{n} x_i = 0$, and is *odd-like* otherwise. The weight of binary even-like vector must be even, and that of binary odd-like vector must be odd. For an $[n, \kappa, d]$ code $\mathcal{C}$ over $\mathrm{GF}(q)$, we call the minimum weight of the even-like codewords, respectively the odd-like codewords, the *minimum even-like weight*, respectively the *minimum odd-like weight,* of the code. Denote the minimum even-like weight by d_e and the minimum odd-like weight by d_o. So $d = \min\{d_e, d_o\}$.

The following theorem can be easily proved.

Theorem 2.1. *Let C be an $[n, \kappa]$ code over $\mathrm{GF}(q)$. Let $C^{(e)}$ be the set of all even-like codewords in C. Then $C^{(e)} = C$ if C does not have odd-like codewords and $C^{(e)}$ is an $[n, \kappa - 1]$ subcode of C otherwise.*

The following theorem characterizes the minimum weight of linear codes and its proof is trivial.

Theorem 2.2. *A linear code has minimum weight d if and only if its parity check matrix has a set of d linearly independent columns but no set of $d - 1$ linearly independent columns.*

Let C be an $[n, \kappa, d]$ code over $\mathrm{GF}(q)$. The *extended code* $\widehat{C}$ of C is defined by

$$\widehat{C} = \left\{ (x_1 x_2 \ldots x_{n+1}) \in \mathrm{GF}(q)^{n+1} : (x_1 x_2 \ldots x_n) \in C \text{ with } \sum_{i=1}^{n+1} x_i = 0 \right\}.$$

It is an easy exercise to prove that $\widehat{C}$ is an $[n + 1, \kappa, \widehat{d}]$ linear code, where $\widehat{d} = d$ or $d + 1$.

When C is an $[n, \kappa, d]$ binary code, the extended code $\widehat{C}$ has only even weight codewords and is an $[n + 1, \kappa, \widehat{d}]$ code, where $\widehat{d}$ equals d if d is even and equals $d + 1$ if d is odd. In the nonbinary case, it is not obvious whether $\widehat{d} = d$ or $\widehat{d} = d + 1$. Recall that d_e and d_o denote the minimum even-like weight and the minimum odd-like weight of C, respectively. If $d_e \leq d_o$, then $\widehat{d} = d_e$. If $d_o < d_e$, then $\widehat{d} = d_o + 1$.

2.1.2 *Equivalences of linear codes*

Two linear codes C_1 and C_2 are *permutation equivalent* if there is a permutation of coordinates which sends C_1 to C_2. If C_1 and C_2 are permutation equivalent, so are $C_1^\perp$ and $C_2^\perp$. Two permutation equivalent linear codes have the same dimension and weight distribution.

The set of coordinate permutations that map a code C to itself forms a group, which is referred to as the *permutation automorphism group* of C and denoted by $\mathrm{PAut}(C)$. If C is a code of length n, then $\mathrm{PAut}(C)$ is a subgroup of the *symmetric group* Sym_n.

A subgroup L of the symmetric group Sym_n is *transitive,* provided that for every ordered pair (i, j), where $1 \leq i, j \leq n$, there is a permutation $\ell \in L$ such that $\ell(i) = j$. When the group $\text{PAut}(\widehat{C})$ is transitive, we have information on the minimum weight of C. A proof of the following theorem can be found in Huffman and Pless (2003)[Theorem 1.6.6].

Theorem 2.3. *Suppose that the group* $\text{PAut}(\widehat{C})$ *is transitive. Then the minimum weight d of C is its minimum odd-like weight* d_o*. Furthermore, every minimum weight codeword of C is odd-like.*

A *monomial matrix* over $\text{GF}(q)$ is a square matrix having exactly one nonzero element of $\text{GF}(q)$ in each row and column. A monomial matrix M can be written either in the form DP or the form PD_1, where D and D_1 are diagonal matrices and P is a permutation matrix.

Let C_1 and C_2 be two linear codes of the same length over $\text{GF}(q)$. Then C_1 and C_2 are *monomially equivalent* if there is a monomial matrix over $\text{GF}(q)$ such that $C_2 = C_1 M$. Monomial equivalence and permutation equivalence are precisely the same for binary codes. If C_1 and C_2 are monomially equivalent, then they have the same weight distribution.

The set of monomial matrices that map C to itself forms the group $\text{MAut}(C)$, which is called the *monomial automorphism group* of C. Clearly, we have $\text{PAut}(C) \subseteq \text{MAut}(C)$. The following theorem will be useful [Huffman and Pless (2003)][Theorem 1.7.13].

Theorem 2.4. *Suppose that the group* $\text{MAut}(\widehat{C})$ *is transitive. Then the minimum weight d of C is its minimum odd-like weight* d_o*. Furthermore, every minimum weight codeword of C is odd-like.*

A more general kind of equivalence of linear codes is treated in Huffman and Pless (2003)[Section 1.7], where more information on equivalences of codes are available. In this monograph, we are interested only in the monomial equivalence as they are the only weight reserving linear transformation on linear codes [Huffman and Pless (2003)][Section 7.9].

2.1.3 *Hamming and simplex codes*

A parity check matrix $H_{q,m}$ of the *Hamming code* $\mathcal{H}_{q,m}$ over $\text{GF}(q)$ is defined by choosing for its columns a nonzero vector from each one-dimensional subspace of $\text{GF}(q)^m$. In terms of finite geometry, the columns of $H_{q,m}$ are the points of the projective geometry $\text{PG}(m - 1, \text{GF}(q))$. Hence $\mathcal{H}_{q,m}$ has length $n = (q^m - 1)/(q - 1)$ and dimension $n - m$. Note that no two columns of $H_{q,m}$ are linearly dependent over $\text{GF}(q)$. The minimum weight of $\mathcal{H}_{q,m}$ is at least 3. Adding two nonzero vectors from two different one-dimensional subspaces gives a nonzero vector from a third one-dimensional space. Therefore, $\mathcal{H}_{q,m}$ has minimum weight 3.

Example 2.2. The Hamming code $\mathcal{H}_{3,3}$ has parameters [13, 10, 3] and generator matrix

$$
G = \begin{bmatrix}
1 & 0 & 0 & 0 & 0 & 0 & 0 & 0 & 0 & 0 & 2 & 1 & 1 \\
0 & 1 & 0 & 0 & 0 & 0 & 0 & 0 & 0 & 0 & 1 & 1 & 0 \\
0 & 0 & 1 & 0 & 0 & 0 & 0 & 0 & 0 & 0 & 0 & 1 & 1 \\
0 & 0 & 0 & 1 & 0 & 0 & 0 & 0 & 0 & 0 & 1 & 2 & 0 \\
0 & 0 & 0 & 0 & 1 & 0 & 0 & 0 & 0 & 0 & 0 & 1 & 2 \\
0 & 0 & 0 & 0 & 0 & 1 & 0 & 0 & 0 & 0 & 2 & 1 & 2 \\
0 & 0 & 0 & 0 & 0 & 0 & 1 & 0 & 0 & 0 & 2 & 0 & 2 \\
0 & 0 & 0 & 0 & 0 & 0 & 0 & 1 & 0 & 0 & 2 & 0 & 1 \\
0 & 0 & 0 & 0 & 0 & 0 & 0 & 0 & 1 & 0 & 1 & 1 & 2 \\
0 & 0 & 0 & 0 & 0 & 0 & 0 & 0 & 0 & 1 & 2 & 2 & 2
\end{bmatrix}. \tag{2.1}
$$

The following theorem is interesting, as it implies that all $[(q^m - 1)/(q - 1), (q^m - 1)/(q - 1) - m, 3]$ codes over GF(q) have the same weight distribution [Huffman and Pless (2003)][Theorem 1.8.2].

Theorem 2.5. *Any* $[(q^m - 1)/(q - 1), (q^m - 1)/(q - 1) - m, 3]$ *code over* GF(q) *is monomially equivalent to the Hamming code* $\mathcal{H}_{q,m}$.

The duals of the Hamming codes $\mathcal{H}_{q,m}$ are called *Simplex codes*, which have parameters $[(q^m - 1)/(q - 1), m, q^{m-1}]$. In fact, we have the following conclusion [Huffman and Pless (2003)][Theorem 1.8.3].

Theorem 2.6. *The nonzero codewords of the* $[(q^m - 1)/(q - 1), m, q^{m-1}]$ *Simplex codes all have weight* q^{m-1}.

Example 2.3. Let $q = 3$ and $m = 3$. Then the Simplex code has parameters [13, 3, 9] and generator matrix

$$
G = \begin{bmatrix}
1 & 0 & 0 & 1 & 2 & 0 & 2 & 0 & 1 & 1 & 1 & 2 & 1 \\
0 & 1 & 0 & 2 & 2 & 2 & 1 & 2 & 2 & 0 & 0 & 2 & 1 \\
0 & 0 & 1 & 2 & 0 & 2 & 0 & 1 & 1 & 1 & 2 & 1 & 1
\end{bmatrix}. \tag{2.2}
$$

2.1.4 Subfield subcodes

Let $\mathcal{C}$ be an $[n, \kappa]$ code over GF(q^t). The *subfield subcode* $\mathcal{C}|_{\text{GF}(q)}$ of $\mathcal{C}$ with respect to GF(q) is the set of codewords in $\mathcal{C}$ each of whose components is in GF(q). Since $\mathcal{C}$ is linear over GF(q^t), $\mathcal{C}|_{\text{GF}(q)}$ is a linear code over GF(q).

The dimension, denoted κ_q, of the subfield subcode $\mathcal{C}|_{GF(q)}$ may not have an elementary relation with that of the code $\mathcal{C}$. However, we have the following lower and upper bounds on κ_q [Huffman and Pless (2003)][Theorems 3.8.3 and 3.8.4].

Theorem 2.7. *Let $\mathcal{C}$ be an $[n, \kappa]$ code over $GF(q^t)$. Then $\mathcal{C}|_{GF(q)}$ is an $[n, \kappa_q]$ code over $GF(q)$, where $\kappa \geq \kappa_q \geq n - t(n - \kappa)$. If $\mathcal{C}$ has a basis of codewords in $GF(q)^n$, then this is also a basis of $\mathcal{C}|_{GF(q)}$ and $\mathcal{C}|_{GF(q)}$ has dimension κ.*

Example 2.4. The Hamming code $\mathcal{H}_{2^2,3}$ over $GF(2^2)$ has parameters $[21, 18, 3]$. The subfield subcode $\mathcal{H}_{2^3,3}|_{GF(2)}$ is a binary $[21, 16, 3]$ with parity check matrix

$$H = \begin{bmatrix} 1 & 0 & 0 & 1 & 1 & 0 & 0 & 1 & 1 & 0 & 0 & 1 & 1 & 1 & 1 & 0 & 0 & 1 & 1 & 0 & 1 \\ 0 & 1 & 0 & 0 & 1 & 0 & 1 & 1 & 0 & 0 & 1 & 1 & 0 & 1 & 0 & 0 & 1 & 1 & 0 & 0 & 1 \\ 0 & 0 & 1 & 1 & 0 & 0 & 1 & 1 & 0 & 0 & 1 & 1 & 0 & 0 & 1 & 1 & 0 & 0 & 1 & 1 & 0 \\ 0 & 0 & 0 & 0 & 0 & 1 & 1 & 1 & 1 & 0 & 0 & 0 & 0 & 0 & 0 & 0 & 0 & 1 & 1 & 1 & 1 \\ 0 & 0 & 0 & 0 & 0 & 0 & 0 & 0 & 1 & 1 & 1 & 1 & 1 & 1 & 1 & 1 & 1 & 0 & 0 & 0 & 0 \end{bmatrix}.$$

In this case, $n = 21$, $\kappa = 18$, and $n - t(n - \kappa) = 15$. Hence $\kappa_q = 16$, which is very close to $n - t(n - \kappa) = 15$.

The *trace* of a vector $\mathbf{c} = (c_1, c_2, \ldots, c_n) \in GF(q^t)^n$ is defined by

$$\text{Tr}_{q^t/q}(\mathbf{c}) = \left(\text{Tr}_{q^t/q}(c_1), \text{Tr}_{q^t/q}(c_2), \ldots, \text{Tr}_{q^t/q}(c_n) \right).$$

The *trace code* of a linear code $\mathcal{C}$ of length n over $GF(q^t)$ is defined by

$$\text{Tr}_{q^t/q}(\mathcal{C}) = \left\{ \text{Tr}_{q^t/q}(\mathbf{c}) : \mathbf{c} \in \mathcal{C} \right\}, \tag{2.3}$$

which is a linear code of length n over $GF(q)$.

The following is called *Delsarte's Theorem*, which exhibits a dual relation between subfield subcodes and trace codes.

Theorem 2.8 (Delsarte's Theorem). *Let $\mathcal{C}$ be a linear code of length n over $GF(q^t)$. Then*

$$(\mathcal{C}|_{GF(q)})^{\perp} = \text{Tr}_{q^t/q}(\mathcal{C}^{\perp}).$$

This theorem is very useful in the design and analysis of linear codes. We will frequently get back to it later.

2.1.5 *Reed–Muller codes*

In this subsection, we briefly introduce binary Reed–Muller codes. A function from $GF(2)^m$ to $GF(2)$ is called a *Boolean function*. Each Boolean function, $f(x) = f(x_1, \ldots, x_m)$, defines a vector $(f(x))_{x \in GF(2)^m} \in GF(2)^{2^m}$, where x ranges over all elements in $GF(2)^m$ in some fixed order. This vector is called the *truth table* of f.

It is easy to prove that any function f from $GF(2)^m$ to $GF(2)$ can be expressed as a linear combination of the following 2^m basis functions:

Basis functions	degree	total number
1	0	$\binom{m}{0}$
$x_i,\ 1 \le i \le m$	1	$\binom{m}{1}$
$x_i x_j,\ 1 \le i < j \le m$	2	$\binom{m}{2}$
$\vdots$	$\vdots$	$\vdots$
$x_{i_1} \cdots x_{i_r},\ 1 \le i_1 < \cdots < i_r \le m$	r	$\binom{m}{r}$
$\vdots$	$\vdots$	$\vdots$
$x_1 x_2 \cdots x_m$	m	$\binom{m}{m}$

The truth tables of these basis functions form a basis of $GF(2)^{2^m}$ over $GF(2)$. The binary Reed–Muller code, denoted by $\mathcal{R}_2(r, m)$, is spanned by the true tables of all the basis functions with degree at most r, where $1 \le r \le m$. Hence $\mathcal{R}$ has dimension

$$\kappa = \sum_{i=0}^{r} \binom{m}{i}.$$

It is known that $\mathcal{R}_2(r, m)$ has minimum weight 2^{m-r} [MacWilliams and Sloane (1977)][Chapter 13].

The parameters of the duals of the Reed–Muller codes are given as follows.

Theorem 2.9. $\mathcal{R}_2(r, m)^{\perp} = \mathcal{R}_2(m - r, m)$.

The error-correcting capability of Reed–Muller codes is not good. But these codes can be decoded efficiently with a majority logic decoding technique. Detailed information on $\mathcal{R}_2(r, m)$ can be found in MacWilliams and Sloane (1977)[Chapters 13–15]. Some linear codes from designs are related to Reed–Muller codes. We will get back to the codes $\mathcal{R}_2(r, m)$ later.

2.2 Bounds on the Size of Linear Codes

The purpose of this section is to collect bounds on the size of both nonlinear and linear codes together. We will not provide a proof for the bounds, but will refer the reader to a reference where a proof can be found.

Recall that an (n, M, d) code C over $GF(q)$ is a code of length n with M codewords whose minimum distance is d. The code C could be either linear or nonlinear. If C is linear, it is an $[n, \kappa, d]$ code over $GF(q)$, where $\kappa = \log_q M$ and d is equal to the minimum weight of C. Let $B_q(n, d)$ (respectively, $A_q(n, d)$) denote the largest number of codewords in a linear (respectively, arbitrary (linear or nonlinear)) code over $GF(q)$ of length n and minimum distance at least d. A code C of length n over $GF(q)$ and minimum distance at least d will be called optimal if it has $A_q(n, d)$ codewords (or $B_q(n, d)$ codewords in the case that C is linear).

The following is a list of targets in the construction of error correcting codes over $GF(q)$.

(1) Given q, n and d, we want to find a code C over $GF(q)$ of length n and minimum distance d with the maximum number of codewords.
(2) Given q, n and M, we wish to find a code C over $GF(q)$ of length n and size M with the largest minimum distance.
(3) Given q, d and M, we wish to find a code C over $GF(q)$ of minimum distance d and size M with the shortest length n.

However, there are constraints on the parameters n, d, q and M. Such constraints define bounds on the parameters of codes.

By definition, we have obviously that $B_q(n, d) \leq A_q(n, d)$ and $B_q(n, d)$ is a nonnegative power of q. The following theorem summarizes basic properties of $B_q(n, d)$ and $A_q(n, d)$ [Huffman and Pless (2003)][Section 2.1].

Theorem 2.10. *Let $d > 1$. Then we have the following.*

(a) $A_q(n, d) \leq A_q(n - 1, d - 1)$ *and* $B_q(n, d) \leq B_q(n - 1, d - 1)$.
(b) *If d is even,* $A_2(n, d) = A_2(n - 1, d - 1)$ *and* $B_2(n, d) = B_2(n - 1, d - 1)$.
(c) *If d is even and $M = A_2(n, d)$, then there exists a binary (n, M, d) code such that all codewords have even weight and the distance between all pairs of codewords is also even.*
(d) $A_q(n, d) \leq q A_q(n - 1, d)$ *and* $B_q(n, d) \leq q B_q(n - 1, d)$.
(e) *If d is even,* $A_2(2d, d) \leq 4d$.
(f) *If d is odd,* $A_2(2d, d) \leq 2d + 2$.
(g) *If d is odd,* $A_2(2d + 1, d) \leq 4d + 4$.

We are now ready to introduce some bounds on the parameters of codes. The following is the Plotkin Bound [Plotkin (1960)], which is useful only when d is close to n.

Theorem 2.11 (Plotkin Bound). *Suppose that* $rn < d$, *where* $r = 1 - q^{-1}$. *Then*

$$A_q(n, d) \leq \left\lfloor \frac{d}{d - rn} \right\rfloor.$$

In the binary case,

$$A_2(n, d) \leq 2 \left\lfloor \frac{d}{2d - n} \right\rfloor$$

provided that $n < 2d$.

The following theorem describes the Singleton Bound [Singleton (1964)], which is simple in format.

Theorem 2.12 (Singleton Bound). *Let* $d \leq n$. *Then*

$$A_q(n, d) \leq q^{n-d+1}.$$

In particular, for any $[n, \kappa, d]$ *linear code over* $\mathrm{GF}(q)$, *we have* $\kappa \leq n - d + 1$.

A code meeting the Singleton Bound is called *maximum distance separable* (MDS). If $\mathcal{C}$ is an MDS linear code, so is $\mathcal{C}^{\perp}$. The weight distribution of MDS codes is given by the following theorem.

Theorem 2.13. *Let* $\mathcal{C}$ *be an* $[n, \kappa]$ *code over* $\mathrm{GF}(q)$ *with* $d = n - \kappa + 1$, *and let the weight enumerator of* $\mathcal{C}$ *be* $1 + \sum_{i=d}^{n} A_i z^i$. *Then*

$$A_i = \binom{n}{i}(q - 1) \sum_{j=0}^{i-d} (-1)^j \binom{i - 1}{j} q^{i-j-d} \text{ for all } d \leq i \leq n.$$

For linear codes over finite fields, we have the following Griesmer Bound [Griesmer (1960)], which is a generalization of the Singleton bound.

Theorem 2.14 (Griesmer Bound). *Let* $\mathcal{C}$ *be an* $[n, \kappa, d]$ *linear code over* $\mathrm{GF}(q)$ *with* $\kappa \geq 1$. *Then*

$$n \geq \sum_{i=0}^{\kappa-1} \left\lceil \frac{d}{q^i} \right\rceil.$$

For certain parameters n, κ, d and q, there may not exist an $[n, \kappa, d]$ linear code over $\mathrm{GF}(q)$ meeting the Griesmer Bound. However, for some other parameters, the Griesmer bound may be achievable. It is quite interesting to construct linear codes meeting the Griesmer Bound. There are a lot of references on this problem. It is clear that the Simplex codes meet the Griesmer Bound.

Binary linear codes over finite fields meeting the Griesmer Bound have the following nice property [Tilborg (1980)].

Theorem 2.15. *Let $\mathcal{C}$ be an $[n, \kappa, d]$ binary code meeting the Griesmer Bound. Then $\mathcal{C}$ has a basis of minimum weight codewords.*

The following is the *sphere packing bound*, also called the *Hamming bound*.

Theorem 2.16 (Sphere Packing Bound).

$$B_q(n, d) \leq A_q(n, d) \leq \frac{q^n}{\sum_{i=0}^{\lfloor \frac{d-1}{2} \rfloor} \binom{n}{i}(q-1)^i}. \qquad (2.4)$$

A code meeting the Sphere Packing Bound is called *perfect*. It is straightforward to verify that the Hamming code $\mathcal{H}_{q,m}$ is perfect.

For a code $\mathcal{C}$ over $\mathrm{GF}(q)$ with minimum distance d, we define its *covering radius* by

$$\rho(\mathcal{C}) = \max_{x \in \mathrm{GF}(q)^n} \min_{\mathbf{c} \in \mathcal{C}} \mathrm{dist}(\mathbf{x}, \mathbf{c}).$$

It is easily seen that

$$\left\lfloor \frac{d-1}{2} \right\rfloor \leq \rho(\mathcal{C})$$

where the equality holds if and only if $\mathcal{C}$ is perfect.

The Gilbert Bound is given in the next theorem [Gilbert (1952)].

Theorem 2.17 (Gilbert Bound).

$$A_q(n, d) \geq B_q(n, d) \geq \frac{q^n}{\sum_{i=0}^{d-1} \binom{n}{i}(q-1)^i}.$$

The Varshamov Bound below is similar to the Gilbert Bound [Varshamov (1957)].

Theorem 2.18 (Varshamov Bound).

$$A_q(n, d) \geq B_q(n, d) \geq q^{n - \left\lceil \log_q \left(1 + \sum_{i=0}^{d-2} \binom{n-1}{i}(q-1)^i \right) \right\rceil}.$$

Our last bound is the Elias Bound documented in the following theorem.

Theorem 2.19 (Elias Bound). *Let $r = 1 - q^{-1}$. Suppose that $w \leq rn$ and $w^2 - 2rnw + rnd > 0$. Then*

$$A_q(n, d) \leq \frac{rnd}{w^2 - 2rnw + rnd} \cdot \frac{q^n}{\sum_{i=0}^{w} \binom{n}{i}(q-1)^i}.$$

In many cases, the linear programming bounds on codes are better than those described before. There are also a number of asymptotic bounds. The reader is referred to Huffman and Pless (2003)[Sections 2.6 and 2.10] for details.

2.3 Bounds on the Size of Constant Weight Codes

A (nonlinear) (n, M, d) code $\mathcal{C}$ over GF(q) is a *constant weight code* if every codeword has the same weight. For example, the codewords of a fixed weight w in a linear code form a constant weight code. The following theorem is easily proved.

Theorem 2.20. *If $\mathcal{C}$ is a constant weight (n, M, d) code with codewords of weight $w > 1$, then $d \leq 2w$.*

Let $A_q(n, d, w)$ denote the maximum number of codewords in a constant weight (n, M) code over GF(q) of length n and minimum distance at least d whose codewords have weight w. Obviously $A_q(n, d, w) \leq A_q(n, d)$. The following is a list of properties of $A_q(n, d, w)$ collected from the literature:

(a) $A_3(n, 3, 3) = \frac{2n(n-1)}{3}$ for $n \equiv 0, 1 \pmod 3$, $n \geq 4$.

(b) $A_3(n, 3, 3) = \frac{2n(n-1)-4}{3}$ for $n \equiv 2 \pmod 3$, $n \geq 5$.

(c) $A_3(n, 3, 4) \geq \lfloor \frac{n^3 - 5n^2 + 6n}{3} \rfloor$, if $n \geq 4$.

(d) $A_3(n, 3, w) \geq \frac{1}{2n+1}\binom{n}{w}2^w$.

(e) $A_3(n, 3, w) \geq \frac{1}{2n}\binom{n}{w}2^w$, if $n \equiv 0, 1 \pmod 4$.

(f) $A_3(2^r - 1, 3, 2^r - 2) = (2^r - 1)2^{2^r - r - 2}$ for $r \geq 2$.

(g) $A_3(2^r - 2, 3, 2^r - 3) = (2^{r-1} - 1)2^{2^r - r - 2}$ for $r \geq 2$.

(h) $A_3(2^r, 3, 2^r - 1) = 2^{2^r - 1}$ for $r \geq 2$.

(i) $A_3\left(q, \frac{q+3}{2}, q - 1\right) = q$, where q is a power of odd prime.

(j) $A_3\left(q\frac{q^m - 1}{q - 1}, q^{m-1}\frac{q+3}{2}, q^m - 1\right) = q^m$, where q is a power of odd prime.

(k) $A_3\left(p^m + 1, \frac{p^m + 3}{2}, p^m\right) = 2p^m + 2$, where $p \geq 3$ is a prime.

(l) $A_q(n, 2, w) = \binom{n}{w}(q - 1)^w$.

(m) $A_q\left(\frac{q^m - 1}{q - 1}, q^{m-1}, q^{m-1}\right) = q^m - 1$, where q is a prime power.

(n) $A_q\left(q + 1, \frac{q+1}{2}, q\right) \geq 2q + 2$, where $q = p^m$.

(o) $A_q(n, d, w) = 1$ if $d > 2w$ and $0 \leq w \leq n$.

(p) $A_q(n, 2w, w) = \lfloor \frac{n}{w} \rfloor$.

(q) $A_q(n, d, n) = A_{q-1}(n, d)$.

The following is the Restricted Johnson Bound for $A_q(n, d, w)$ Huffman and Pless (2003)[Theorem 2.3.4].

Theorem 2.21 (Restricted Johnson Bound for $A_q(n, d, w)$).

$$A_q(n, d, w) \leq \left\lfloor \frac{n(q-1)d}{qw^2 - 2(q-1)nw + n(q-1)d} \right\rfloor,$$

provided that $qw^2 - 2(q-1)nw + n(q-1)d > 0$.

This bound is restricted because of the condition $qw^2 - 2(q-1)nw + n(q-1)d > 0$. The following is another bound without such a condition [Huffman and Pless (2003)][Theorem 2.3.6].

Theorem 2.22 (Unrestricted Johnson Bound for $A_q(n, d, w)$). *If* $2w \geq d$ *and* $d \in \{2e - 1, 2e\}$, *then*

$$A_q(n, d, w) \leq \left\lfloor \frac{n\hat{q}}{w} \left\lfloor \frac{(n-1)\hat{q}}{w-1} \left\lfloor \cdots \left\lfloor \frac{(n-w+e)\hat{q}}{w-1} \right\rfloor \cdots \right\rfloor \right\rfloor \right\rfloor,$$

where $\hat{q} = q - 1$.

2.4 Cyclic Codes Over GF(q)

An $[n, \kappa]$ code $\mathcal{C}$ over GF(q) is called *cyclic* if $\mathbf{c} = (c_0, c_1, \ldots, c_{n-1}) \in \mathcal{C}$ implies $(c_{n-1}, c_0, c_1, \ldots, c_{n-2}) \in \mathcal{C}$.

As a subclass of linear codes, cyclic codes have applications in consumer electronics, data storage systems, and communication systems as they have efficient encoding and decoding algorithms. In this section, we introduce the basic theory of cyclic codes over finite fields without providing proofs. We refer the reader to Huffman and Pless (2003)[Chapter 4] for proofs.

2.4.1 *Factorization of $x^n - 1$ over* GF(q)

To deal with cyclic codes of length n over GF(q), we have to study the canonical factorization of $x^n - 1$ over GF(q). To this end, we need to introduce q-cyclotomic cosets modulo n. Note that $x^n - 1$ has no repeated factors over GF(q) if and only if $\gcd(n, q) = 1$. Throughout Section 2.4, we assume that $\gcd(n, q) = 1$.

Recall that $\mathbb{Z}_n$ denotes the set $\{0, 1, 2, \ldots, n - 1\}$. Let s be an integer with $0 \leq s < n$. The *q-cyclotomic coset of s modulo n* is defined by

$$C_s = \{s, sq, sq^2, \ldots, sq^{\ell_s - 1}\} \bmod n \subseteq \mathbb{Z}_n,$$

where ℓ_s is the smallest positive integer such that $s \equiv sq^{\ell_s} \pmod{n}$, and is the size of the q-cyclotomic coset. The smallest integer in C_s is called the *coset leader* of C_s. Let $\Gamma_{(n,q)}$ be the set of all the coset leaders. We have then $C_s \cap C_t = \emptyset$ for any two distinct elements s and t in $\Gamma_{(n,q)}$, and

$$\bigcup_{s \in \Gamma_{(n,q)}} C_s = \mathbb{Z}_n. \tag{2.5}$$

Hence, the distinct q-cyclotomic cosets modulo n partition $\mathbb{Z}_n$.

Let $m = \mathrm{ord}_n(q)$, and let α be a generator of $\mathrm{GF}(q^m)^*$. Put $\beta = \alpha^{(q^m-1)/n}$. Then β is a primitive nth root of unity in $\mathrm{GF}(q^m)$. In Section 1.3.1, we defined the minimal polynomial $\mathbb{M}_{\beta^s}(x)$ of β^s over $\mathrm{GF}(q)$. It is now straightforward to prove that this polynomial is given by

$$\mathbb{M}_{\beta^s}(x) = \prod_{i \in C_s} (x - \beta^i) \in \mathrm{GF}(q)[x], \tag{2.6}$$

which is irreducible over $\mathrm{GF}(q)$. It then follows from (2.5) that

$$x^n - 1 = \prod_{s \in \Gamma_{(n,q)}} \mathbb{M}_{\beta^s}(x) \tag{2.7}$$

which is the factorization of $x^n - 1$ into irreducible factors over $\mathrm{GF}(q)$. This canonical factorization of $x^n - 1$ over $\mathrm{GF}(q)$ is crucial for the study of cyclic codes.

Example 2.5. Let $q = 3$ and $n = 11$. Then $\mathrm{ord}_{11}(3) = 5$. It is easily checked that $\Gamma_{(11,3)} = \{0, 1, 2\}$ and

$$C_0 = \{0\}, \quad C_1 = \{1, 3, 4, 5, 9\}, \quad C_2 = \{2, 6, 7, 8, 10\}.$$

Let α be a generator of $\mathrm{GF}(3^{11})^*$ with $\alpha^5 + 2\alpha + 1 = 0$. Then $\beta = \alpha^{22}$ and

$$\mathbb{M}_{\beta^0}(x) = x + 2,$$
$$\mathbb{M}_{\beta^1}(x) = x^5 + x^4 + 2x^3 + x^2 + 2,$$
$$\mathbb{M}_{\beta^2}(x) = x^5 + 2x^3 + x^2 + 2x + 2.$$

The following result will be useful and is not hard to prove [Huffman and Pless (2003)][Theorem 4.1.4].

Theorem 2.23. *The size ℓ_s of each q-cyclotomic coset C_s is a divisor of $\mathrm{ord}_n(q)$, which is the size ℓ_1 of C_1.*

2.4.2 *Generator and parity check polynomials*

Recall that an $[n, \kappa]$ code $\mathcal{C}$ over $\mathrm{GF}(q)$ is called *cyclic* if $\mathbf{c} = (c_0, c_1, \ldots, c_{n-1}) \in \mathcal{C}$ implies $(c_{n-1}, c_0, c_1, \ldots, c_{n-2}) \in \mathcal{C}$. Put

$$\mathcal{R}_{(n,q)} = \mathrm{GF}(q)[x]/(x^n - 1),$$

which is the residue class ring. By identifying any vector $(c_0, c_1, \ldots, c_{n-1}) \in \mathrm{GF}(q)^n$ with

$$c_0 + c_1 x + c_2 x^2 + \cdots + c_{n-1} x^{n-1} \in \mathcal{R}_{(n,q)},$$

any code $\mathcal{C}$ of length n over $\mathrm{GF}(q)$ corresponds to a subset of the residue class ring $\mathcal{R}_{(n,q)}$. One can easily prove that the linear code $\mathcal{C}$ is cyclic if and only if the corresponding subset in $\mathcal{R}_{(n,q)}$ is an ideal of the ring $\mathcal{R}_{(n,q)}$. We identify the cyclic code $\mathcal{C}$ with the corresponding subset in $\mathcal{R}_{(n,q)}$.

It is well-known that $\mathcal{R}_{(n,q)}$ is a principal ideal ring. Hence, for every cyclic code $\mathcal{C}$ of length n over $\mathrm{GF}(q)$, there is a unique monic polynomial $g(x) \in \mathrm{GF}(q)[x]$ of the smallest degree such that $\mathcal{C} = \langle g(x) \rangle$. This polynomial $g(x)$ must be a divisor of $x^n - 1$, and is called the *generator polynomial* of $\mathcal{C}$. Therefore, there is a one-to-one correspondence between the set of all cyclic codes of length n over $\mathrm{GF}(q)$ and the set of all monic divisors of $x^n - 1$ with degree at least one over $\mathrm{GF}(q)$. Hence, the total number of cyclic codes of length n over $\mathrm{GF}(q)$ is $2^t - 1$, where t is the total number of distinct q-cyclotomic cosets modulo n, i.e., $t = |\Gamma_{(n,q)}|$. We usually do not consider the zero code $\{\bar{\mathbf{0}}\}$ of length n over $\mathrm{GF}(q)$.

Let $\mathcal{C}$ be an $[n, \kappa]$ cyclic code over $\mathrm{GF}(q)$ with generator polynomial $g(x)$. By definition, $\kappa = n - \deg(g(x))$ and $\{g(x), xg(x), \ldots, x^{\kappa-1}g(x)\}$ is a basis for $\mathcal{C}$. Let $g(x) = \sum_{i=0}^{n-\kappa} g_i x^i$, where $g_{n-\kappa} = 1$. Then the following is a generator matrix of $\mathcal{C}$:

$$G = \begin{bmatrix} g_0 & g_1 & g_2 & \cdots & g_{n-\kappa} & 0 & \cdots & 0 & 0 \\ 0 & g_0 & g_1 & g_2 & \cdots & g_{n-\kappa} & \cdots & 0 & 0 \\ \vdots & \vdots & \vdots & \vdots & \vdots & \vdots & \vdots & \vdots & \vdots \\ 0 & 0 & \cdots & 0 & g_0 & g_1 & g_2 & \cdots & g_{n-\kappa} \end{bmatrix}. \tag{2.8}$$

Let $\mathcal{C}$ be an $[n, \kappa]$ cyclic code over $\mathrm{GF}(q)$ with generator polynomial $g(x)$. Let $h(x) = (x^n - 1)/g(x) = \sum_{i=0}^{\kappa} h_i x^i$. The polynomial $h(x)$ is called the *parity check polynomial* of $\mathcal{C}$. It is straightforward to verify that the dual $\mathcal{C}^\perp$ of $\mathcal{C}$ is also cyclic, and the generator polynomial $g^\perp(x) = x^\kappa h(x^{-1})/h(0)$. Furthermore, a generator matrix for $C^\perp$, and hence a parity check matrix for C is

$$
\begin{bmatrix}
h_\kappa & h_{\kappa-1} & h_{\kappa-2} & \cdots & h_0 & 0 & \cdots & 0 & 0 \\
0 & h_\kappa & h_{\kappa-1} & h_{\kappa-2} & \cdots & h_0 & \cdots & 0 & 0 \\
\vdots & \vdots & \vdots & \vdots & \vdots & \vdots & \vdots & \vdots & \vdots \\
0 & 0 & \cdots & 0 & h_\kappa & h_{\kappa-1} & h_{\kappa-2} & \cdots & h_0
\end{bmatrix}. \tag{2.9}
$$

It should be noticed that the cyclic code generated by $h(x)$ is in general different from $C^\perp$, but has the same parameters and weight distribution as $C^\perp$.

Example 2.6. Let $q = 3$ and $n = 11$. The cyclic code of length n over GF(q) with generator polynomial $g(x) = x^5 + x^4 + 2x^3 + x^2 + 2$ has parameters $[11, 6, 5]$ and parity check polynomial $h(x) = x^6 + 2x^5 + 2x^4 + 2x^3 + x^2 + 1$. Its dual code has generator polynomial $x^6 + x^4 + 2x^3 + 2x^2 + 2x + 1$.

The conclusions in the following two theorems are straightforward.

Theorem 2.24. *If C is a cyclic code over* GF(q^t), *then $C|_{\text{GF}(q)}$ is also cyclic.*

Theorem 2.25. *Let C_1 and C_2 be two cyclic codes with generator polynomials $g_1(x)$ and $g_2(x)$, respectively. Then $C_1 \subseteq C_2$ if and only if g_2 divides $g_1(x)$.*

2.4.3 *Idempotents of cyclic codes*

An element e in a ring $\mathcal{R}$ is called an *idempotent* if $e^2 \triangleq e$. The ring $\mathcal{R}_{(n,q)}$ has in general quite a number of idempotents. Every cyclic code C over GF(q) can be produced with its generator polynomial. In fact, many polynomials can generate C. Let C be a cyclic code over GF(q) with generator polynomial $g(x)$. It can be proved that a polynomial $f(x) \in$ GF$(q)[x]$ generates C if and only if $\gcd(f(x), x^n - 1) = g(x)$.

If an idempotent $e(x) \in$ GF$(q)[x]$ generates a cyclic code C, it is then unique and called the *generating idempotent*. Given the generator polynomial of a cyclic code, one can compute its generating idempotent with the following theorem [Huffman and Pless (2003)][Theorem 4.3.3].

Theorem 2.26. *Let C be a cyclic code of length n over* GF(q) *with generator polynomial $g(x)$. Let $h(x) = (x^n - 1)/g(x)$. Then $\gcd(g(x), h(x)) = 1$. Employing the Extended Euclidean Algorithm, one can compute two polynomials $a(x) \in$ GF$(q)[x]$ and $b(x) \in$ GF(q) such that $1 = a(x)g(x) + b(x)h(x)$. Then $e(x) = a(x)g(x) \bmod (x^n - 1)$ is the generating idempotent of C.*

Given the generating idempotent of a cyclic code, one can obtain the generator polynomial of this code as follows [Huffman and Pless (2003)][Theorem 4.3.3].

Theorem 2.27. *Let C be a cyclic code over* GF(q) *with generating idempotent* $e(x)$. *Then the generator polynomial of C is given by* $g(x) = \gcd(e(x), x^n - 1)$ *computed in* GF$(q)[x]$.

Example 2.7. Let $q = 3$ and $n = 11$. The cyclic code C of length n over GF(q) with generator polynomial $g(x) = x^5 + x^4 + 2x^3 + x^2 + 2$ has parameters $[11, 6, 5]$ and parity check polynomial $h(x) = x^6 + 2x^5 + 2x^4 + 2x^3 + x^2 + 1$.

Let $a(x) = 2x^5 + x^4 + x^2$ and $b(x) = x^4 + x^3 + 1$. It is then easily verified that $1 = a(x)g(x) + b(x)h(x)$. Hence

$$e(x) = a(x)g(x) \bmod (x^n - 1) = 2x^{10} + 2x^8 + 2x^7 + 2x^6 + 2x^2$$

is the generating idempotent of C. On the other hand, we have $g(x) = \gcd(e(x), x^n - 1)$.

A generator matrix of a cyclic code can be derived from its generating idempotent as follows [Huffman and Pless (2003)][Theorem 4.3.6].

Theorem 2.28. *Let C be an $[n, \kappa]$ cyclic code with generating idempotent $e(x) = \sum_{i=0}^{n-1} e_i x^i$. The the following $\kappa \times n$ matrix*

$$\begin{bmatrix} e_0 & e_1 & e_2 & \cdots & e_{n-2} & e_{n-1} \\ e_{n-1} & e_0 & e_1 & \cdots & e_{n-3} & e_{n-2} \\ & & & \vdots & & \\ e_{n-\kappa+1} & e_{n-\kappa+2} & e_{n-\kappa+3} & \cdots & e_{n-\kappa-1} & e_{n-\kappa} \end{bmatrix}$$

is a generator matrix of C.

The *sum* of two cyclic codes C_1 and C_2 of length n over GF(q) is denoted by $C_1 + C_2$, and defined by

$$C_1 + C_2 = \{\mathbf{c}_1 + \mathbf{c}_2 : \mathbf{c}_1 \in C_1, \ \mathbf{c}_2 \in C_2\}.$$

Both the sum and intersection of two cyclic codes over GF(q) are also cyclic codes. Their generator polynomial and generating idempotent are given in the following theorem [Huffman and Pless (2003)][Theorem 4.3.7].

Theorem 2.29. *Let C_1 and C_2 be two cyclic codes of length n over* GF(q) *with generator polynomials $g_1(x)$ and $g_2(x)$, and generating idempotents $e_1(x)$ and $e_2(x)$, respectively. Then*

(i) *the intersection code $C_1 \cap C_2$ has generator polynomial* $\mathrm{lcm}(g_1(x), g_2(x))$ *and generating idempotent $e_1(x)e_2(x)$, and*

(ii) *the sum code $\mathcal{C}_1 + \mathcal{C}_2$ has generator polynomial* $\gcd(g_1(x), g_2(x))$ *and generating idempotent* $e_1(x) + e_2(x) - e_1(x)e_2(x)$.

The ring $\mathcal{R}_{(n,q)}$ has a special set of idempotents, called primitive idempotents, which can be used to produce all idempotents in $\mathcal{R}_{(n,q)}$ and therefore all cyclic codes of length n over GF(q). We now introduce them. Before doing this, we recall that an ideal $\mathcal{I}$ in a ring $\mathcal{R}$ is a *minimal ideal* if there is no proper ideal between $\{0\}$ and $\mathcal{I}$.

Let $x^n - 1 = f_1(x)f_2(x)\cdots f_t(x)$, where $f_i(x)$ is irreducible over GF(q) for $1 \le i \le t$. Since we always assume that $\gcd(n, q) = 1$, all $f_i(x)$'s are distinct. Define

$$\tilde{f}_i(x) = \frac{x^n - 1}{f_i(x)}$$

for all i with $1 \le i \le t$. It is known that all the ideals $\langle \tilde{f}_i(x) \rangle$ of $\mathcal{R}_{(n,q)}$ are minimal, and called *minimal cyclic codes* and *irreducible cyclic codes* of length n over GF(q). Let $\tilde{e}_i(x)$ denote the generating idempotent of $\langle \tilde{f}_i(x) \rangle$ for all i. These $\tilde{e}_i(x)$ are the *primitive idempotents* of $\mathcal{R}_{(n,q)}$.

The following theorem lists basic properties of the primitive idempotents [Huffman and Press (2003)][Theorem 4.3.8].

Theorem 2.30. *Let symbols and notation be the same as before. The following statements hold in* $\mathcal{R}_{(n,q)}$.

(a) $\mathcal{R}_{(n,q)}$ *is the vector space direct sum of* $\langle \tilde{f}_i(x) \rangle$ *for* $1 \le i \le t$.
(b) *For every pair of distinct i and j, we have* $\tilde{e}_i(x)\tilde{e}_j(x) = 0$ *in* $\mathcal{R}_{(n,q)}$.
(c) $\sum_{i=1}^{t} \tilde{e}_i(x) = 1$ *in* $\mathcal{R}_{(n,q)}$.
(d) *If $e(x)$ is a nonzero idempotent in* $\mathcal{R}_{(n,q)}$, *then there is a subset T of* $\{1, 2, \ldots, t\}$ *such that*

$$e(x) = \sum_{i \in T} \tilde{e}_i(x) \quad and \quad \langle e(x) \rangle = \sum_{i \in T} \langle \tilde{f}_i(x) \rangle.$$

Part (d) of this theorem says that every nonzero idempotent in $\mathcal{R}_{(n,q)}$ is the sum of some primitive idempotents, and every nonzero cyclic code of length n over GF(q) is the sum of some minimal cyclic codes of length n over GF(q). It is also known that any minimal ideal of $\mathcal{R}_{(n,q)}$ is an extension field of GF(q) [Huffman and Press (2003)][Theorem 4.3.9].

We will need the following result later [Huffman and Press (2003)][Corollary 4.3.15].

Theorem 2.31. *Let C be a cyclic code of length n over* GF(q) *with generating idempotent $e(x) = \sum_{i=0}^{n-1} e_i(x)$. Then*

(i) *$e_i = e_j$ if i and j in the same q-cyclotomic coset modulo n;*
(ii) *There is a subset T of $\mathbb{Z}_n$ such that*

$$e(x) = \sum_{j \in T} a_j \sum_{i \in C_j} x^i,$$

where each $a_i \in$ GF$(q)^*$.
If $q = 2$, the set T is a set of coset leaders and $a_j = 1$ for all $j \in T$.

2.4.4 *Zeros of cyclic codes*

Let C be a cyclic code of length n over GF(q) with generator polynomial $g(x)$, and let β be a primitive nth root of unity over GF(q^m), where $m = \text{ord}_n(q)$. It then follows from (2.7) that

$$g(x) = \prod_{t \in T} \mathbb{M}_{\beta^t}(x) = \prod_{t \in T} \prod_{i \in C_t} (x - \beta^i),$$

where T is a set of coset leaders, i.e., $T \subset \Gamma_{(n,q)}$. The set $\bigcup_{t \in T} \{\beta^i : i \in C_t\}$ is called the *zeros* of C, and the set $\bigcup_{t \in T} C_t$ is referred to as the *defining set* of C. By definition, $c(x)$ is a codeword of C if and only if $c(\beta^i) = 0$ for all $i \in \bigcup_{t \in T} C_t$.

If C is a code of length n over GF(q), then a *complement* of C, is a code C^c such that $C + C^c = $ GF$(q)^n$ and $C \cap C^c = \{\bar{0}\}$. If C is cyclic, then C^c is unique and cyclic. We have the following information on the code C^c.

Theorem 2.32. *Let C be a cyclic code of length n over* GF(q) *with generator polynomial $g(x)$, generating idempotent $e(x)$, and defining set S. We have then the following conclusions about the complement C^c of C:*

(1) *$h(x) = (x^n - 1)/g(x)$ is the generator polynomial and $1 - e(x)$ is its generating idempotent of C^c.*
(2) *C^c is the sum of the minimal ideals of $\mathcal{R}_{(n,q)}$ not contained in C.*
(3) *$\mathbb{Z}_n \setminus S$ is the defining set of C^c.*

Let $f(x) = \sum_{i=0}^{\ell} f_i x^i$ be a polynomial of degree ℓ. Then its *reciprocal* is defined by

$$f^*(x) = x^{\ell} f(x^{-1}) = \sum_{i=0}^{\ell} f_{\ell-i} x^{\ell}.$$

The following result is fundamental and follows from Huffman and Pless (2003)[Theorem 4.4.9].

Theorem 2.33. *Let C be a cyclic code of length n over GF(q) with generator polynomial $g(x)$. Let $h(x) = (x^n - 1)/g(x)$ be the parity-check polynomial of C. Then the following statements are true:*

(1) $h^*(x)/h(0)$ *is the generator polynomial of $C^\perp$.*
(2) $C^\perp$ *and C^c are permutation equivalent. Hence they have the same dimension and weight distribution.*

Self-orthogonal codes are an interesting class of codes. The next theorem gives a characterization of self-orthogonal cyclic codes [Huffman and Pless (2003)][Corollary 4.4.10].

Theorem 2.34. *Let C be a cyclic code of length n over GF(q) with generator polynomial $g(x)$. Let $h(x) = (x^n - 1)/g(x)$ be the parity-check polynomial of C. Then C is self-orthogonal if and only if $h^*(x)|g(x)$.*

Minimal cyclic codes (also called irreducible cyclic codes) have the following trace representation [Huffman and Pless (2003)][Theorem 4.4.19]. It is the direct consequence of Delsarte's Theorem.

Theorem 2.35 (Trace representation of irreducible cyclic codes). *Let $g(x)$ be an irreducible factor of $x^n - 1$ over GF(q). Suppose that $g(x)$ has degree m. Let $\gamma \in$ GF(q^m) be a root of $g(x)$. Then*

$$\mathcal{C}(\gamma) = \left\{ \sum_{i=0}^{n-1} \mathrm{Tr}_{q^m/q}(a\gamma^i)x^i : a \in \mathrm{GF}(q^m) \right\}$$

is the $[n, m]$ irreducible cyclic code with parity check polynomial

$$h(x) = \prod_{i=0}^{m-1} \left(x - \gamma^{q^i} \right) \in \mathrm{GF}(q)[x].$$

2.4.5 *Lower bounds on the minimum distance*

It is usually very difficult to determine the minimum distance of cyclic codes over finite fields. However, due to the cyclicity of cyclic codes, we have some lower bounds on their minimum distance. The first one is described in the following theorem [Bose and Ray-Chaudhuri (1960)] and [Hocquenghem (1959)].

Theorem 2.36 (BCH Bound). *Let C be a cyclic code of length n over* GF(q) *with defining set S and minimum distance d. Assume S contains $\delta - 1$ consecutive elements for some integer δ. Then the $d \geq \delta$.*

Note that the BCH bound depends on the primitive nth root of unity that defines the cyclic code. It may yield a very bad lower bound on the minimum distance sometimes. In this case, the lower bound given in the following theorem may be much better. It was discovered by Hartmann and Tzeng (1972). To introduce this bound, we define

$$A + B = \{a + b : a \in A, \, b \in B\},$$

where A and B are two subsets of the ring $\mathbb{Z}_n$, n is a positive integer, and $+$ denotes the integer addition modulo n.

Theorem 2.37. *Let C be a cyclic code of length n over* GF(q) *with defining set S and minimum distance d. Let A be a set of $\delta - 1$ consecutive elements of S and $B(b, s) = \{jb \bmod n : 0 \leq j \leq s\}$, where $\gcd(b, n) < \delta$. If $A + B(b, s) \subseteq S$ for some b and s, then the $d \geq \delta + s$.*

The following theorem is also a generalization of the BCH bound [Huffman and Pless (2003)][Corollary 4.5.11].

Theorem 2.38. *Let C be a cyclic code of length n over* GF(q). *Suppose that $f(x)$ is a codeword such that $f(\beta^b) = f(\beta^{b+1}) = \cdots = f(\beta^{b+w-1}) = 0$ but $f(\beta^{b+w}) \neq 0$, where β is a primitive nth root of unity in an extension field of* GF(q). *Then* $\mathrm{wt}(f(x)) \geq w + 1$.

When $s = 0$, The Hartmann and Tzeng Bound becomes the BCH bound. There are also other bounds. The Roos Bounds, which are generalizations of the Hartmann Tzeng Bound, can be found in Roos (1982a) and Roos (1982b). A number of techniques for finding a lower bound on the minimum distance of cyclic codes are given in Lint and Wilson (1986).

2.4.6 *BCH codes*

Let δ be an integer with $2 \leq \delta \leq n$. A *BCH code* C over GF(q) of length n and *design distance* δ is a cyclic code with defining set

$$S(b, \delta) = C_b \cup C_{b+1} \cup \cdots C_{b+\delta-2}, \tag{2.10}$$

where C_i is the q-cyclotomic coset modulo n containing i and b is an nonnegative integer. It follows from Theorem 2.36 that a cyclic code with design distance δ has minimum weight at least δ. It is possible that the actual minimum distance is equal to the design distance. Sometimes the actual minimum weight is much larger than the design distance.

It may happen that $S(b_1, \delta_1) = S(b_2, \delta_2)$ for two distinct pairs (b_1, δ_1) and (b_2, δ_2). The *maximum design distance* of a BCH code is defined to be the largest δ such that the set $S(b, \delta)$ in (2.10) defines the code for some $b \geq 0$. The maximum design distance of a BCH code is also called the *Bose distance*.

When $b = 1$, the code $\mathcal{C}$ with defining set in (2.10) is called a *narrow-sense* BCH code. If $n = q^m - 1$, then $\mathcal{C}$ is referred to as a *primitive* BCH code.

The dimension of the BCH code $\mathcal{C}$ with defining set $S(b, \delta)$ in (2.10) depends on the size of the defining set. However, it may not have a clear relation with n, q, b and δ, and thus cannot be given exactly in terms of these parameters. The best we can do in general is to develop tight lower bounds on the dimension of BCH codes. The next theorem introduces such bounds [Huffman and Pless (2003)][Theorem 5.1.7].

Theorem 2.39. *Let $\mathcal{C}$ be an $[n, \kappa]$ BCH code over $\mathrm{GF}(q)$ of design distance δ. Then the following statements hold:*

(i) $\kappa \geq n - \mathrm{ord}_n(q)(\delta - 1)$.
(ii) *If $q = 2$ and $\mathcal{C}$ is a narrow-sense BCH code, then δ can be assumed odd; furthermore if $\delta = 2w + 1$, then $\kappa \geq n - \mathrm{ord}_n(q)w$.*

The bounds in Theorem 2.39 may not be improved for the general case, as demonstrated by the following theorem. However, in some special cases, they could be improved.

Example 2.8. Let $q = 2$ and $n = 15$. Then $m = \mathrm{ord}_{15}(2) = 4$ and the 2-cyclotomic cosets modulo 15 are the following:

$$C_0 = \{0\}, \quad C_1 = \{1, 2, 4, 8\}, \quad C_3 = \{3, 6, 9, 12\},$$
$$C_5 = \{5, 10\}, \quad C_7 = \{7, 11, 13, 14\}.$$

Let α be a generator of $\mathrm{GF}(2^4)^*$ with $\alpha^4 + \alpha + 1 = 0$ and let $\beta = \alpha^{(q^m - 1)/n} = \alpha$ be the primitive nth root of unity.

When $(b, \delta) = (0, 3)$, the defining set $S(b, \delta) = \{0, 1, 2, 4, 8\}$, and the binary cyclic code has parameters $[15, 10, 4]$ and generator polynomial $x^5 + x^4 + x^2 + 1$. In this case, the actual minimum weight is more than the design distance, and the dimension is larger than the first bound in Theorem 2.39.

When $(b, \delta) = (1, 3)$, the defining set $S(b, \delta) = \{1, 2, 4, 8\}$, and the binary cyclic code has parameters $[15, 11, 3]$ and generator polynomial $x^4 + x + 1$. It is a narrow-sense BCH code. In this case, the actual minimum weight is equal to the design distance, and the dimension reaches the second bound in Theorem 2.39.

When $(b, \delta) = (2, 3)$, the defining set $S(b, \delta) = \{1, 2, 3, 4, 6, 8, 9, 12\}$, and the binary cyclic code has parameters $[15, 7, 5]$ and generator polynomial $x^8 + x^7 + x^6 + x^4 + 1$. In this case, the actual minimum weight is more than the design distance, and the dimension achieves the first bound in Theorem 2.39.

When $(b, \delta) = (1, 5)$, the defining set $S(b, \delta) = \{1, 2, 3, 4, 6, 8, 9, 12\}$, and the binary cyclic code has parameters $[15, 7, 5]$ and generator polynomial $x^8 + x^7 + x^6 + x^4 + 1$. In this case, the actual minimum weight is equal to the design distance, and the dimension is larger than the first bound in Theorem 2.39. Note that the three pairs $(b_1, \delta_1) = (2, 3)$, $(b_2, \delta_2) = (2, 4)$ and $(b_3, \delta_3) = (1, 5)$ define the same binary cyclic code with generator polynomial $x^8 + x^7 + x^6 + x^4 + 1$. Hence the maximum design distance of this $[15, 7, 5]$ cyclic code is 5.

When $(b, \delta) = (3, 4)$, the defining set $S(b, \delta) = \{1, 2, 3, 4, 5, 6, 8, 9, 10, 12\}$, and the binary cyclic code has parameters $[15, 5, 7]$ and generator polynomial $x^{10} + x^8 + x^5 + x^4 + x^2 + x + 1$. In this case, the actual minimum weight is more than the design distance, and dimension is larger than the first bound in Theorem 2.39.

Let $\mathcal{C}$ be a primitive narrow-sense BCH code of length $n = q^m - 1$ over $\mathrm{GF}(q)$ with design distance δ. The defining set is then $S(1, \delta) = C_1 \cup C_2 \cup \cdots \cup C_{\delta-1}$. The following theorem provides useful information on the minimum weight of primitive narrow-sense BCH codes [Huffman and Pless (2003)][Theorem 5.1.9].

Theorem 2.40. *Let $\mathcal{C}$ be a primitive narrow-sense BCH code of length $n = q^m - 1$ over $\mathrm{GF}(q)$ with design distance δ. Then the minimum weight of $\mathcal{C}$ is its minimum odd-like weight.*

We dealt with Hamming codes in Section 2.1.3, which may not be equivalent to a cyclic code. Now, we are ready to tell when they are cyclic codes [Huffman and Pless (2003)][Theorem 5.1.4].

Theorem 2.41. *Let $n = (q^m - 1)/(q - 1)$ with $\gcd(m, q - 1) = 1$. Let $\mathcal{C}$ be the narrow-sense BCH code with defining set C_1. Then $\mathcal{C}$ is the Hamming code $\mathcal{H}_{q,m}$.*

After introducing BCH codes in general, we are now ready to describe an important subclass of BCH codes, which are referred to as *Reed–Solomon codes* [Reed and Solomon (1960)].

Let $n = q - 1$. Then $\mathrm{ord}_n(q) = 1$, and all the q-cyclotomic cosets modulo n are of size 1. Therefore, the canonical factorization of $x^n - 1$ over $\mathrm{GF}(q)$ is given by

$$x^n - 1 = \prod_{i=0}^{n-1} (x - \alpha^i),$$

where α is a generator of $GF(q)^*$.

A Reed–Solomon code over $GF(q)$, denoted $RS_{(q,\kappa)}$, is of length $n = q - 1$ with design distance $n - \kappa + 1$. It is straightforward to prove the following theorem [Huffman and Pless (2003)][Theorem 5.2.1].

Theorem 2.42. *Let $n = q - 1$. Then $RS_{(q,\kappa)}$ has defining set $S(b, n - \kappa + 1)$ for some integer $b \geq 0$ and parameters $[n, \kappa, n - \kappa + 1]$. Furthermore, it is MDS.*

The following theorem gives another way to define the narrow-sense Reed–Solomon code $RS_{(q,\kappa)}$.

Theorem 2.43. *Let α be a generator of $GF(q)^*$ and let κ be an integer with $0 \leq \kappa \leq n = q - 1$. Then*

$$\mathcal{C} = \{(f(1), f(\alpha), f(\alpha^2), \ldots, f(\alpha^{n-1})) : f \in \mathcal{P}_{(q,\kappa)}\}$$

is the narrow-sense $[n, \kappa, n - \kappa + 1]$ Reed–Solomon code $RS_{(q,\kappa)}$, where $\mathcal{P}_{(q,\kappa)}$ denotes all the polynomials of degree less than κ over $GF(q)$.

2.4.7 Quadratic residue codes

Throughout this subsection, let n be an odd prime, and let γ be a primitive element of $GF(n)$. Recall that the cyclotomic classes of order two with respect to $GF(n)$ are defined by

$$C_i^{(2,n)} = \gamma^i \langle \gamma^2 \rangle,$$

where $i \in \{0, 1\}$. The elements in $C_0^{(2,n)}$ and $C_1^{(2,n)}$ are quadratic residues and quadratic nonresidues modulo n, respectively.

Put $m = \mathrm{ord}_n(q)$. Let α be a generator of $GF(q^m)^*$ and let $\beta = \alpha^{(q^m-1)/n}$. Then β is a primitive nth root of unity in $GF(q^m)$. Define

$$g_0(x) = \prod_{i \in C_0^{(2,n)}} (x - \beta^i) \quad \text{and} \quad g_1(x) = \prod_{i \in C_1^{(2,n)}} (x - \beta^i).$$

When $q \in C_0^{(2,n)}$, it is straightforward to verify that $g_i(x) \in GF(q)[x]$ for all $i \in \{0, 1\}$. In the reminder of this subsection, we always assume that $q \in C_0^{(2,n)}$. Let $QRC_i^{(n,q)}$ and $\overline{QRC}_i^{(n,q)}$ denote the cyclic code over $GF(q)$ of length n with generator polynomial $g_i(x)$ and $(x-1)g_i(x)$, respectively, for each $i \in \{0,1\}$. It then follows that $\mathrm{QRC}_i^{(n,q)}$ and $\overline{\mathrm{QRC}}_i^{(n,q)}$ have dimension $(n+1)/2$ and $(n-1)/2$, respectively, for each i.

Since $q \in C_0^{(2,n)}$, each $C_i^{(2,n)}$ is the union of some q-cyclotomic cosets modulo n. Therefore, $\mathrm{QRC}_i^{(n,q)}$ and $\overline{\mathrm{QRC}}_i^{(n,q)}$ are cyclic codes with defining set $C_i^{(2,n)}$ and $\{0\} \cup C_i^{(2,n)}$, respectively. The four codes $\mathrm{QRC}_i^{(n,q)}$ and $\overline{\mathrm{QRC}}_i^{(n,q)}$ are called *quadratic residue codes*. The two codes $\mathrm{QRC}_i^{(n,q)}$ are called odd-like QR codes, and the two codes $\overline{\mathrm{QRC}}_i^{(n,q)}$ are even-like QR codes.

Note that the two codes $\mathrm{QRC}_0^{(n,q)}$ and $\mathrm{QRC}_1^{(n,q)}$ depend on the choice of the primitive nth root of unity. They have the same parameters and weight distribution. The following theorem provides information on the minimum weight of quadratic residue codes [Huffman and Pless (2003)][Theorem 6.6.22].

Theorem 2.44. *Let d_i and $\bar{d}_i$ denote the minimum weight of $\mathrm{QRC}_i^{(n,q)}$ and $\overline{\mathrm{QRC}}_i^{(n,q)}$, respectively. Then $d_0 = d_1$ and $\bar{d}_0 = \bar{d}_1$. Furthermore, $d_i = \bar{d}_i - 1$ and $d_i^2 \geq n$. If $n \equiv 3 \pmod 4$, then $d_i^2 - d_i + 1 \geq n$. Additionally, every minimum weight codeword of $\mathrm{QRC}_i^{(n,q)}$ is odd-like. If $\mathrm{QRC}_i^{(n,q)}$ is binary, d_i is odd; and if in addition, $n \equiv -1 \pmod 8$, then $d_i \equiv 3 \pmod 4$.*

Example 2.9. Let $n = 23$ and $q = 2$. Then $m = \mathrm{ord}_n(q) = 11$. Let α be a generator of $\mathrm{GF}(2^{11})^*$ with $\alpha^{11} + \alpha^2 + 1 = 0$, and let $\beta = \alpha^{(q^m-1)/n}$. Then $\mathrm{QRC}_0^{(n,q)}$ and $\mathrm{QRC}_1^{(n,q)}$ have parameters $[23, 12, 7]$ and generator polynomials

$$g_0(x) = x^{11}+x^9+x^7+x^6+x^5+x+1 \text{ and } g_1(x) = x^{11}+x^{10}+x^6+x^5+x^4+x^2+1.$$

The two codes $\overline{\mathrm{QRC}}_0^{(n,q)}$ and $\overline{\mathrm{QRC}}_1^{(n,q)}$ have parameters $[23, 11, 8]$ and generator polynomials $(x-1)g_0(x)$ and $(x-1)g_1(x)$.

Example 2.10. Let $n = 13$ and $q = 3$. Then $m = \mathrm{ord}_n(q) = 3$. Let α be a generator of $\mathrm{GF}(3^3)^*$ with $\alpha^3 + 2\alpha + 1 = 0$, and let $\beta = \alpha^{(q^m-1)/n}$. Then $\mathrm{QRC}_0^{(n,q)}$ and $\mathrm{QRC}_1^{(n,q)}$ have parameters $[13, 7, 5]$ and generator polynomials

$$g_0(x) = x^6 + 2x^4 + 2x^3 + 2x^2 + 1 \text{ and } g_1(x) = x^6 + x^5 + 2x^4 + 2x^2 + x + 1.$$

The two codes $\overline{\mathrm{QRC}}_0^{(n,q)}$ and $\overline{\mathrm{QRC}}_1^{(n,q)}$ have parameters $[13, 6, 6]$ and generator polynomials $(x-1)g_0(x)$ and $(x-1)g_1(x)$.

The following theorem demonstrates further relations among the four quadratic residue codes [Assmus and Key (1992a)][Theorem 2.10.1].

Theorem 2.45. *If* $n \equiv -1$ *(mod 4), then* $(\text{QRC}_0^{(n,q)})^\perp = \overline{\text{QRC}}_0^{(n,q)}$. *If* $n \equiv 1$ *(mod 4), then* $(\text{QRC}_0^{(n,q)})^\perp = \overline{\text{QRC}}_1^{(n,q)}$.

Recall that $\widehat{\text{QRC}}_i^{(n,q)}$ denotes the extended code of $\text{QRC}_0^{(n,q)}$. The relations among the two extended codes and their duals are documented in the following theorem [Assmus and Key (1992a)][Corollary 2.10.1].

Theorem 2.46. *If* $n \equiv -1$ *(mod 4), both* $\widehat{\text{QRC}}_i^{(n,q)}$ *are self-dual* $[n+1, (n+1)/2]$ *codes.*

If $n \equiv 1$ *(mod 4), then* $\left(\widehat{\text{QRC}}_0^{(n,q)} \right)^\perp = \widehat{\text{QRC}}_1^{(n,q)}$ *and* $\left(\widehat{\text{QRC}}_1^{(n,q)} \right)^\perp = \widehat{\text{QRC}}_0^{(n,q)}$.

The format of the generating idempotent of QR codes is known and described in the following theorem [Huffman and Pless (2003)][Theorem 6.6.3].

Theorem 2.47. *If* C *is a quadratic residue code over* $\text{GF}(q)$ *with generating idempotent* $e(x)$. *Then*

$$e(x) = a_0 + a_1 \sum_{i \in C_0^{(2,n)}} x^i + a_2 \sum_{i \in C_1^{(2,n)}} x^i,$$

for some $a_1, a_1,$ *and* a_2 *in* $\text{GF}(q)$.

The following theorem provides more detailed information on the generating idempotents of binary quadratic residue codes [Huffman and Pless (2003)] [Theorem 6.6.5].

Theorem 2.48. *Let* $n \equiv \pm 1$ *(mod 8) be a prime. Then the following hold.*

(a) *The binary codes* $\overline{\text{QRC}}_i^{(n,2)}$ *have generating idempotents*

$$\delta + \sum_{i \in C_0^{(2,n)}} x^i \quad \text{and} \quad \delta + \sum_{i \in C_1^{(2,n)}} x^i,$$

where $\delta = 1$ *if* $n \equiv -1$ *(mod 8) and* $\delta = 0$ *if* $n \equiv 1$ *(mod 8).*

(b) *The binary codes* $\text{QRC}_i^{(n,2)}$ *have generating idempotents*

$$\epsilon + \sum_{i \in C_0^{(2,n)}} x^i \quad \text{and} \quad \epsilon + \sum_{i \in C_1^{(2,n)}} x^i,$$

where $\epsilon = 0$ *if* $n \equiv -1$ *(mod 8) and* $\epsilon = 1$ *if* $n \equiv 1$ *(mod 8).*

For further information about the generating idempotents of quadratic residue codes over other finite fields, the reader is referred to Huffman and Pless (2003)[Section 6.6]. More information on the minimum weight of quadratic residue codes can be found in Newhart (1988).

2.4.8 *Duadic codes*

In the previous subsection, we described quadratic residue codes briefly. In this subsection, we introduce a family of cyclic codes that are generalizations of the quadratic residue codes. Binary duadic codes were defined in Leon, Masley and Pless (1984) and were generalized to arbitrary finite fields in Pless (1986; 1993); Rushanan (1986).

As usual, let n be a positive integer and q a prime power with $\gcd(n, q) = 1$. Let S_1 and S_2 be two subsets of $\mathbb{Z}_n$ such that

- $S_1 \cap S_2 = \emptyset$ and $S_1 \cup S_2 = \mathbb{Z}_n \setminus \{0\}$, and
- both S_1 and S_2 are a union of some q-cyclotomic cosets modulo n.

If there is a unit $\mu \in Z_n$ such that $S_1\mu = S_2$ and $S_2\mu = S_1$, then (S_1, S_2, μ) is called a *splitting* of $\mathbb{Z}_n$.

Let $m = \mathrm{ord}_n(q)$ and let β be an nth root of unity in $\mathrm{GF}(q^m)$. Let (S_1, S_2, μ) be a splitting of $\mathbb{Z}_n$, Define

$$g_i(x) = \prod_{i \in S_i}(x - \beta^i) \quad \text{and} \quad \bar{g}_i(x) = (x - 1)g_i(x)$$

for $i \in \{1, 2\}$. The pair of cyclic codes $\mathcal{C}_1$ and $\mathcal{C}_2$ of length n over $\mathrm{GF}(q)$ with generator polynomials $g_1(x)$ and $g_2(x)$ are called *odd-like duadic codes*, and the pair of cyclic codes $\bar{\mathcal{C}}_1$ and $\bar{\mathcal{C}}_2$ of length n over $\mathrm{GF}(q)$ with generator polynomials $\bar{g}_1(x)$ and $\bar{g}_2(x)$ are called *even-like duadic codes*.

By definition, $\mathcal{C}_1$ and $\mathcal{C}_2$ have parameters $[n, (n + 1)/2]$ and $\bar{\mathcal{C}}_1$ and $\bar{\mathcal{C}}_2$ have parameters $[n, (n - 1)/2]$. For odd-like duadic codes, we have the following result [Huffman and Pless (2003)][Theorem 6.5.2].

Theorem 2.49 (Square Root Bound). *Let $\mathcal{C}_1$ and $\mathcal{C}_2$ be a pair of odd-like duadic codes of length n over $\mathrm{GF}(q)$. Let d_o be their (common) minimum odd-like weight. Then the following hold*:

(a) $d_o^2 \geq n$.

(b) *If the splitting defining the duadic codes is given by $\mu = -1$, then $d_o^2 - d_o + 1 \geq n$.*

(c) *Suppose $d_o^2 - d_o + 1 = n$, where $d_o > 2$, and assume that the splitting defining the duadic codes is given by $\mu = -1$. Then d_o is the minimum weight of both $\mathcal{C}_1$ and $\mathcal{C}_2$.*

It is easily seen that quadratic residue codes are duadic codes. Duadic codes can be defined in terms of their generating idempotents. For further information on the existence, constructions, and properties of duadic codes, the reader is referred to Huffman and Pless (2003)[Chapter 6], Ding, Lam and Xing (1999) and Ding and Pless (1999).

Example 2.11. Let $(n, q) = (49, 2)$. Define

$$S_1 = \{1, 2, 4, 8, 9, 11, 15, 16, 18, 22, 23, 25, 29, 30, 32, 36, 37, 39, 43, 44, 46\}$$
$$\cup \{7, 14, 28\}$$

and

$$S_2 = \{1, 2, \ldots, 48\} \backslash S_1.$$

It is easily seen that $(S_1, S_2, -1)$ is a splitting of $\mathbb{Z}_{48}$. The pair of odd-like duadic codes $\mathcal{C}_1$ and $\mathcal{C}_2$ defined by this splitting have parameters $[49, 25, 4]$ and generator polynomials

$$x^{24} + x^{22} + x^{21} + x^{10} + x^8 + x^7 + x^3 + x + 1$$

and

$$x^{24} + x^{23} + x^{21} + x^{17} + x^{16} + x^{14} + x^3 + x^2 + 1,$$

respectively. The minimum weight of the two codes is even.

2.4.9 *Bounds on weights in irreducible cyclic codes*

Let $N > 1$ be an integer dividing $r - 1$, and put $n = (r-1)/N$. Let α be a primitive element of $GF(r)$ and let $\theta = \alpha^N$. Recall that the set

$$\mathcal{C}(r, N) = \{(\text{Tr}_{r/q}(\beta), \text{Tr}_{r/q}(\beta\theta), \ldots, \text{Tr}_{r/q}(\beta\theta^{n-1})) : \beta \in GF(r)\} \quad (2.11)$$

is an irreducible cyclic $[n, m_0]$ code over $GF(q)$, where m_0 is the multiplicative order of q modulo n and m_0 divides m.

Irreducible cyclic codes have been an interesting subject of study for many years. The celebrated Golay code is an irreducible cyclic code and was used on the Mariner Jupiter–Saturn Mission. They form a special class of codes and are interesting in theory as they are minimal cyclic codes.

To study the weights in irreducible cyclic codes, we need the following lemma.

Lemma 2.1. *Let e_1 be a positive divisor of $r - 1$ and let i be any integer with $0 \leq i < e_1$. We have the following multiset equality:*

$$\left\{ xy : y \in \mathrm{GF}(q)^*, \; x \in C_i^{(e_1,r)} \right\}$$

$$= \frac{(q-1)\gcd((r-1)/(q-1), e_1)}{e_1} * C_i^{(\gcd((r-1)/(q-1),e_1),r)}, \quad (2.12)$$

where $\frac{(q-1)\gcd((r-1)/(q-1),e_1)}{e_1} * C_i^{(\gcd((r-1)/(q-1),e_1),r)}$ denotes the multiset in which each element in the set $C_i^{(\gcd((r-1)/(q-1),e_1),r)}$ appears in the multiset with multiplicity $\frac{(q-1)\gcd((r-1)/(q-1),e_1)}{e_1}$.

Proof.   We just prove the conclusion for $i = 0$. The proof is similar for $i \neq 0$ since

$$C_i^{(\gcd((r-1)/(q-1),e_1),r)} = \alpha^i C_0^{(\gcd((r-1)/(q-1),e_1),r)}.$$

Note that every $y \in \mathrm{GF}(q)^*$ can be expressed as $y = \alpha^{\frac{r-1}{q-1}\ell}$ for an unique ℓ with $0 \leq \ell < q - 1$ and every $x \in C_0^{(e_1,r)}$ can be expressed as $x = \alpha^{e_1 j}$ for an unique j with $0 \leq j < (r-1)/e_1$. Then, we have

$$xy = \alpha^{\frac{r-1}{q-1}\ell + e_1 j}.$$

It follows that

$$xy = \alpha^{\frac{r-1}{q-1}\ell + e_1 j}$$

$$= \left(\alpha^{\gcd((r-1)/(q-1),e_1)} \right)^{\frac{r-1}{(q-1)\gcd((r-1)/(q-1),e_1)}\ell + \frac{e_1}{\gcd((r-1)/(q-1),e_1)}j}.$$

Note that

$$\gcd\left(\frac{r-1}{(q-1)\gcd((r-1)/(q-1),e_1)}, \frac{e_1}{\gcd((r-1)/(q-1),e_1)} \right) = 1.$$

When ℓ ranges over $0 \leq \ell < q - 1$ and j ranges over $0 \leq j < (r-1)/e_1$, xy takes on the value 1 exactly $\frac{q-1}{e_1}\gcd((r-1)/(q-1), e_1)$ times.

Let $x_{i_1} \in C_0^{(e_1,r)}$ for $i_1 = 1$ and $i_1 = 2$, and let $y_{i_2} \in \mathrm{GF}(q)^*$ for $i_2 = 1$ and $i_2 = 2$. Then $\frac{x_1}{x_2} \in C_0^{(e_1,r)}$ and $\frac{y_1}{y_2} \in \mathrm{GF}(q)^*$. Note that $x_1 y_1 = x_2 y_2$ if and only if $\frac{x_1}{x_2}\frac{y_1}{y_2} = 1$. Then the conclusion of the lemma for the case $i = 0$ follows from the discussions above. $\square$

Let $N > 1$ be an integer dividing $r - 1$, and put $n = (r - 1)/N$. Let α be a primitive element of $\text{GF}(r)$ and let $\theta = \alpha^N$. Let $Z(r, a)$ denote the number of solutions $x \in \text{GF}(r)$ of the equation $\text{Tr}_{r/q}(ax^N) = 0$. Let $\zeta_p = e^{2\pi\sqrt{-1}/p}$, and $\chi(x) = \zeta_p^{\text{Tr}_{r/p}(x)}$, where $\text{Tr}_{r/p}$ is the trace function from $\text{GF}(r)$ to $\text{GF}(p)$. Then χ is an additive character of $\text{GF}(r)$. We have then by Lemma 2.1:

$$
\begin{aligned}
Z(r, a) &= \frac{1}{q} \sum_{y \in \text{GF}(q)} \sum_{x \in \text{GF}(r)} \zeta_p^{\text{Tr}_{q/p}\left(y\text{Tr}_{r/q}(ax^N)\right)} \\
&= \frac{1}{q} \sum_{y \in \text{GF}(q)} \sum_{x \in \text{GF}(r)} \chi(yax^N) \\
&= \frac{1}{q}\left[q + r - 1 + \sum_{y \in \text{GF}(q)^*} \sum_{x \in \text{GF}(r)^*} \chi(yax^N) \right] \\
&= \frac{1}{q}\left[q + r - 1 + N \sum_{y \in \text{GF}(q)^*} \sum_{x \in C_0^{(N,r)}} \chi(yax) \right] \\
&= \frac{1}{q}\left[q + r - 1 + (q-1)\gcd\left(\frac{r-1}{q-1}, N\right) \cdot \sum_{z \in C_0^{\left(\gcd\left(\frac{r-1}{q-1}, N\right), r\right)}} \chi(az) \right].
\end{aligned}
$$

$$(2.13)$$

Then the Hamming weight of the codeword

$$
\left(\text{Tr}_{r/q}(\beta), \text{Tr}_{r/q}(\beta\theta), \ldots, \text{Tr}_{r/q}(\beta\theta^{n-1})\right)
$$

in the irreducible cyclic code of (2.11) is equal to

$$
\begin{aligned}
n &- \frac{Z(r, \beta) - 1}{N} \\
&= \frac{(q-1)\left(r - 1 - \gcd\left(\frac{r-1}{q-1}, N\right)\eta_k^{\left(\gcd\left(\frac{r-1}{q-1}, N\right), r\right)}\right)}{qN}.
\end{aligned}
$$

$$(2.14)$$

The weight expression of (2.14) is the key observation and proves that the determination of the weight distribution of an irreducible cyclic code is equivalent to that of the Gaussian periods of order $N_1 = \gcd((r - 1)/(q - 1), N)$. McEliece (1974) gave a different proof of (2.14) by Gaussian sums. From (1.9), we know that the weights of an irreducible cyclic code can be expressed as a linear combination of Gaussian sums.

Theorem 2.50. *Let $N_1 = \gcd((r-1)/(q-1), N)$. Then for all i with $0 \le i \le N_1 - 1$, we have*

(i) $\eta_i^{(N_1,r)} \in \mathbb{Z}$;
(ii) $N_1 \eta_i^{(N_1,r)} + 1 \equiv 0 \pmod{q}$; and
(iii) $\left| \eta_i^{(N_1,r)} + \frac{1}{N_1} \right| \le \left\lfloor \frac{(N_1-1)\sqrt{r}}{N_1} \right\rfloor$.

Proof.   The conclusions of Parts (i) and (ii) follow from (2.14) directly, and that of Part (iii) follows from (1.5) and (1.9).   □

Theorem 2.50 is an interesting result in the theory of cyclotomy. The next theorem is a divisibility result of irreducible cyclic codes.

Theorem 2.51 (Ding–Yang). *Let $N_1 = \gcd((r-1)/(q-1), N)$. Then the Hamming weight of every codeword in the irreducible cyclic code of (2.11) is divisible by*

$$\frac{(q-1)}{\gcd(q-1, N/N_1)}.$$

Proof.   By (2.14), the Hamming weight of every nonzero codeword is equal to

$$\frac{q-1}{\gcd(q-1, N/N_1)} \frac{r - (1 + N_1 \eta_k)}{q^{\frac{N}{\gcd(q-1,N/N_1)}}}.$$

The desired conclusion then follows from the fact that

$$\gcd\left(q - 1, q \frac{N}{\gcd(q-1, N/N_1)} \right) = 1.$$
                                                                                        □

Particularly, when N divides $(r-1)/(q-1)$, the Hamming weight of every codeword in the irreducible cyclic code of (2.11) is divisible by $q - 1$.

Example 2.12. Let $q = 5$. $m = 4$, $N = 4$. Then the irreducible cyclic code of (2.11) over $\mathrm{GF}(q)$ has length 156, dimension 4, and the following weight distribution:

$$1 + 156x^{112} + 156x^{124} + 156x^{128} + 156x^{136}.$$

So by Theorem 2.51, 4 is a common divisor of all nonzero weights. Note that

$$\gcd(112, 124, 128, 136) = 4.$$

Example 2.13. Let $q = 3$. $m = 4$, $N = 2$. Then the irreducible cyclic code of (2.11) over GF(q) has length 40, dimension 4, and the following weight distribution:

$$1 + 40x^{24} + 40x^{30}.$$

So by Theorem 2.51, 2 is a common divisor of all nonzero weights. Note that $\gcd(24, 30) = 6$.

The weight distribution of irreducible cyclic codes is known in some special cases, but open in general [Ding and Yang (2013)]. Since it is notoriously hard to determine the weight distributions of the irreducible cyclic codes, it would be interesting to develop tight bounds on the weights in irreducible cyclic codes. Such tight bounds can give information on the error-correcting capability of this class of cyclic codes. The following bounds were deduced in Ding and Yang (2013).

Theorem 2.52. *Let N be a positive divisor of $r - 1$ and define $N_1 = \gcd((r - 1)/ (q - 1), N)$. Let m_0 be the multiplicative order of q modulo n. Then the set $C(r, N)$ in (2.11) is a $[(q^m - 1)/N, m_0]$ cyclic code over GF(q) in which the weight w of every nonzero codeword satisfies that*

$$w_H(c(\beta)) \geq (q - 1) \left\lceil \frac{r - \lfloor (N_1 - 1)\sqrt{r} \rfloor}{qN} \right\rceil,$$

$$w_H(c(\beta)) \leq (q - 1) \left\lfloor \frac{r + \lfloor (N_1 - 1)\sqrt{r} \rfloor}{qN} \right\rfloor.$$

In particular, if $N_1(N_1 - 1) < r$, then $m_0 = m$.

Proof. The results of this theorem follow from Theorem 2.50 and (2.14). □

The lower bound of Theorem 2.52 is tight when $\gcd((r - 1)/(q - 1), N)$ is small, but it can be negative in some other cases. In fact, when $\gcd((r - 1)/(q - 1), N) = 1$, the lower and upper bounds of Theorem 2.52 are the same, and they are indeed achieved as the code in this case is a constant-weight code. Table 2.1 lists some experimental data, where n, k, d are the length, dimension and minimum nonzero weight of the code.

At the end of this section, we derive bounds on the weights in two more classes of cyclic codes that are closely related to the irreducible cyclic codes.

Let $\gcd(n, q) = 1$ and let $k = \mathrm{ord}_n(q)$ denote the order of q modulo n. Define $r = q^k$. Let $N > 1$ be an integer dividing $r - 1$, and put $n = (r - 1)/N$. Let α be

Table 2.1   The lower bound of Theorem 2.52.

n	k	d	q	lower bound of Theorem 2.52	$\frac{r-1}{q-1} \bmod N$
5	4	2	2	2	0
21	6	8	2	8	0
21	3	12	2^2	12	0
85	4	64	2^2	64	1
13	3	9	3	9	1
40	4	24	3	24	0
121	5	81	3	81	1
312	4	240	5	236	0

a primitive element of GF(r) and let $\theta = \alpha^N$. The set

$$\overline{C}(r, N) = \left\{ (\mathrm{Tr}_{r/q}(a+b), \mathrm{Tr}_{r/q}(a\theta+b), \ldots, \mathrm{Tr}_{r/q}(a\theta^{n-1}+b)) : a, b \in \mathrm{GF}(r) \right\}$$

is a cyclic $[n, k+1]$ code over GF(q), where $\mathrm{Tr}_{r/q}$ is the trace function from GF(r) onto GF(q).

Using Delsarte's Theorem, one can prove that the code $\overline{C}(r, N)$ is the cyclic code with parity check polynomial $(x - 1)m_{\theta^{-1}}(x)$, where $m_{\theta^{-1}}(x)$ is the minimal polynomial of θ^{-1} over GF(q) and is irreducible over GF(q).

Theorem 2.53. *Let N be a positive divisor of $r - 1$ and define $N_1 = \gcd((r-1)/(q-1), N)$. Let k be the multiplicative order of q modulo n. Then the set $\overline{C}(r, N)$ in (2.15) is a $[(q^m - 1)/N, k + 1, d]$ cyclic code over GF(q), where*

$$d \geq \min \left\{ \begin{array}{c} (q-1)\left\lceil \frac{r - \lfloor (N_1-1)\sqrt{r} \rfloor}{qN} \right\rceil, \\ \frac{(q-1)(r-1)-1}{qN} - \frac{q-1}{q}\left\lfloor \frac{(N-1)\sqrt{r}}{N} \right\rfloor \end{array} \right\}.$$

Proof.   Let $\zeta_p = e^{2\pi\sqrt{-1}/p}$, and $\chi(x) = \zeta_p^{\mathrm{Tr}_{r/p}(x)}$, where $\mathrm{Tr}_{r/p}$ is the trace function from GF(r) to GF(p). Then χ is an additive character of GF(r).

Let $b \in \mathrm{GF}(r)$. We have

$$\sum_{y \in \mathrm{GF}(q)} \chi(-by) = \sum_{y \in \mathrm{GF}(q)} \zeta_p^{\mathrm{Tr}_{q/p}(\mathrm{Tr}_{r/q}(-by))} = \begin{cases} 0 & \text{if } \mathrm{Tr}_{r/q}(b) \neq 0, \\ q & \text{if } \mathrm{Tr}_{r/q}(b) = 0. \end{cases} \quad (2.15)$$

Note that the code $\overline{C}(r, N)$ of (2.15) contains the code $C(r, N)$ of (2.11) as a subcode. Define

$$\mathbf{c}_{(a,b)} = (\mathrm{Tr}_{r/q}(a+b), \mathrm{Tr}_{r/q}(a\theta+b), \ldots, \mathrm{Tr}_{r/q}(a\theta^{n-1}+b)),$$

where $a, b \in \mathrm{GF}(r)$.

If $\mathrm{Tr}_{r/q}(b) = 0$, $\mathbf{c}_{(a,b)}$ is a codeword of the code $\mathcal{C}(r, N)$ of (2.11) and the Hamming weight of this codeword satisfies the bounds of Theorem 2.52. If $a = 0$ and $\mathrm{Tr}_{r/q}(b) \neq 0$, the Hamming weight of $\mathbf{c}_{(a,b)}$ is equal to n.

We now consider the weight of $\mathbf{c}_{(a,b)}$ for the case that $a \neq 0$ and $\mathrm{Tr}_{r/q}(b) \neq 0$. Let $Z(r, a, b)$ denote the number of solutions $x \in \mathrm{GF}(r)$ of the equation $\mathrm{Tr}_{r/q}(ax^N - b) = 0$. It then follows from (2.15) that

$$
\begin{aligned}
Z(r, a, b) &= \frac{1}{q} \sum_{y \in \mathrm{GF}(q)} \sum_{x \in \mathrm{GF}(r)} \zeta_p^{\mathrm{Tr}_{q/p}(y\mathrm{Tr}_{r/q}(ax^N - b))} \\
&= \frac{1}{q} \sum_{y \in \mathrm{GF}(q)} \sum_{x \in \mathrm{GF}(r)} \chi(y(ax^N - b)) \\
&= \frac{1}{q} \left[\sum_{y \in \mathrm{GF}(q)} \sum_{x \in \mathrm{GF}(r)^*} \chi(y(ax^N - b)) + \sum_{y \in \mathrm{GF}(q)} \chi(-by) \right] \\
&= \frac{1}{q} \left[r - 1 + \sum_{y \in \mathrm{GF}(q)^*} \sum_{x \in \mathrm{GF}(r)^*} \chi(yax^N - b) \right] \\
&= \frac{1}{q} \left[r - 1 + N \sum_{y \in \mathrm{GF}(q)^*} \sum_{x \in C_0^{(N,r)}} \chi(y(ax - b)) \right]. \quad (2.16)
\end{aligned}
$$

Then the Hamming weight w of the codeword $\mathbf{c}_{(a,b)}$ is then given by

$$
w = n - \frac{Z(r, a, b)}{N} = \frac{(q - 1)(r - 1) - N \sum_{y \in \mathrm{GF}(q)^*} \sum_{x \in C_0^{(N,r)}} \chi(y(ax - b))}{qN}.
$$

It then follows that

$$
w - \frac{(q - 1)(r - 1) - 1}{qN} = -\frac{\sum_{y \in \mathrm{GF}(q)^*} \chi(-by) \left(\sum_{x \in C_0^{(N,r)}} \chi(ayx) + \frac{1}{N} \right)}{q}.
$$

By Theorem 2.50, we have then

$$
\left| w - \frac{(q - 1)(r - 1) - 1}{qN} \right| \leq \frac{q - 1}{q} \left\lfloor \frac{(N - 1)\sqrt{r}}{N} \right\rfloor.
$$

Whence,

$$
w \geq \frac{(q - 1)(r - 1) - 1}{qN} - \frac{q - 1}{q} \left\lfloor \frac{(N - 1)\sqrt{r}}{N} \right\rfloor. \quad (2.17)
$$

Combining the lower bound of (2.17) and that of Theorem 2.52 proves the conclusions of this theorem.   □

2.5   A Combinatorial Approach to Cyclic Codes

In this section, we present a combinatorial approach to the construction of cyclic codes over finite fields. It will be employed in later chapters.

Let D be any nonempty subset of $\mathbb{Z}_n$. The polynomial

$$D(x) := \sum_{i=0}^{n-1} s(D)_i x^i = \sum_{i \in D} x^i \tag{2.18}$$

is called the *Hall polynomial* of the set D, and can be viewed as a polynomial over any finite field.

The cyclic code over $\mathrm{GF}(q)$ with generator polynomial

$$g(x) = \gcd(x^n - 1, D(x)) \tag{2.19}$$

is called the cyclic code of the set D, where $\gcd(x^n - 1, D(x))$ is computed over $\mathrm{GF}(q)$.

It is clear that every linear code over a finite field can be produced in this way. The key question about this approach is how to choose a subset D of $\mathbb{Z}_n$ so that the cyclic code $\mathcal{C}_{\mathrm{GF}(q)}(D)$ has good parameters. Intuitively, if the subset D has good combinatorial structures, its code $\mathcal{C}_{\mathrm{GF}(q)}(D)$ may be optimal. In some of the subsequent chapters, we study the cyclic code $\mathcal{C}_{\mathrm{GF}(q)}(D)$ for some combinatorial designs D.

Note that the correspondence between $s(D)^\infty$ and $D(x)$ is one-to-one. This approach is also a sequence approach to the construction of cyclic codes.

Chapter 3

Designs and Their Codes

3.1 Incidence Structures

Many combinatorial designs are special incidence structures. The objective of this section is to give a short introduction to incidence structures.

3.1.1 *Definitions*

An *incidence structure* is a triple $\mathbb{D} = (\mathcal{P}, \mathcal{B}, \mathcal{I})$, where $\mathcal{P}$ is a set of elements called *points* and $\mathcal{B}$ is a set of elements called *blocks* (lines), and $\mathcal{I} \subseteq \mathcal{P} \times \mathcal{B}$ is a binary relation, called *incidence relation*. The elements of $\mathcal{I}$ are called *flags*.

In this monograph, we use upper case Latin letters to denote blocks and lower case Latin letters to denote points, and allow that $\mathcal{B}$ be a multiset. In other words, the set $\mathcal{B}$ may contain a block B more than once. When distinct blocks of a given incidence structure have distinct point sets, we shall often identify each of its blocks with the point set incident with it for convenience. If a point p is incident with a block B, we can write either $p\mathcal{I}B$ or $p \in B$ and use geometric languages such as p lies on B, B passes through p, and B contains p.

An incidence structure $\mathbb{D} = (\mathcal{P}, \mathcal{B}, \mathcal{I})$ is called a *finite incidence structure* if both $\mathcal{P}$ and $\mathcal{B}$ are finite sets. In this monograph, we consider only finite incidence structures. A finite incidence structure with equally many points and blocks is called *square*.

Example 3.1. Let $\mathcal{P} = \{1, 2, 3, 4, 5, 6, 7\}$ and put

$$\mathcal{B} = \{\{1, 2, 3\}, \{1, 4, 5\}, \{1, 6, 7\}, \{2, 4, 7\}, \{2, 5, 6\}, \{3, 4, 6\}, \{3, 5, 7\}\}.$$

Define $\mathcal{I} = \mathcal{P} \times \mathcal{B}$. Then $(\mathcal{P}, \mathcal{B}, \mathcal{I})$ is an incidence structure obtained from the Fano plane.

3.1.2 *Incidence matrices*

Let $\mathbb{D} = (\mathcal{P}, \mathcal{B}, \mathcal{I})$ be an incidence structure with $v \geq 1$ points and $b \geq 1$ blocks. The points of $\mathcal{P}$ are usually indexed with $p_1, p_2, \ldots, p_v$, and the blocks of $\mathcal{B}$ are normally denoted by $B_1, B_2, \ldots, B_b$. The *incidence matrix* $M_{\mathbb{D}} = (m_{ij})$ of $\mathbb{D}$ is a $b \times v$ matrix where $m_{ij} = 1$ if p_j is on B_i and $m_{ij} = 0$ otherwise. It is clear that the incidence matrix $M_{\mathbb{D}}$ depends on the labeling of the points and blocks of $\mathbb{D}$, but is unique up to row and column permutations. Conversely, every $(0, 1)$-matrix (entries are 0 or 1) determines an incidence structure. Our definition of the incidence matrix follows [Assmus and Key (1992a)][p. 12]. In other books, the transpose of the matrix $M_{\mathbb{D}}$ above is defined as the incidence matrix.

Example 3.2. Consider the incidence structure $\mathbb{D} = (\mathcal{P}, \mathcal{B}, \mathcal{I})$ of Example 3.1. Let the labeling of the points and blocks of $\mathbb{D}$ be in the order of their appearance in $\mathcal{P}$ and $\mathcal{B}$. Then the incidence matrix of $\mathbb{D}$ is

$$
M_{\mathbb{D}} = \begin{bmatrix} 1 & 1 & 1 & 0 & 0 & 0 & 0 \\ 1 & 0 & 0 & 1 & 1 & 0 & 0 \\ 1 & 0 & 0 & 0 & 0 & 1 & 1 \\ 0 & 1 & 0 & 1 & 0 & 0 & 1 \\ 0 & 1 & 0 & 0 & 1 & 1 & 0 \\ 0 & 0 & 1 & 1 & 0 & 1 & 0 \\ 0 & 0 & 1 & 0 & 1 & 0 & 1 \end{bmatrix}. \tag{3.1}
$$

Let p be a prime. The *p-rank* of an incidence structure $\mathbb{D} = (\mathcal{P}, \mathcal{B}, \mathcal{I})$ is the rank of an incidence matrix over $\mathrm{GF}(p)$, and will be useful later.

3.1.3 *Isomorphisms and automorphisms*

An *isomorphism* γ from an incidence structure $\mathbb{D}_1$ onto an incidence structure $\mathbb{D}_2$ is a one-to-one mapping from the points of $\mathbb{D}_1$ onto the points of $\mathbb{D}_2$ and from the blocks of $\mathbb{D}_1$ onto the blocks of $\mathbb{D}_2$ such that p is on B if and only if $\gamma(p)$ is on $\gamma(B)$.

If there is an isomorphism between $\mathbb{D}_1$ and $\mathbb{D}_2$, then we say that the two incidence structures are *isomorphic*. Let M_1 and M_2 be the incidence matrix of $\mathbb{D}_1$ and $\mathbb{D}_2$, respectively. Then $\mathbb{D}_1$ and $\mathbb{D}_2$ are isomorphic if and only if there are permutation matrices P and Q such that

$$
P M_1 Q = M_2.
$$

An *automorphism* of a given incidence structure $\mathbb{D}$ is an isomorphism of $\mathbb{D}$ onto itself. Obviously, the set of automorphisms of $\mathbb{D}$ forms a group, which is called the *full automorphism group* of $\mathbb{D}$ and denoted by Aut($\mathbb{D}$). Any subgroup of Aut($\mathbb{D}$) will be called an *automorphism group* of $\mathbb{D}$.

3.1.4 *Linear codes of incidence structures*

Let $\mathbb{D} = (\mathcal{P}, \mathcal{B}, \mathcal{I})$ be an incidence structure with $v \geq 1$ points and $b \geq 1$ blocks, and let $M_{\mathbb{D}}$ be its incidence matrix with respect to a labeling of the points and blocks of $\mathbb{D}$. The *linear code* of $\mathbb{D}$ over a field F is the subspace $\mathcal{C}_F(\mathbb{D})$ of F^v spanned by the row vectors of the incidence matrix $M_{\mathbb{D}}$. By definition $\mathcal{C}_F(\mathbb{D})$ is a linear code over F with length v and dimension at most b.

The code $\mathcal{C}_F(\mathbb{D})$ depends on the labeling of the points and blocks of the incidence structure $\mathbb{D} = (\mathcal{P}, \mathcal{B}, \mathcal{I})$. Whatever the labeling, these codes are all equivalent.

Example 3.3. Consider the incidence structure of Example 3.1 and the labeling of the points and blocks in Example 3.2. Then the code $\mathcal{C}_F(\mathbb{D})$ over GF(2) is the binary Hamming code with parameters $[7, 4, 3]$, and has the generator matrix $M_{\mathbb{D}}$ of (3.1).

The linear code of an incidence structure may be very bad, as an incidence structure may not have good combinatorial properties. Hence, we will consider the code of special designs in the sequel.

3.2 *t*-Designs and Their Codes

Let $\mathbb{D} = (\mathcal{P}, \mathcal{B}, \mathcal{I})$ be an incidence structure with $v \geq 1$ points and $b \geq 1$ blocks. Let t and λ be two positive integers. Then $\mathbb{D}$ is said to be *t-balanced* with parameter λ if and only if every subset of t points of $\mathcal{P}$ is incident with exactly λ blocks of $\mathcal{B}$. If every block of $\mathbb{D}$ is also of the same size k, then $\mathbb{D}$ is called a *t-(v, k, λ) design*, or simply *t-design*. The integers t, v, k, λ are referred to as the *parameters* of the design. It is possible for a design to have repeated blocks. But in this monograph, we usually consider designs without repeated blocks, which are called *simple t-designs*. A *t*-design is called *symmetric* if $v = b$.

It is clear that *t*-designs with $k = t$ or $k = v$ always exist. Such *t*-designs are *trivial*. A 1-design is referred to as a *tactical configuration*. A nontrivial 2-design is called a *balanced incomplete block design*. A *t-(v, k, λ) design* is referred to as a *Steiner system* if $t \geq 2$ and $\lambda = 1$.

Example 3.4. The incidence structure $(\mathcal{P}, \mathcal{B}, \mathcal{I})$ of Example 3.1 is a 2-$(7, 3, 1)$ design, and also a Steiner system.

Given a t-(v, k, λ_t) design $\mathbb{D}$, it is not difficult to show that $\mathbb{D}$ is also an s-(v, k, λ_s) design for any $s < t$. Precisely, the parameter λ_s (the number of blocks containing an s-set) equals

$$\lambda_t \frac{\binom{v-s}{t-s}}{\binom{k-s}{t-s}},$$

where λ_t is the number of blocks containing a t-set (see [Beth, Jungnickel and Lenz (1999)][Theorem 3.2] for a proof).

In a t-(v, k, λ) design,

$$b = \lambda \frac{\binom{v}{t}}{\binom{k}{t}} \tag{3.2}$$

and every point is incident with (contained in) exactly bk/v blocks. Therefore, the incidence matrix of the t-design $\mathbb{D}$ has a constant row sum which is denoted by r. The *order* of t-(v, k, λ) is defined to be $r - \lambda$.

When $t = 2$, we have

$$r = \lambda(v - 1)/(k - 1) \tag{3.3}$$

and $r > \lambda$ if $k < v$.

When $t = 2$, we can easily prove the following necessary and sufficient condition for $\mathbb{D}$ to be a 2-design [Assmus and Key (1992a)][Theorem 1.4.1].

Theorem 3.1. *Let M be an incidence matrix of an incidence structure $\mathbb{D}$. Then $\mathbb{D}$ is a 2-(v, k, λ) design if and only if*

$$M^T M = (r - \lambda)I_v + \lambda J_v, \tag{3.4}$$

where I_v is the $v \times v$ identity matrix and J_v is the all-one $v \times v$ matrix.

Note that the matrix J_v has the eigenvector $(1, 1, \ldots, 1)$ with eigenvalue v and the following $v - 1$ eigenvectors

$$(1, -1, 0, 0, \ldots, 0), (1, 0, -1, 0, \ldots, 0), \cdots, (1, 0, 0, 0, \ldots, 0, -1)$$

with eigenvalue 0. The matrix $M^T M$ satisfying (3.4) has one eigenvalue $r - \lambda + \lambda v$ and $v - 1$ eigenvalues $r - \lambda$. Hence, we have the following theorem [Assmus and Key (1992a)][Theorem 1.4.1].

Theorem 3.2. *Let M be a $v \times v$ matrix with $M^T M$ satisfying (3.4). Then*

$$\det(M^T M) = (r - \lambda)^{v-1}(v\lambda - \lambda + r).$$

If M is an incidence matrix of a 2-(v, k, λ) design with $v > k$, then by (3.3)

$$\det(M^T M) = (r - \lambda)^{v-1} rk \neq 0.$$

A symmetric 2-$(v, k, 1)$ design is in fact a projective plane of order $k - 1$.

As a corollary of Theorem 3.2, we have the following result due to Fisher [Assmus and Key (1992a)][Corollary 1.4.1].

Corollary 3.1. *If $\mathbb{D}$ is a nontrivial t-design with $t \geq 2$, then there are at least as many blocks as points, i.e., $b \geq v$.*

A symmetric 2-(v, k, λ) design with $\lambda < k < v - 1$ is simply called a (v, k, λ)-*design*. We have the following conclusions about (v, k, λ)-designs [Assmus and Key (1992a)][Theorem 4.2.1].

Theorem 3.3. *Let $\mathbb{D} = (\mathcal{P}, \mathcal{B}, \mathcal{I})$ be a (v, k, λ)-design. Then*

(i) *$r = k$ and the order of $\mathbb{D}$ is $k - \lambda$;*
(ii) *$\lambda(v - 1) = k(k - 1)$; and*
(iii) *λ divides $(k - \lambda)(k - \lambda - 1)$*

For the codes of 2-designs, we can say much more. The following theorem gives information on the codes of 2-designs [Assmus and Key (1992a)][Theorem 2.4.1].

Theorem 3.4. *Let $\mathbb{D} = (\mathcal{P}, \mathcal{B}, \mathcal{I})$ be a nontrivial 2-(v, k, λ) design with order $r - \lambda$, where r is defined in* (3.3). *Let p be a prime and let F be a field of characteristic p where p does not divide $r - \lambda$. Then*

$$\text{rank}_p(\mathbb{D}) \geq v - 1$$

with equality if and only if p divides k.

If $\text{rank}_p(\mathbb{D}) = v - 1$, then $(\mathcal{C}_F(\mathbb{D}))^{\perp}$ is generated by the all-one vector. If $\text{rank}_p(\mathbb{D}) = v$, then $\mathcal{C}_F(\mathbb{D}) = F^b$.

Example 3.5. As a demonstration of Theorem 3.4, we consider the code of the 2-$(7, 3, 1)$ design of Example 3.4. In this case, $r = 3$ and the order of this design $\mathbb{D}$ is $r - \lambda = 2$. We consider now the code $\mathcal{C}_{\text{GF}(p)}(\mathbb{D})$ whose generator matrix is given in (3.1).

When $p = 3$, p does not divide $r - \lambda$ and p divides $k = 3$. In this case, we have

$$\text{rank}_3(\mathbb{D}) = v - 1 = 6.$$

Hence, the dimension of the code $\mathcal{C}_{GF(3)}(\mathbb{D})$ is 6. The ternary code $\mathcal{C}_{GF(3)}(\mathbb{D})^{\perp}$ has parameters $[7, 1, 7]$ and generator matrix $[1111111]$.

When $p = 5$, p does not divide $r - \lambda$ and p does not divide $k = 3$. In this case, we have

$$\text{rank}_3(\mathbb{D}) = v = 7.$$

Hence, the dimension of the code $\mathcal{C}_{GF(5)}(\mathbb{D})$ is 7. The code $\mathcal{C}_{GF(3)}(\mathbb{D})$ is $GF(5)^7$.

For symmetric 2-designs $\mathbb{D}$, we have $r = k$ and the order of $\mathbb{D}$ becomes $r - \lambda = k - \lambda$. In this case, Theorem 3.4 can be refined as follows [Huffman and Pless (2003)][Theorem 8.5.2].

Theorem 3.5. *Let $\mathbb{D} = (\mathcal{P}, \mathcal{B}, \mathcal{I})$ be a symmetric 2-(v, k, λ) design with order $k - \lambda$. Let p be a prime and let F be a field of characteristic p. Then we have the following.*

(a) *If p divides $(k - \lambda)$ and p divides k, then $\mathcal{C}_F(\mathbb{D})$ is self-orthogonal and has dimension at most $v/2$.*

(b) *If p does not divide $(k - \lambda)$ but p divides k, then $\mathcal{C}_F(\mathbb{D})$ has dimension $v - 1$.*

(c) *If p does not divide $(k - \lambda)$ and p does not divide k, then $\mathcal{C}_F(\mathbb{D})$ has dimension v.*

The following theorem is due to Klemm (1986) (see also [Assmus and Key (1992a)][Theorem 2.4.2]).

Theorem 3.6. *Let $\mathbb{D} = (\mathcal{P}, \mathcal{B}, \mathcal{I})$ be a 2-(v, k, λ) design with order $r - \lambda$, where r is defined in (3.3). Let p be a prime dividing $r - \lambda$. Then*

$$\text{rank}_p(\mathbb{D}) \leq \frac{|\mathcal{B}| + 1}{2};$$

moreover, if p does not divide λ and p^2 does not divide $r - \lambda$, then

$$\mathcal{C}_F(\mathbb{D})^{\perp} \subseteq \mathcal{C}_F(\mathbb{D})$$

and $\text{rank}_p(\mathbb{D}) \geq v/2$, where F is a finite field with characteristic p.

Example 3.6. As a demonstration of Theorem 3.6, we consider the code of the 2-$(7, 3, 1)$ design of Example 3.4. In this case, $r = 3$ and the order of this design $\mathbb{D}$ is $r - \lambda = 2$. We consider the case that $p = 2$ and the code $\mathcal{C}_{GF(p)}(\mathbb{D})$. It is clear that p divides $r - \lambda$ and

$$\text{rank}_p(\mathbb{D}) = \frac{|\mathcal{B}| + 1}{2} = 4.$$

Hence, the dimension of the code $\mathcal{C}_{GF(p)}(\mathbb{D})$ is 4. The binary code $\mathcal{C}_{GF(p)}(\mathbb{D})$ was treated in Example 3.2, and has parameters $[7, 4, 3]$.

Clearly, p does not divide λ and p^2 does not divide $r - \lambda$. The dual code $\mathcal{C}_{\mathrm{GF}(p)}(\mathbb{D})^{\perp}$ has the following generator matrix

$$\begin{bmatrix} 1 & 0 & 1 & 0 & 1 & 1 & 0 \\ 0 & 1 & 1 & 0 & 0 & 1 & 1 \\ 0 & 0 & 0 & 1 & 1 & 1 & 1 \end{bmatrix}.$$

Note that the row vectors of this matrix are pairwise orthogonal. In this case, we have indeed

$$\mathcal{C}_{\mathrm{GF}(p)}(\mathbb{D})^{\perp} \subseteq \mathcal{C}_{\mathrm{GF}(p)}(\mathbb{D}).$$

For symmetric 2-designs, Theorem 3.6 can be refined as follows [Huffman and Pless (2003)][Theorem 8.5.3].

Theorem 3.7. *Let* $\mathbb{D} = (\mathcal{P}, \mathcal{B}, \mathcal{I})$ *be a symmetric* 2-(v, k, λ) *design with order* $k - \lambda$. *Let* p *be the characteristic of a finite field* F. *Assume that* p *divides* $k - \lambda$, *but* p^2 *does not divide* $k - \lambda$. *Then* v *is odd, and we have the following.*

(a) *If* p *divides* k, *then* $\mathcal{C}_F(\mathbb{D})$ *is self-orthogonal and has dimension* $(v - 1)/2$.
(b) *If* p *does not divide* k, *then* $\mathcal{C}_F(\mathbb{D})^{\perp} \subset \mathcal{C}_F(\mathbb{D})$ *and* $\mathcal{C}_F(\mathbb{D})$ *has dimension* $(v + 1)/2$.

In Sections 1.6.3 and 1.6.4, we introduced the projective space $\mathrm{PG}(m, \mathrm{GF}(q))$ and the affine space $\mathrm{AG}(m, \mathrm{GF}(q))$, defined their points and k-flats. If we define the incidence relation as the natural containment of sets, we can derive designs from the finite geometries whose blocks are the k-flats. The theorem below documents these designs.

Theorem 3.8. *Let notation and symbols be the same as in Sections* 1.6.3 *and* 1.6.4 *as well as above.*

(a) *The points and t-dimensional flats in* $\mathrm{PG}(m, \mathrm{GF}(q))$ *form a 2-design with parameters*

$$\left(\begin{bmatrix} m + 1 \\ 1 \end{bmatrix}_q, \begin{bmatrix} t + 1 \\ 1 \end{bmatrix}_q, \begin{bmatrix} m - 1 \\ t - 1 \end{bmatrix}_q \right).$$

(b) *The points and t-dimensional flats of* $\mathrm{AG}(m, \mathrm{GF}(q))$ *form a 2-design with parameters*

$$\left(q^m, q^t, \begin{bmatrix} m - 1 \\ t - 1 \end{bmatrix}_q \right).$$

If $q = 2$, *then it is a 3-design, with* $\lambda = \begin{bmatrix} m-1 \\ t-1 \end{bmatrix}_q$.

So far, we have constructed codes from designs. Conversely, we can obtain t-designs from linear codes. Let C be an code of length n. Consider all the codewords of weight w in C. Let $\mathbf{c} = (c_1, c_2, \ldots, c_n)$ be a codeword of weight w in C. The *support* of $\mathbf{c}$ is defined by

$$\text{suppt}(\mathbf{c}) = \{1 \leq i \leq n : c_i \neq 0\} \subseteq \{1, 2, 3, \ldots, n\}.$$

Two different codewords of weight w may have the same support. Let $\mathcal{P} = \{1, 2, \ldots, n\}$ and $\mathcal{B}$ be the set of the supports of the codewords of weight w in C, where no repeated blocks are allowed. Let the incidence relation $\mathcal{I}$ be the usual containment of sets. Then it is possible that $(\mathcal{P}, \mathcal{B}, \mathcal{I})$ is a t-design for some t. In this case, we say that the codewords of weight w in C hold a t-design.

Example 3.7. The binary $[7, 4, 3]$ Hamming code has the weight enumerator $1 + 7x^3 + 7x^4 + x^7$. It has 7 codewords of weight 3. The 7 codewords are exactly the row vectors of the matrix in (3.1). The supports of 7 codewords form the set $\mathcal{B}$ in Example 3.1. Hence, the codewords of weight 3 in the binary $[7, 4, 3]$ Hamming code hold a 2-$(7, 3, 1)$ design, which is the Fano plane.

The Assmus–Mattson Theorem below describes t-designs from linear codes [Assmus and Mattson (1969)].

Theorem 3.9 (Assmus–Mattson). *Let C be an $[n, k, d]$ code over $\text{GF}(q)$. Let $d^\perp$ denote the minimum distance of $C^\perp$. Let w be the largest integer satisfying $w \leq n$ and*

$$w - \left\lfloor \frac{w + q - 2}{q - 1} \right\rfloor < d.$$

Define $w^\perp$ analogously using $d^\perp$. Let $(A_i)_{i=0}^n$ and $(A_i^\perp)_{i=0}^n$ denote the weight distribution of C and $C^\perp$, respectively. Fix a positive integer t with $t < d$, and let s be the number of i with $A_i^\perp \neq 0$ for $0 \leq i \leq n - t$. Suppose $s \leq d - t$. Then

(1) *the codewords of weight i in C hold a t-design provided $A_i \neq 0$ and $d \leq i \leq w$, and*

(2) *the codewords of weight i in $C^\perp$ hold a t-design provided $A_i^\perp \neq 0$ and $d^\perp \leq i \leq \min\{n - t, w^\perp\}$.*

The Assmus–Mattson Theorem applied to C is most useful when $C^\perp$ has only a few nonzero weights. The Assmus–Mattson Theorem has been the main tool in discovering designs in codes. We refer the reader to Huffman and Pless (2003)[Section 8.4] for more information.

3.3 *t*-Adesigns and Their Codes

Let $\mathbb{D} = (\mathcal{P}, \mathcal{B}, \mathcal{I})$ be an incidence structure with $v \geq 1$ points and $b \geq 1$ blocks, where every block has size k. If every subset of t points of $\mathcal{P}$ is incident with either λ or $\lambda + 1$ blocks of $\mathcal{B}$, then $\mathbb{D}$ is called a t-(v, k, λ) *adesign*, or simply t-*adesign*. The integers t, v, k, λ are referred to as the *parameters* of the adesign. A t-adesign is called *symmetric* if $v = b$.

Example 3.8. Let $\mathcal{P} = Z_{13} = \{0, 1, 2, \ldots, 12\}$ and $\mathcal{B}$ consists of the following blocks of size 6:

$$
\begin{aligned}
B_0 &= \{1, 3, 4, 9, 10, 12\}, \\
B_1 &= \{0, 2, 4, 5, 10, 11\}, \\
B_2 &= \{1, 3, 5, 6, 11, 12\}, \\
B_3 &= \{0, 2, 4, 6, 7, 12\}, \\
B_4 &= \{0, 1, 3, 5, 7, 8\}, \\
B_5 &= \{1, 2, 4, 6, 8, 9\}, \\
B_6 &= \{2, 3, 5, 7, 9, 10\}, \\
B_7 &= \{3, 4, 6, 8, 10, 11\}, \\
B_8 &= \{4, 5, 7, 9, 11, 12\}, \\
B_9 &= \{0, 5, 6, 8, 10, 12\}, \\
B_{10} &= \{0, 1, 6, 7, 9, 11\}, \\
B_{11} &= \{1, 2, 7, 8, 10, 12\}, \\
B_{12} &= \{0, 2, 3, 8, 9, 11\}.
\end{aligned}
$$

Let $\mathcal{I}$ be the membership of sets. Then the incidence structure $(\mathcal{P}, \mathcal{B}, \mathcal{I})$ is a 2-$(13, 6, 2)$ adesign.

The incidence matrix, p-rank, and the code of t-adesigns are defined in the same way as those of t-designs. However, the theory of t-designs may not work for t-adesigns in many cases. It is open how to develop general theory for t-adesigns.

Chapter 4

Difference Sets

In this chapter, we introduce the basic theory of difference sets and present some important families of difference sets for applications later. We will not provide proofs of the basic theory and the difference set property of the difference sets, but will provide the reader with a reference, where a proof is available.

4.1 Fundamentals of Difference Sets

In this monograph, we consider only difference sets in finite abelian groups A. To make the notation of the order of the abelian group consistent with the length of a linear code, we use n to denote the order of abelian groups A that contain a difference set. We also write the operation of an abelian group additively as $+$, and call the identity element of A the zero element.

A subset D of size k in an abelian group $(A, +)$ with order n is called an (n, k, λ) *difference set* in $(A, +)$ if the multiset $\{a_1 - a_2 : a_1 \in A, a_2 \in A\}$ contains every nonzero element of A exactly λ times.

For convenience later, we define the *difference function* of a subset D of $(A, +)$ as

$$\text{diff}_D(x) = |D \cap (D + x)|, \qquad (4.1)$$

where $D + x = \{y + x : y \in D\}$.

In terms of the difference function, a subset D of size k in an abelian group $(A, +)$ with order n is called an (n, k, λ) *difference set* in $(A, +)$ if the difference function $\text{diff}_D(x) = \lambda$ for every nonzero $x \in A$. A difference set D in $(A, +)$ is called *cyclic* if the abelian group A is so. The *order* of an (n, k, λ) difference set is defined to be $k - \lambda$.

By definition, if an (n, k, λ) difference set exists, then

$$k(k - 1) = (n - 1)\lambda. \qquad (4.2)$$

If D is an (n, k, λ) difference set in $(A, +)$, its *complement*, $D^c = A \backslash D$, is an $(n, n - k, n - 2k + \lambda)$ difference set in $(A, +)$.

Example 4.1. Let $n = 7$ and let $D = \{1, 2, 4\}$. Then D is a $(7, 3, 1)$ difference set in $(\mathbb{Z}_7, +)$. Its complement, $D^c = \{0, 3, 5, 6\}$, is a $(7, 4, 2)$ difference set in $(\mathbb{Z}_7, +)$.

Recall that a symmetric 2-(v, k, λ) design with $\lambda < k < v - 1$ is simply called a (v, k, λ)-design. The order of any (v, k, λ)-design is $k - \lambda$ and $\lambda(v - 1) = k(k - 1)$.

Let D be an (n, k, λ) difference set in an abelian group $(A, +)$. We associate D with an incidence structure $\mathbb{D}$, called the *development* of D, by defining $\mathbb{D} = (\mathcal{P}, \mathcal{B}, \mathcal{I})$, where $\mathcal{P}$ is the set of the elements in A,

$$\mathcal{B} = \{a + D : a \in A\},$$

and the incidence $\mathcal{I}$ is the membership of sets. Each block $a + D = \{a + x : x \in D\}$ is called a *translate* of D.

It is an easy exercise to prove the following result [Assmus and Key (1992a)][Theorem 4.4.1].

Theorem 4.1. *Let $\mathbb{D}$ be the development of an (n, k, λ) difference set in a group A. Then $\mathbb{D}$ is an (n, k, λ) design, and $A \subseteq \mathrm{Aut}(\mathbb{D})$ with A acting regularly on points and blocks of $\mathbb{D}$.*

The development $\mathbb{D}$ of a difference set D is called a *difference set design* and also the *translate design* of D.

Example 4.2. The development of the $(7, 3, 1)$ difference set in $(\mathbb{Z}_7, +)$ given in Example 4.1 is the projective plane of order 2.

The converse of Theorem 4.1 is also true and is stated as follows [Assmus and Key (1992a)][Theorem 4.4.2].

Theorem 4.2. *If D is an (n, k, λ) design with a group of automorphisms acting regularly on points, then D is the development of a difference set for that group.*

Let D be a difference set in $(A, +)$ and let $\sigma \in \mathrm{Aut}(A)$. Then $\sigma(D) = \{\sigma(d) : d \in D\}$ is also a difference set in $(A, +)$. Furthermore, since $\sigma(D + a) = \sigma(D) + \sigma(a)$, the automorphism σ will induce an automorphism of the development of D when $\sigma(D) = D + a$ for some element $a \in A$. We are now ready to introduce the important concept of multipliers of difference sets.

If D is an (n, k, λ) difference set in $(A, +)$ and $\mu \in \mathrm{Aut}(A)$, then μ is a *multiplier* of D if $\mu(D) = D + a$ for some $a \in A$. It is obvious that the set of multipliers of a difference set D in a group A form a subgroup of $\mathrm{Aut}(A)$.

In this monograph, we are much more interested in difference sets in $(\mathbb{Z}_n, +)$ as they give cyclic codes. The group $(\mathbb{Z}_n^*, \times)$, i.e., the group of invertible integers in $\mathbb{Z}_n$ with the integer multiplication, acts on the group $(\mathbb{Z}_n, +)$ by multiplication. In this case, $\ell \in \mathbb{Z}_n^*$ is a multiplier of a difference set D in $(\mathbb{Z}_n, +)$ if $\ell D = D + a$ for some $a \in A$, where $\ell D = \{\ell d \bmod n : d \in D\}$. Such a multiplier is called a *numerical multiplier*. It is well-known that $\mathrm{Aut}((Z_n, +))$ is isomorphic to $(\mathbb{Z}_n^*, \times)$.

Example 4.3. The set $D = \{1, 3, 4, 5, 9\}$ is a $(11, 5, 2)$ difference set in $(\mathbb{Z}_{11}, +)$. Its multiplier group is $(\mathbb{Z}_{11}^*, \times)$.

The following is a result about multipliers [Beth, Jungnickel and Lenz (1999)][Theorem 4.4].

Theorem 4.3. *Let D be an abelian (n, k, λ) difference set, and let p be a prime dividing $k - \lambda$ but not n. If $p > \lambda$, then p is a multiplier of D.*

It is conjectured that Theorem 4.3 holds without the restriction $p > \lambda$, i.e., every prime divisor of $k - \lambda$ is a multiplier.

The following results is due to Bruck–Ryser–Chowla Theorem.

Theorem 4.4. *If $\ell \equiv 1, 2 \pmod 4$, and the square part of ℓ is divisible by a prime $p \equiv 3 \pmod 4$, then no difference set of order ℓ exists.*

Studying the group of numerical multipliers is useful for proving the nonexistence of abelian planar difference sets. McFarland and Rice proved the following [McFarland and Rice (1978)].

Theorem 4.5. *Let D be an abelian (n, k, λ) difference set in $(A, +)$, and let M be the group of numerical multipliers of D. Then there exists a translate of D that is fixed by every element of M.*

A lot of results about multipliers have been developed. The reader is referred to Beth, Jungnickel and Lenz (1999)[Chapter VI] for further information.

Difference sets can also be characterized with group characters. For any character χ of an abelian finite group $(A, +)$, recall that its conjugate $\bar{\chi}$ is defined $\bar{\chi}(a) = \overline{\chi(a)}$. For any subset D of A, we define

$$\chi(D) = \sum_{d \in D} \chi(d).$$

We have then the following result [Beth, Jungnickel and Lenz (1999)] [Lemma 3.12].

Theorem 4.6. *A subset D of a finite abelian group $(A, +)$ of order n is an (n, k, λ) difference set if and only if for every nontrivial complex character χ of $(A, +)$,*

$$|\chi(D)|^2 = \chi(D)\bar{\chi}(D) = \chi(D)\chi(-D) = k - \lambda.$$

This theorem is important as it can be employed to prove the difference set property in many cases, while other tools and methods may not work easily.

4.2 Divisible and Relative Difference Sets

Let $(A, +)$ be a group of order mn and $(N, +)$ a subgroup of A of order n. A k-subset D of A is an $(m, n, k, \lambda_1, \lambda_2)$ *divisible difference set* if the multiset $\{*d_1 - d_2 : d_1, d_2 \in D, d_1 \neq d_2*\}$ contain every nonidentity element of N exactly λ_1 times and every element of $A \setminus N$ exactly λ_2 times. If $\lambda_1 = 0$, D is called an (m, n, k, λ_2) *relative difference set*, and N is called the *forbidden subgroup*.

Example 4.4. The set $D = \{0, 1, 3\}$ is a $(4, 2, 3, 0, 1)$ divisible difference set in $(\mathbb{Z}_8, +)$, and also a relative difference set in $(\mathbb{Z}_8, +)$ relative to the subgroup $\{0, 4\}$.

There is a huge amount of references on divisible and relative difference sets. The reader is referred to Beth, Jungnickel and Lenz (1999)[Chapter VI] for information. In this monograph, we will not study divisible and relative difference sets, but will need the concepts later.

4.3 Characteristic Sequence of Difference Sets in $\mathbb{Z}_n$

Let D be any subset of $(\mathbb{Z}_n, +)$, where $n \geq 2$ and n is a positive integer. The *characteristic sequence* of D, denoted by $s(D)^{\infty}$, is a binary sequence of period n, where

$$s(D)_i = \begin{cases} 1 & \text{if } i \bmod n \in D, \\ 0 & \text{otherwise.} \end{cases} \tag{4.3}$$

Difference sets in $(\mathbb{Z}_n, +)$ can be characterized with its characteristic sequence $s(D)^{\infty}$ as follows.

Theorem 4.7. *Let D be any subset of $(\mathbb{Z}_n, +)$ with size k. Then D is an (n, k, λ) difference set in $(\mathbb{Z}_n, +)$ if and only if*

$$\mathrm{AC}_{s(D)}(w) = n - 4(k - \lambda) \tag{4.4}$$

for any nonzero w in $\mathbb{Z}_n$, where $\mathrm{AC}_{s(D)}(w)$ is the autocorrelation function of the periodic binary sequence $s(D)^{\infty}$ defined in Section 1.8.2.

It is straightforward to prove this theorem. Hence, cyclic difference sets can be defined with the language of sequences.

The following result follows from Theorem 4.7.

Theorem 4.8. *Let $s(D)^\infty$ denote the characteristic sequence of a subset D of $\mathbb{Z}_n$. Let $n \equiv 3 \pmod 4$. Then $\mathrm{AC}_{s(D)}(w) = -1$ for all $w \not\equiv 0 \pmod n$ if and only if D is an $(n, (n+1)/2, (n+1)/4)$ or $(n, (n-1)/2, (n-3)/4)$ difference set in $\mathbb{Z}_n$.*

Theorem 4.8 shows that the characteristic sequence of certain cyclic difference sets has optimal (ideal) autocorrelation.

4.4 Characteristic Functions of Difference Sets

Let D be any subset of an abelian group $(A, +)$ of order n. The *characteristic function* of D, denoted by ξ_D, is defined by

$$\xi_D(x) = \begin{cases} 1 & \text{if } x \in D, \\ 0 & \text{otherwise.} \end{cases} \tag{4.5}$$

Difference sets in $(A, +)$ can be characterized with its characteristic function ξ_D as follows.

Theorem 4.9. *Let D be any k-subset of an abelian group $(A, +)$ with order n. Then D is an (n, k, λ) difference set in $(A, +)$ if and only if*

$$\mathrm{AC}_{\xi_D}(w) := \sum_{a \in A} (-1)^{\xi_D(a+w) - \xi_D(a)} = n - 4(k - \lambda) \tag{4.6}$$

for any nonzero w in A.

It is also easy to prove this theorem. Thus, difference sets can be defined in terms of the characteristic function ξ_D. In many cases, the characteristic function of a difference set has optimum nonlinearity [Carlet and Ding (2004)].

4.5 Cyclotomic Difference Sets

We need to recall the cyclotomic classes in $\mathrm{GF}(r)$ dealt with in Section 1.4.2. Let r be a power of a prime p. Let $r - 1 = nN$ for two positive integers $n > 1$ and $N > 1$, and let α be a fixed primitive element of $\mathrm{GF}(r)$. Define $C_i^{(N,r)} = \alpha^i \langle \alpha^N \rangle$ for $i = 0, 1, \ldots, N - 1$, where $\langle \alpha^N \rangle$ denotes the subgroup of $\mathrm{GF}(r)^*$ generated by α^N. The cosets $C_i^{(N,r)}$ are the cyclotomic classes of order N in $\mathrm{GF}(r)$. The cyclotomic numbers of order N are defined by

$$(i, j)^{(N,r)} = \left| (C_i^{(N,r)} + 1) \cap C_j^{(N,r)} \right|$$

for all $0 \le i \le N - 1$ and $0 \le j \le N - 1$.

The following is an existence result [Jungnickel (1992)].

Lemma 4.1. *Let $r = nN + 1$ be an odd prime. If a union of cyclotomic classes of order N forms a difference set, then n is odd and N is even.*

Cyclotomic classes in $\mathrm{GF}(r)$ are very useful building blocks for constructing difference sets. The following theorem plays an important role in this direction [Lehmer (1953)].

Theorem 4.10 (Lehmer). *Let r be a prime power. Then $C_0^{(N,r)}$ is a difference set in $(\mathrm{GF}(r), +)$ with parameters $(r, n, (n - 1)/N)$ if and only if*

$$(i, 0)^{(N,r)} = (n - 1)/N \text{ for all } i \in \{0, 1, \dots, N - 1\}.$$

Similarly, then $C_0^{(N,r)} \cup \{0\}$ is a difference set in $(\mathrm{GF}(r), +)$ with parameters $(r, n + 1, (n + 1)/N$ if and only if

$$1 + (0, 0)^{(N,r)} = (i, 0)^{(N,r)} = (n + 1)/N \text{ for all } i \in \{1, \dots, N - 1\}.$$

In either case, the only multipliers of the difference set are the elements of $C_0^{(N,r)}$.

Proof.   We prove the conclusion of the first part only. The desired conclusion of the second part can be proved similarly. Let $D = C_0^{(N,r)}$. Let $a \in C_i^{(N,r)}$ for some i. Then $a^{-1} \in C_{N-i}^{(N,r)}$ and

$$\begin{aligned}
\mathrm{diff}_D(a) &= \left| \left(C_0^{(N,r)} + a \right) \cap C_0^{(N,r)} \right| \\
&= \left| \left(a^{-1} C_0^{(N,r)} + 1 \right) \cap a^{-1} C_0^{(N,r)} \right| \\
&= \left| \left(C_{N-i}^{(N,r)} + 1 \right) \cap C_{N-i}^{(N,r)} \right| \\
&= (N - i, N - i)^{(N,r)} \\
&= (i, 0)^{(N,r)},
\end{aligned}$$

where the last equality follows from Equation (B) in Theorem 1.17. Hence, D is a (r, n, λ) difference set if and only if $(i, 0)^{(N,r)} = \lambda$ for all i. When $(i, 0)^{(N,r)} = \lambda$ for all i, it follows from Equation (D) in Theorem 1.17 that $\lambda = (n - 1)/N$. $\square$

Plugging the cyclotomic numbers of specific orders documented in Section 1.4.2 and in Storer (1967) into Theorem 4.10, we obtain the following theorem.

Theorem 4.11. *Below is a list of cyclotomic difference sets in* $(GF(r), +)$.

(1) $C_0^{(2,r)}$ *with parameters* $\left(r, \frac{r-1}{2}, \frac{r-3}{4}\right)$, *where* $r \equiv 3 \pmod 4$ *[Paley (1933)]*.

(2) $C_0^{(4,r)}$ *with parameters* $\left(r, \frac{r-1}{4}, \frac{r-5}{16}\right)$, *where* $r = 4t^2 + 1$ *and* t *is odd [Lehmer (1953)]*.

(3) $C_0^{(4,r)} \cup \{0\}$ *with parameters* $\left(r, \frac{r+3}{4}, \frac{r+3}{16}\right)$, *where* $r = 4t^2 + 9$ *and* t *is odd [Lehmer (1953)]*.

(4) $C_0^{(8,r)}$ *with parameters* $\left(r, \frac{r-1}{8}, \frac{r-9}{64}\right)$, *where* $r = 8t^2 + 1 = 64u^2 + 9$ *for odd* t *and odd* u *[Lehmer (1953)]*.

(5) $C_0^{(8,r)} \cup \{0\}$ *with parameters* $\left(r, \frac{r+7}{8}, \frac{r+7}{64}\right)$, *where* $r = 8t^2 + 49 = 64u^2 + 441$ *for odd* t *and even* u *[Lehmer (1953)]*.

Example 4.5. Let $r = 37$. Then $C_0^{(4,r)} = \{1, 7, 9, 10, 12, 16, 26, 33, 34\}$, which is a $(37, 9, 2)$ difference set in $(\mathbb{Z}_{37}, +)$.

Example 4.6. Let $r = 13$. Then $C_0^{(4,r)} \cup \{0\} = \{0, 1, 3, 9\}$, which is a $(13, 4, 1)$ difference set in $(\mathbb{Z}_{13}, +)$.

Example 4.7. Let $r = 73$. Then $C_0^{(8,r)} = \{1, 2, 4, 8, 16, 32, 37, 55, 64\}$, which is a $(73, 9, 1)$ difference set in $(\mathbb{Z}_{73}, +)$.

The following construction is due to Hall (1956).

Theorem 4.12 (Hall). *The set* $C_0^{(6,r)} \cup C_1^{(6,r)} \cup C_3^{(6,r)}$ *is a difference set in* $(GF(r), +)$ *with parameters* $\left(r, \frac{r-1}{2}, \frac{r-3}{4}\right)$, *where* $r = 4t^2 + 27$ *and* $\gcd(3, t) = 1$.

Note that under the conditions of Theorem 4.12, the set $C_0^{(6,r)} \cup C_1^{(6,r)} \cup C_3^{(6,r)}$ is a difference set only when the primitive element α of $GF(r)$ employed to define the cyclotomic classes is properly chosen. Different choices of the α may lead to a permutation among the cyclotomic classes $C_i^{(6,r)}$ for $1 \leq i \leq 5$. The following two examples demonstrate this fact.

Example 4.8. Let $r = 31 = 4 \times 1^2 + 27$, and let the primitive element $\alpha = 3$. Then

$$D = C_0^{(6,r)} \cup C_1^{(6,r)} \cup C_3^{(6,r)}$$

$$= \{1, 2, 3, 4, 6, 8, 12, 15, 16, 17, 23, 24, 27, 29, 30\}$$

is a $(31, 15, 7)$ difference set in $(\mathbb{Z}_{31}, +)$.

Example 4.9. Let $r = 31 = 4 \times 1^2 + 27$, and let the primitive element $\alpha = 11$. Then

$$D = C_0^{(6,r)} \cup C_1^{(6,r)} \cup C_3^{(6,r)}$$
$$= \{1, 2, 4, 8, 11, 13, 15, 16, 21, 22, 23, 26, 27, 29, 30\},$$

which is not a difference set as the multiset $D - D$ is given by

$$\{*0^{15}, 1^6, 2^6, 3^8, 4^6, 5^7, 6^8, 7^8, 8^6, 9^7, 10^7, 11^7, 12^8, 13^7, 14^8, 15^6, 16^6,$$
$$17^8, 18^7, 19^8, 20^7, 21^7, 22^7, 23^6, 24^8, 25^8, 26^7, 27^6, 28^8, 29^6, 30^6*\}.$$

Note that the Hall difference sets have the same parameters as the Paley difference sets. One would ask if further difference sets can be constructed by combining a number of cyclotomic classes together. The answer is positive. In Section 4.9, we will demonstrate a class of difference sets that are a union of cyclotomic classes of higher-orders.

The following is due to Hall (1956).

Theorem 4.13. *Let D be a difference set in $(GF(r), +)$, where $r \equiv 1$ (mod 6) and is a prime power. If D admits the sextic residues as multipliers, then we have (up to equivalence) one of the following two cases:*

- $r \equiv 3$ (mod 4) *and D consists of the quadratic residues.*
- r *is of the form $r = 4x^2 + 27$ and $D = C_0^{(6,r)} \cup C_1^{(6,r)} \cup C_3^{(6,r)}$.*

Except the Paley and Hall difference sets, one can prove that no other unions of cyclotomic classes of order six can form a difference set in $(GF(r), +)$. For cyclotomic classes of order 10, we have the following result due to Hayashi (1965).

Theorem 4.14. *Let D be a cyclic difference set in $(GF(r), +)$, where r is a prime congruent to 1 modulo 10, which admits the 10th-powers as multipliers. Then we have (up to equivalence) one of the following two cases:*

- $r \equiv 3$ (mod 4) *and D consists of the quadratic residues.*
- $r = 31$ *and $D = C_0^{(10,r)} \cup C_1^{(10,r)}$.*

There are also other sporadic cyclotomic difference sets. The reader is referred to Jungnickel (1992) for information.

4.6 Twin-Prime Difference Sets

Let n_1 and n_2 be two distinct odd primes, and let $n = n_1 n_2$. Put $N = \gcd(n_1 - 1, n_2 - 1)$. Recall the generalized cyclotomic classes $W_{2i}^{(N)}$ of order N introduced in

Section 1.5. Define

$$D_0^{(2)} = \bigcup_{i=0}^{(N-2)/2} W_{2i}^{(N)} \quad \text{and} \quad D_1^{(2)} = \bigcup_{i=0}^{(N-2)/2} W_{2i+1}^{(N)}.$$

Clearly $D_0^{(2)}$ is a subgroup of $\mathbb{Z}_n^*$ and $D_1^{(2)} = \varrho D_0^{(2)}$, where ϱ was defined in Section 1.5.

Theorem 4.15. *Let*

$$D = D_1^{(2)} \cup \{n_1, 2n_1, \ldots, (n_2 - 1)n_1\}. \tag{4.7}$$

If $n_2 - n_1 = 2$, then D is a difference set in $(\mathbb{Z}_{n_1(n_1+2)}, +)$ with parameters

$$\left(n_1(n_1 + 2), \frac{(n_1 + 1)^2}{2}, \frac{(n_1 + 1)^2}{4} \right).$$

The difference set D and its complement D^c are called the twin-prime difference sets, and were discovered in Whiteman (1962).

Example 4.10. Let $(n_1, n_2) = (5, 7)$. Then

$$D = D_1^{(2)} \cup \{n_1, 2n_1, \ldots, (n_2 - 1)n_1\}$$

$$= \{2, 5, 6, 8, 10, 15, 18, 19, 20, 22, 23, 24, 25, 26, 30, 31, 32, 34\},$$

which is a $(35, 18, 9)$ difference set in $(\mathbb{Z}_{35}, +)$.

The twin-prime difference set D above can be generalized as follows [Stanton and Sprott (1958)].

Theorem 4.16. *Let r and $r + 2$ both be prime powers. Define*

$$D = \left[C_0^{(2,r)} \times C_1^{(2,r+2)} \right] \cup \left[C_1^{(2,r)} \times C_0^{(2,r+2)} \right] \cup [\{0\} \times \mathrm{GF}(r + 2)]. \tag{4.8}$$

Then D is a difference set in $(\mathrm{GF}(r) \times \mathrm{GF}(r + 2), +)$ with parameters

$$\left(r^2 + 2r, \frac{(r + 1)^2}{2}, \frac{(r + 1)^2}{4} \right).$$

4.7 McFarland Difference Sets and Variations

McFarland discovered a construction of difference sets in 1973, which is described in the next theorem [McFarland (1973)].

Theorem 4.17 (McFarland). *Let q be a prime power and m a positive integer. Let A be an abelian group of order $q^{m+1}(q^m + \cdots + q^2 + q + 2)$ which contains an elementary abelian subgroup E of order q^{m+1}. View E as the additive group of $\mathrm{GF}(q)^{m+1}$. Put $r = (q^{m+1} - 1)/(q - 1)$ and let $H_1, H_2, \ldots, H_r$ be the hyperplanes (i.e., the linear subspaces of dimension m) of E. If $g_1, g_2, \ldots, g_r$ are distinct coset representatives of E in A, then*

$$D = (g_1 + H_1) \cup (g_2 + H_2) \cup \cdots \cup (g_r + H_r)$$

is a difference set with parameters

$$\left(q^{m+1} \left(1 + \frac{q^{m+1} - 1}{q - 1} \right), q^m \frac{q^{m+1} - 1}{q - 1}, q^m \frac{q^m - 1}{q - 1} \right).$$

When $q = 2$, the parameters of the McFarland difference sets become

$$\left(2^{2t+2}, 2^{2t+1} - 2^t, 2^{2t} - 2^t \right). \tag{4.9}$$

Difference sets with such parameters are a special type of difference sets, named *Hadamard difference sets*. There are a number of constructions of Hadamard difference sets in $(\mathrm{GF}(2)^{2t+2}, +)$, which are equivalent to bent functions. We will introduce them and their codes in Section 6.2.

Dillon gave variations of the McFarland difference sets [Dillon (1985)]. Another variation of the McFarland difference sets is outlined in the following theorem [Spence (1977)].

Theorem 4.18 (Spence). *Let E be the elementary abelian group of order 3^{m+1}, and let A be a group of order $3^{m+1}(3^{m+1} - 1)/2$ containing E. Put $r = (3^{m+1} - 1)/2$ and let $H_1, H_2, \ldots, H_r$ be the subgroups of E of order 3^m. If $g_1, g_2, \ldots, g_r$ are distinct coset representatives of E in A, then*

$$D = (g_1 + (E \backslash H_1)) \cup (g_2 + (E \backslash H_2)) \cup \cdots \cup (g_r + (E \backslash H_r))$$

is a difference set with parameters

$$\left(3^{m+1} \frac{3^{m+1} - 1}{2}, 3^m \frac{3^{m+1} + 1}{2}, 3^m \frac{3^m + 1}{2} \right).$$

4.8 Menon Difference Sets

Difference sets D with parameters

$$(4u^2, 2u^2 \pm u, u^2 \pm u) \tag{4.10}$$

are called *Menon difference sets* ([Menon (1960)] and [Menon (1962)]).

In 1974, Dillion proposed the following construction of Menon difference sets ([Dillon (1974)] and [Dillon (1975)]).

Theorem 4.19. *Let A be a group of order $4u^2$ that contains u pairwise disjoint subgroups of order $2u$, denoted by $A_1, A_2, \ldots, A_u$. Then*

$$D = A_1 \cup A_2 \cup \cdots \cup A_u$$

is a difference set with parameters of (4.10).

For more information on the existence and construction of Menon difference sets, the reader is referred to Jungnickel (1992).

4.9 Skew Hadamard Difference Sets

Difference sets with parameters $(\ell, (\ell+1)/2, (\ell+1)/4)$ and $(\ell, (\ell-1)/2, (\ell-3)/4)$ are called *Paley–Hadamard difference sets*. The Paley and Hall difference sets are then Paley–Hadamard difference sets.

A difference set D in an abelian group $(A, +)$ is called a *skew difference set* if A is the disjoint union of D, $-D$, and $\{0\}$. Skew difference sets must have parameters $(4\ell - 1, 2\ell - 1, \ell - 1)$, and are called *skew Hadamard difference sets* in general.

For many years the only known example of skew Hadamard difference sets was the classical Paley–Hadamard difference sets formed by the nonzero quadratic residues of a finite field and described by Paley in 1933 [Paley (1933)]. It was proved that the only skew difference sets in cyclic groups are the Paley difference sets in groups of prime order [Kelly (1954)] (see also Johnson (1966)). It was conjectured that no other skew Hadamard difference sets exist. Unexpectedly, a family of new skew Hadamard difference sets was discovered in 2006 [Ding and Yuan (2006)]. Afterwards, several new constructions have been developed. In this section, we will introduce the newly constructed skew Hadamard difference sets.

4.9.1 *Properties of skew Hadamard difference sets*

The following result follows from Theorem 4.6, and was employed to prove the difference set property in several papers.

Theorem 4.20. *A subset D of a finite abelian group $(A, +)$ of order n is an $(n, (n-1)/2, (n-3)/4)$ skew Hadamard difference set if and only if for every nontrivial complex character χ of $(A, +)$,*

$$|\chi(D)|^2 = \chi(D)\chi(-D) = \frac{n-1}{4}.$$

The next lemma is useful in dealing with the equivalence of skew Hadamard difference sets [Weng and Hu (2009)].

Lemma 4.2. *Let D be a skew Hadamard difference set in an abelian group $(A, +)$ with $|A| > 3$, and let $a \in A$ and $\sigma \in \mathrm{Aut}(A)$. Then $\sigma(D) + a$ is also a skew Hadamard difference set if and only if $a = 0$.*

This lemma implies that there must exists an automorphism $\sigma \in \mathrm{Aut}(A)$ such that $D_2 = \sigma(D_1)$ if D_1 and D_2 are equivalent skew Hadamard difference sets.

4.9.2 *Construction with planar functions and presemifields*

In this subsection, we introduce the construction of skew Hadamard difference sets in $(\mathrm{GF}(3^m), +)$ for odd m with a class of planar functions [Ding and Yuan (2006)] and its generalizations [Weng, Qiu, Wang and Xiang (2007)].

Before doing this, we need to recall Dickson polynomials of order 5 over $\mathrm{GF}(3^m)$, which are defined as

$$D_5(x, u) = x^5 + ux^3 - u^2x \in \mathrm{GF}(3^m)[x].$$

We then define, for any $u \in \mathrm{GF}(3^m)$, the polynomial

$$g_u(x) = D_5(x^2, u) = x^{10} + ux^6 - u^2x^2.$$

These polynomials $g_u(x)$ turn out to be planar functions [Ding and Yuan (2006)].

Theorem 4.21. *For any $u \in \mathrm{GF}(3^m)$, $g_u(x)$ is a planar function from $\mathrm{GF}(3^m)$ to $\mathrm{GF}(3^m)$, where m is odd.*

Proof. Since m is odd, -1 is a quadratic nonresidue in $\mathrm{GF}(3^m)$. For any $a \neq 0$, we have $g_u(x + a) - g_u(x) = ax^9 + ua^3x^3 + (a^9 + u^2a)x + g_u(a)$. This is a permutation polynomial if and only if the linearized polynomial $L_{a,u}(x) = x^9 + ua^2x^3 + (a^8 + u^2)x$ is a permutation polynomial. Since $L_{a,u}(x)$ is a GF(3)-linear mapping defined over $\mathrm{GF}(3^m)$, it then follows from Lidl and Niederreiter (1997a)[pp. 107–124] that $g_u(x + a) - g_u(x)$ is a permutation polynomial if and only if $L_{a,u}(x) \neq 0$ for all $x \neq 0$. Suppose there is an $x \neq 0$ such that $L_{a,u}(x) = 0$. So, we have $x^8 + a^8 + ua^2x^2 + u^2 = 0$. This is equivalent to $(x^4 + a^4)^2 + (a^2x^2 - u)^2 = 0$. Since $x^4 + a^4 \neq 0$ (otherwise $-1 = (\frac{a^2}{x^2})^2$ is a quadratic residue), we have $-1 = (\frac{a^2x^2 - u}{x^4 + a^4})^2$. This is contrary to the fact that -1 is a quadratic nonresidue. The conclusion then follows. □

Notice that the family of planar functions g_u contains the Coulter–Matthew planar function $x^{10} + x^6 - x^2$ [Coulter and Matthews (1997)].

To describe the family of skew Hadamard difference sets, we still need to prove a few lemmas below.

Lemma 4.3. *For any $u \in \mathrm{GF}(3^m)$, $D_5(x, -u)$ is a permutation polynomial of $\mathrm{GF}(3^m)$, where m is odd.*

Proof. Note that $\gcd(5, 3^{2m} - 1) = 1$ because m is odd. The conclusion follows from Theorem 1.7. □

Lemma 4.4. *Let $m > 0$ be odd. Then $|\mathrm{Im}(g_u)| = (3^m + 1)/2$ for each $u \in \mathrm{GF}(3^m)$, where $\mathrm{Im}(g_u) = \{g_u(x) : x \in \mathrm{GF}(3^m)\}$.*

Proof. For any $u \neq 0$, the conclusion follows from the fact that $g_u(0) = 0$, $g_u(x) = g_u(-x) = D_5(x^2, -u)$ for any $x \in \mathrm{GF}(q)$, and Lemma 4.3. □

Now, we investigate the structure of $\mathrm{Im}(g_u)$ and prove the following result.

Lemma 4.5. *Let $m > 0$ be odd, $u \in \mathrm{GF}(3^m)$, and $x, y \in \mathrm{GF}(3^m)$. Then $g_u(x) + g_u(y) = 0$ if and only if $(x, y) = (0, 0)$.*

Proof. Let $s = x + y$ and $t = x - y$. It is easy to show that $x^4 + y^4 = -(s^4 + t^4)$. Then, we have

$$
\begin{aligned}
g_u(x) &+ g_u(y) \\
&= x^{10} - ux^6 - u^2x^2 + y^{10} - uy^6 - u^2y^2 \\
&= (x^2 + y^2)[(x^8 - x^6y^2 + x^4y^4 - x^2y^6 + y^8) - u(x^4 - x^2y^2 + y^4) - u^2] \\
&= (x^2 + y^2)[x^8 + y^8 + x^4y^4 - x^6y^2 - x^2y^6 - u(x^4 - x^2y^2 + y^4) - u^2] \\
&= (x^2 + y^2)[(x^4 + y^4)^2 - x^2y^2(x^4 + x^2y^2 + y^4) - u(x^4 - x^2y^2 + y^4) - u^2] \\
&= (x^2 + y^2)[(x^4 + y^4)^2 - x^2y^2(x^2 - y^2)^2 - u(x^2 + y^2)^2 - u^2] \\
&= -(s^2 + t^2)[(s^4 + t^4)^2 - (s^2 - t^2)^2s^2t^2 - u(s^2 + t^2)^2 - u^2] \\
&= -(s^2 + t^2)[s^8 + t^8 - s^4t^4 - (s^4 + t^4 + s^2t^2)s^2t^2 - u(s^4 + t^4 - s^2t^2) - u^2] \\
&= -(s^2 + t^2)[s^8 + t^8 - s^6t^2 - s^2t^6 + s^4t^4 - us^4 - ut^4 + us^2t^2 - u^2].
\end{aligned}
$$

Let $\beta \in \mathrm{GF}(3^2)$ such that $\beta^2 = -1$. Then $\beta \notin \mathrm{GF}(3^m)$ but $\beta \in \mathrm{GF}(3^{2m})$. Then, it is easy to verify

$$
\begin{aligned}
g_u(x) + g_u(y) = -(s^2 + t^2)&[(s^4 + t^4 + s^2t^2 + u) + \beta(s^2t^2 + u)] \\
&\times [(s^4 + t^4 + s^2t^2 + u) - \beta(s^2t^2 + u)].
\end{aligned} \tag{4.11}
$$

Suppose that

$$(s^4 + t^4 + s^2t^2 + u) + \beta(s^2t^2 + u) = 0 \tag{4.12}$$

for a pair $(s, t) \in \mathrm{GF}(3^m)^2$. If $s^2t^2 + u \neq 0$, then $\beta = -(s^4 + t^4 + s^2t^2 + u)/(s^2t^2 + u) \in \mathrm{GF}(3^m)$, which is contrary to the fact that $\beta \notin \mathrm{GF}(3^m)$. So, we have $s^2t^2 + u = 0$, and consequently $s^4 + t^4 = 0$. Then we must have $s = t = 0$, for otherwise $s \neq 0$, $t \neq 0$, and $-1 = (\frac{s^2}{t^2})^2$, which is contrary to the fact that -1 is a quadratic nonresidue in $\mathrm{GF}(3^m)$. Hence $(s^4 + t^4 + s^2t^2 + u) + \beta(s^2t^2 + u) \neq 0$ for all pairs (s, t).

Similarly, we can prove that $(s^4 + t^4 + s^2t^2 + u) - \beta(s^2t^2 + u) \neq 0$ for all pairs (s, t). It then follows from (4.11) that $g_u(x) + g_u(y) = 0$ if and only if $s^2 + t^2 = 0$.

Again because -1 is a quadratic nonresidue, $s^2 + t^2 = 0$ if and only if $(s, t) = (0, 0)$. By definition, $(x, y) = (0, 0)$ if and only if $(s, t) = (0, 0)$. Whence, $g_u(x) + g_u(y) = 0$ if and only if $(x, y) = (0, 0)$. $\qquad \square$

Lemma 4.6. *Let m be odd, and $u \in \mathrm{GF}(3^m)$. For any $b \in \mathrm{GF}(3^m)^*$, one and only one of the equations $g_u(x) = b$ and $g_u(x) = -b$ has a solution $x \in \mathrm{GF}(3^m)^*$.*

Proof. It follows from Lemma 4.5 that at most one of the two equations has a solution. The conclusion then follows from Lemma 4.4. $\qquad \square$

Note that Lemma 4.6 is not true for all planar functions. One example is the perfect nonlinear function $g(x) = x^{10} + x^6 - x^2 + x$ from $\mathrm{GF}(3^3)$ to $\mathrm{GF}(3^3)$.

The following follows from Lemma 4.6.

Corollary 4.1. *Let m be odd, and $u \in \mathrm{GF}(3^m)$. Then*

$$\mathrm{GF}(3^m) = (\mathrm{Im}(g_u) \backslash \{0\}) \cup [-(\mathrm{Im}(g_u) \backslash \{0\})] \cup \{0\}.$$

Now, we are ready to describe the skew Hadamard difference sets [Ding and Yuan (2006)].

Theorem 4.22 (Ding–Yuan). *Let m be odd, and $u \in \mathrm{GF}(3^m)$. The set*

$$\mathrm{DY}(u) := \mathrm{Im}(g_u) \backslash \{0\} = \{g_u(x) : x \in \mathrm{GF}(3^m)^*\}$$

is a $(3^m, (3^m - 1)/2, (3^m - 3)/4)$ skew Hadamard difference set in the abelian group $(\mathrm{GF}(3^m), +)$.

Proof. By Lemma 4.4, $|\mathrm{DY}(u)| = (3^m - 1)/2$. We now consider the number of solutions $(x, y) \in \mathrm{GF}(3^m)^2$ to the equation $g_u(x) - g_u(y) = b$ for any nonzero

$b \in \mathrm{GF}(3^m)$. By the definition of perfect nonlinearity and Lemma 4.21, for any $a \neq 0$,

$$x - y = a, \quad g_u(x) - g_u(y) = b$$

have exactly one solution (x, y). Therefore $g_u(x) - g_u(y) = b$ has $3^m - 1$ solutions (x, y) for any nonzero $b \in \mathrm{GF}(3^m)$.

By Lemma 4.6, $g_u(x) - g_u(y) = b$ has exactly two solutions of the forms $(0, y)$ and $(x, 0)$ for any nonzero $b \in \mathrm{GF}(3^m)$. Thus, the total number of solutions (x, y) of $g_u(x) - g_u(y) = b$ with $xy \neq 0$ is equal to $3^m - 1 - 2$. Since $g_u(x) = g_u(-x)$, we have then

$$|\mathrm{DY}(u) \cap (\mathrm{DY}(u) + b)| = \frac{3^m - 3}{4}$$

for any nonzero $b \in \mathrm{GF}(3^m)$. This proves that $\mathrm{DY}(u)$ is a $(3^m, (3^m - 1)/2, (3^m - 3)/4)$ difference set in $(\mathrm{GF}(3^m), +)$. It then follows from Corollary 4.1 that this difference set is skew. □

Example 4.11. When $m = 3$, $\mathrm{DY}(-1)$ is a $(27, 13, 6)$ skew Hadamard difference set in $(\mathrm{GF}(27), +)$ and

$$\mathrm{DY}(-1) = \left\{ w^{18}, w^{23}, w^2, w^3, w^{25}, w^{17}, w^7, w^{21}, w^6, w, w^{11}, w^9, 1 \right\},$$

where w is a primitive element of $\mathrm{GF}(27)$ satisfying $w^3 - w + 1 = 0$. In this difference set, only four elements are quadratic residues.

It is straightforward to prove that $\mathrm{DY}(u)$ and $\mathrm{DY}(v)$ are equivalent if both are squares or nonsquares in $\mathrm{GF}(3^m)^*$.

Let $(A, +)$ and $(B, +)$ be two abelian groups of the same order n. A planar function from A to B is said to be *2-to-1* if $|f^{-1}(y)| = 0$ or 2 for every nonzero y in B.

The construction of Theorem 4.22 is generalized as follows in Weng, Qiu, Wang and Xiang (2007).

Theorem 4.23. *Let $(A, +)$ and $(B, +)$ be two abelian groups of the same order n. Let f be a 2-to-1 planar function from A to B, and let $D = \mathrm{Im}(f) \backslash \{0\}$. If $n \equiv 3 \pmod{4}$, then D is a skew Hadamard difference set in $(B, +)$.*

Let $(A, +)$ and $(B, +)$ be two abelian groups. A function from A to B is called *even* if $f(-x) = f(x)$ for every $x \in A$ and $f(0) = 0$. As a consequence of Theorem 4.23, we have the following [Weng, Qiu, Wang and Xiang (2007)].

Corollary 4.2. *Let $(A, +)$ and $(B, +)$ be two abelian groups of the same order n. Let f be a planar function from A to B, and let $D = \text{Im}(f) \backslash \{0\}$. If $n \equiv 3 \pmod 4$ and f is even, then D is a skew Hadamard difference set in $(B, +)$.*

In Section 1.7.3, we discussed the connections between planar functions and presemifields. In view of the correspondence between planar functions and commutative presemifields, the later naturally gives skew Hadamard difference sets. The theorem below was proved in Weng, Qiu, Wang and Xiang (2007).

Theorem 4.24. *Let $(K, +, \times)$ be an odd order presemifield with commutative multiplication. Then the set $\{x \times x : x \in K, x \neq 0\}$ is a skew Hadamard difference set in $(K, +)$ if $|K| \equiv 3 \pmod 4$.*

4.9.3 Construction with Ree-Tifts slice sympletic spreads

It was proved in Ball and Zieve (2004) that

$$f_a(x) = x^{2 \times 3^{(m+1)/2} + 3} + (ax)^{3^{(m+1)/2}} - a^2 x \tag{4.13}$$

is permutation polynomial over $\text{GF}(3^m)$ for any odd $m \geq 3$ and $a \in \text{GF}(3^m)$. This class of permutation polynomials $f_a(x)$ are derived from the Ree-Tifts slice sympletic spreads and were employed to construct skew Hadamard difference sets in Ding, Wang and Xiang (2007).

Define

$$\text{RT}(a) = \{f_a(x^2) : x \in \text{GF}(3^m)^*\}. \tag{4.14}$$

We have then the following [Ding, Wang and Xiang (2007)].

Theorem 4.25 (Ding–Wang–Xiang). *Let m be odd. Then $\text{RT}(a)$ is a skew Hadamard difference set in $(\text{GF}(3^m), +)$.*

The reader is referred to Ding, Wang and Xiang (2007) for a very technical proof of Theorem 4.25. It is also easy to see that $\text{RT}(u)$ and $\text{RT}(v)$ are equivalent if both are squares or nonsquares in $\text{GF}(3^m)^*$. Notice that $f_a(x^2)$ is not a planar polynomial on $\text{GF}(3^m)$.

4.9.4 Cyclotomic constructions

Traditionally, cyclotomy is a powerful tool for constructing difference sets. The Paley skew Hadamard difference sets are cyclotomic classes of order 2. In 2012, Feng and Xiang revisited this approach and constructed new skew Hadamard difference sets using cyclotomic classes of higher-orders [Feng and Xiang (2012)].

Theorem 4.26 (Feng–Xiang). *Let p_1 be a prime, $N \equiv 7$ (mod 8), and let $p \equiv 3$ (mod 4) be a prime such that $f := \mathrm{ord}_N(p) = \phi(N)/2$. Let s be any odd integer, I any subset of $\mathbb{Z}_N$ such that $\{i \pmod{p_1^m} : i \in I\} = \mathbb{Z}_{p_1^m}$ and let $D = \bigcup_{i \in I} C_i^{(N,q)}$, where $q = p^{fs}$. Then D is a skew Hadamard difference set in $(\mathrm{GF}(q), +)$.*

The proof of the difference set property makes use of the evaluation of index 2 Gauss sums developed in Yang and Xia (2010).

The construction of Feng and Xiang was generalized in Momihara (2013) and is described as follows.

Theorem 4.27. *Let p_1 be a prime, $N = 2p_1^m$, and let $p \equiv 3$ (mod 4) be a prime such that $2 \in \langle p \rangle$ (mod p_1^m), $\gcd(p_1, p-1) = 1$, and $f := \mathrm{ord}_N(p)$ is odd. Let s be any odd integer, I any subset of $\mathbb{Z}_N$ such that $\{i \pmod{p_1^m} : i \in I\} = \mathbb{Z}_{p_1^m}$ and let $D = \bigcup_{i \in I} C_i^{(N,q)}$, where $q = p^{fs}$. Then D is a skew Hadamard difference set in $(\mathrm{GF}(q), +)$.*

4.9.5 Construction with Dickson polynomials of order 7

Recall that the Dickson polynomials of order 7 over $\mathrm{GF}(3^m)$ are given by

$$D_7(x, u) = x^7 - ux^5 - u^2x^3 - u^3x \in \mathrm{GF}(3^m)[x].$$

Similar to the construction of Section 4.9.2, the Dickson polynomials of order 7 were used to construct skew Hadamard difference sets in Ding, Pott and Wang (2013). We briefly introduce this construction below.

For each $u \in \mathrm{GF}(3^m)^*$, define

$$D_u := \{D_7(x^2, u) : x \in \mathrm{GF}(3^m)^*\}. \tag{4.15}$$

Theorem 4.28 ([Ding, Pott and Wang (2013)]). *Let $u \in \mathrm{GF}(3^m)^*$ and D_u be defined as in (4.15). If m is odd and $m \not\equiv 0$ (mod 3), D_u is a skew Hadamard difference set in $(\mathrm{GF}(q), +)$.*

The proof of Theorem 4.28 is very technical and can be found in Ding, Pott and Wang (2013). This construction is different from that of Section 4.9.2 in the sense that $D_7(x^2, u)$ is not planar.

4.9.6 Equivalence of skew Hadamard difference sets

In many cases, the p-rank and the Smith normal form can be hired to distinguish inequivalent difference sets (see Xiang (2005) for the definitions and a survey). However, skew Hadamard difference sets with the same parameters have the same p-rank [Jungnickel (1992)][pp. 297–299] and the same Smith normal form [Michael and Wallis (1998)]. Thus they cannot be employed to distinguish inequivalent skew Hadamard difference sets.

A coding-theoretical technique for testing the equivalence of skew Hadamard difference sets was developed in Ding and Yuan (2006). But it is open if this approach is really useful.

For a difference set D in $(\mathrm{GF}(q), +)$, define

$$\mathrm{T}\{a, b\} := |D \cap (D + a) \cap (D + b)|,$$

where $a, b \in \mathrm{GF}(q)^*$. These numbers $\mathrm{T}\{a, b\}$ are called the *triple intersection numbers*. It is clear that two equivalent difference sets have the same distribution of triple intersection numbers.

When $m = 5$, the distributions of the triple intersection numbers of some skew Hadamard difference sets in $(\mathrm{GF}(3^m), +)$ are summarized in Table 4.1. Note that the exponents denote the multiplicities of the corresponding triple intersection numbers.

It is easily checked that the distributions of all these seven classes of skew Hadamard difference sets are pairwise distinct. Therefore, we conclude that all these seven classes of skew Hadamard difference sets are pairwise inequivalent for $m = 5$.

When $m = 7$, we only need to check the maximum and the minimum triple intersection numbers of these difference sets in $(\mathrm{GF}(3^m), +)$, as listed in Table 4.2. Therefore, for $m = 7$, all these seven classes of skew Hadamard difference sets are also pairwise inequivalent. Consequently, the following conjecture was made in Ding, Pott and Wang (2013).

Conjecture 4.1. *For all odd* $m > 7$ *with* $m \not\equiv 0$ (mod 3), *the seven skew Hadamard difference sets,* Paley, DY(1), DY(−1), RT(1), RT(−1), D_1 *and* D_{-1}, *are pairwise inequivalent.*

Table 4.1   Triple intersection numbers of skew difference sets.

DS	Triple intersection numbers with multiplicities
Paley	$26^{1815} 27^{3630} 28^{1815} 29^{7260} \ldots 33^{1815}$
DY(1)	$23^{15} 24^{30} 25^{285} 26^{1245} \ldots 35^{45}$
DY(−1)	$24^{75} 25^{435} 26^{1155} 27^{2385} \ldots 35^{120}$
RT(1)	$24^{75} 25^{330} 26^{1155} 27^{2535} \ldots 35^{105}$
RT(−1)	$24^{90} 25^{330} 26^{1095} 27^{2655} \ldots 35^{120}$
D_1	$23^{30} 24^{60} 25^{390} 26^{1110} \ldots 36^{45}$
D_{-1}	$23^{15} 24^{75} 25^{330} 26^{1005} \ldots 36^{15}$

Table 4.2 Maximum and minimum triple intersection numbers.

DS	MIN	MAX
Paley	261	284
$DY(1)$	246	300
$DY(-1)$	248	297
$RT(1)$	250	295
$RT(-1)$	249	296
D_1	244	301
D_{-1}	246	299

Momihara introduced the triple intersection numbers modulo p and employed them to prove the inequivalence between the Paley and the Feng–Xiang skew difference sets [Momihara (2013)].

Note that there are many planar functions and commutative presemifields. Hence, there are many skew Hadamard difference sets. The classification of them into equivalence classes is an important and difficult problem.

Problem 4.1. *Classify the known skew Hadamard difference sets into equivalence classes.*

4.9.7 *Other constructions*

It would be worthy to inform the reader that Feng gave the first construction of nonabelian skew Hadamard difference sets [Feng (2011)]. Inspired by Feng's work, Muzychuk gave a prolific construction of skew Hadamard difference sets in an elementary abelian group of order q^3 and showed that his skew Hadamard difference sets are inequivalent to the Paley difference sets [Muzychuk (2010)].

4.10 Difference Sets with Twin-Prime Power Parameters

The complement of the difference sets in Theorem 4.16 is stated as follows.

Theorem 4.29. *Let q and $q + 2$ be odd prime powers. Then the set*

$$D = \{(x, y) : x \in GF(q)^*, y \in GF(q+2)^*, \chi(x) = \chi(y)\} \cup \{(x, 0) : x \in GF(q)\}$$

is a $(4\ell - 1, 2\ell - 1, \ell - 1)$ difference set in $(GF(q), +) \times (GF(q+2), +)$, where $\ell = \frac{(q+1)^2}{4}$ and χ is the quadratic character.

A variation of the above construction is the following [Ding, Wang and Xiang (2007)].

Theorem 4.30. *Let q and $q + 2$ be prime powers, and let $q \equiv 3$ (mod 4). Let E be a skew Hadamard difference set in $(\mathrm{GF}(q), +)$. Then the set*

$$D = \{(x, y) : x \in E, \ y \in \mathrm{GF}(q + 2)^*, \ \chi(y) = 1\}$$
$$\cup \{(x, y) : x \in -E, \ y \in \mathrm{GF}(q + 2)^*, \ \chi(y) = -1\}$$
$$\cup \{(x, 0) : x \in \mathrm{GF}(q)\}$$

is a $(4\ell - 1, 2\ell - 1, \ell - 1)$ difference set in $(\mathrm{GF}(q), +) \times (\mathrm{GF}(q + 2), +)$, where $\ell = \frac{(q+1)^2}{4}$.

In view of the fact that there exist inequivalent skew Hadamard difference sets in $(\mathrm{GF}(q), +)$, the following theorem is of interest [Ding, Wang and Xiang (2007)].

Theorem 4.31. *Let q and $q + 2$ be prime powers, and let $q \equiv 3$ (mod 4). Let E and F be inequivalent skew Hadamard difference sets in $(\mathrm{GF}(q), +)$. Then the two difference sets*

$$D = \{(x, y) : x \in E, \ y \in \mathrm{GF}(q + 2)^*, \ \chi(y) = 1\}$$
$$\cup \{(x, y) : x \in -E, \ y \in \mathrm{GF}(q + 2)^*, \ \chi(y) = -1\}$$
$$\cup \{(x, 0) : x \in \mathrm{GF}(q)\}$$

and

$$D' = \{(x, y) : x \in F, \ y \in \mathrm{GF}(q + 2)^*, \ \chi(y) = 1\}$$
$$\cup \{(x, y) : x \in -F, \ y \in \mathrm{GF}(q + 2)^*, \ \chi(y) = -1\}$$
$$\cup \{(x, 0) : x \in \mathrm{GF}(q)\}$$

are inequivalent.

4.11 Difference Sets with Singer Parameters

A difference set in an abelian group A is said to have Singer parameters if their parameters are of the form

$$\left(\frac{\ell^m - 1}{\ell - 1}, \frac{\ell^{m-1} - 1}{\ell - 1}, \frac{\ell^{m-2} - 1}{\ell - 1} \right) \quad \text{or} \quad \left(\frac{\ell^m - 1}{\ell - 1}, \ell^{m-1}, \ell^{m-2}(\ell - 1) \right),$$

where m and ℓ are positive integers. There are a lot of constructions of difference sets with Singer parameters. In this section, we will briefly introduce such differences sets. Further information about these difference sets can be found in Xiang (1999).

4.11.1 *The Singer construction*

The first class of difference sets with Singer parameters was discovered by Singer (1938), and is described in the following theorem.

Theorem 4.32 (Singer difference sets). *Let q be a prime power and let $m \geq 3$ be a positive integer. Let α be a generator of* $GF(q^m)^*$. *Put $n = (q^m - 1)/(q - 1)$. Define*

$$D = \{0 \leq i < n : \text{Tr}_{q^m/q}(\alpha^i) = 0\} \subset \mathbb{Z}_n. \tag{4.16}$$

Then D is a difference set in $(\mathbb{Z}_n, +)$ with parameters

$$\left(\frac{q^m - 1}{q - 1}, \frac{q^{m-1} - 1}{q - 1}, \frac{q^{m-2} - 1}{q - 1} \right). \tag{4.17}$$

Example 4.12. Let $(q, m) = (3, 3)$ and let α be a generator of $GF(q^m)^*$ with $\alpha^3 + 2\alpha + 1 = 0$. Then $n = 13$ and the Singer difference set $D = \{0, 1, 3, 9\}$.

4.11.2 *The HKM and Lin constructions*

In this subsection, we introduce two constructions of cyclic difference sets with the Singer parameters

$$\left(\frac{3^m - 1}{3 - 1}, \frac{3^{m-1} - 1}{3 - 1}, \frac{3^{m-2} - 1}{3 - 1} \right). \tag{4.18}$$

The first is due to Helleseth, Kumar and Martinsen (2001) and is described in the following theorem.

Theorem 4.33. *Let $m = 3k \geq 3$ for some positive integer k and $d = 3^{2k} - 3^k + 1$. Define $n = (3^m - 1)/2$ and*

$$D = \left\{ t : \text{Tr}_{3^m/3}(\alpha^t + \alpha^{td}) = 0, \ 0 \leq t \leq n - 1 \right\}, \tag{4.19}$$

where α is a generator of $GF(3^m)^*$. Then D is a difference set with parameters of (4.18) in $(\mathbb{Z}_n, +)$.

A generalization of the construction of Theorem 4.33 can be found in Helleseth and Gong (2002). The second construction is stated in the next theorem.

Theorem 4.34. *Let $m = 2h+1 > 3$ for some positive integer h and $d = 2 \times 3^h + 1$. Define $n = (3^m - 1)/2$ and*

$$D = \left\{ t : \text{Tr}_{3^m/3}(\alpha^t + \alpha^{td}) = 0, \ 0 \leq t \leq n - 1 \right\}, \tag{4.20}$$

where α is a generator of $\mathrm{GF}(3^m)^*$. Then D is a difference set with parameters of (4.18) in $(\mathbb{Z}_n, +)$.

This family of difference sets was conjectured in Lin (1998). A proof of the difference set property was announced in Arasu, Dillion and Player (2004) and further proof outlines were given in Arasu (2011). Another proof was recently developed in Hu, Shao, Gong and Helleseth (2013).

Example 4.13. Let $m = 5$ and let α be a generator of $\mathrm{GF}(3^5)^*$ with $\alpha^5 + 2\alpha + 1 = 0$. Then

$$D = \{2, 5, 6, 8, 10, 13, 14, 15, 18, 24, 28, 30, 31, 34, 37, 39, 41, 42, 43, 45,$$

$$54, 64, 71, 72, 76, 79, 84, 85, 90, 91, 92, 93, 95, 102, 106, 107, 109,$$

$$111, 116, 117\}$$

which is a $(121, 40, 13)$ difference set in $(\mathbb{Z}_{121}, +)$.

4.11.3 *The Maschietti construction*

An *h-arc* in the projective plane $\mathrm{PG}(2, q)$, with q a prime power, is a set of h-points such that no three of them are collinear. The maximum value for h is $q + 1$ if q is odd, and $q + 2$ if q is even. If q is odd, $(q + 1)$-arcs are called *ovals*. If q is even, $(q + 2)$-arcs are called *hyperovals*.

In 1998, Maschietti discovered a connection between hyperoval sets and difference sets and proved the following result [Maschietti (1998)].

Theorem 4.35 (Maschietti). *Let m be odd and let $n = 2^m - 1$. Then*

$$M_\rho := \log_\alpha \left[\mathrm{GF}(2^m) \backslash \{x^\rho + x : x \in \mathrm{GF}(2^m)\} \right]$$

is a difference set with Singer parameters in $(\mathbb{Z}_n, +))$ if $x \mapsto x^\rho$ is a permutation on $\mathrm{GF}(2^m)$ and the mapping $x \mapsto x^\rho + x$ is two-to-one, where α is a generator of $\mathrm{GF}(2^m)^*$.

In particular, the following ρ yields difference sets:

- $\rho = 2$ *(Singer case).*
- $\rho = 6$ *(Segre case).*
- $\rho = 2^\sigma + 2^\pi$ *with $\sigma = (m + 1)/2$ and $4\pi \equiv 1 \bmod m$ (Glynn I case).*
- $\rho = 3 \cdot 2^\sigma + 4$ *with $\sigma = (m + 1)/2$ (Glynn II case).*

The following is a list of examples of the hyperoval difference sets for all the four cases.

Example 4.14 (Singer Case). Let $m = 5$ and let α be a generator of $GF(2^m)^*$ with $\alpha^5 + \alpha^2 + 1 = 0$. Then

$$M_2 = \{0, 3, 5, 6, 9, 10, 11, 12, 13, 17, 18, 20, 21, 22, 24, 26\}$$

is a $(31, 16, 8)$ difference set in $(\mathbb{Z}_{31}, +)$.

Example 4.15 (Segre Case). Let $m = 5$ and let α be a generator of $GF(2^m)^*$ with $\alpha^5 + \alpha^2 + 1 = 0$. Then

$$M_6 = \{0, 1, 2, 4, 5, 7, 8, 9, 10, 14, 16, 18, 19, 20, 25, 28\}$$

is a $(31, 16, 8)$ difference set in $(\mathbb{Z}_{31}, +)$.

Example 4.16 (Glynn I Case). Let $m = 5$ and let α be a generator of $GF(2^m)^*$ with $\alpha^5 + \alpha^2 + 1 = 0$. Then

$$M_{24} = \{0, 1, 2, 4, 5, 8, 9, 10, 15, 16, 18, 20, 23, 27, 29, 30\}$$

is a $(31, 16, 8)$ difference set in $(\mathbb{Z}_{31}, +)$.

Example 4.17 (Glynn II Case). Let $m = 5$ and let α be a generator of $GF(2^m)^*$ with $\alpha^5 + \alpha^2 + 1 = 0$. Then

$$M_{28} = \{0, 1, 2, 4, 8, 11, 13, 15, 16, 21, 22, 23, 26, 27, 29, 30\}$$

is a $(31, 16, 8)$ difference set in $(\mathbb{Z}_{31}, +)$.

4.11.4 *Another construction*

Let $m \not\equiv 0 \pmod 3$ be a positive integer. Define $\delta_k(x) = x^d + (x + 1)^d \in GF(2^m)[x]$, where $d = 4^k - 2^k + 1$ and $k = (m \pm 1)/3$. Put

$$N_k = \begin{cases} \delta_k(GF(2^m)) & \text{if } m \text{ is odd,} \\ GF(2^m) \backslash \delta_k(GF(2^m)) & \text{if } m \text{ is even.} \end{cases}$$

Theorem 4.36. *Let $n = 2^m - 1$ and α be a generator of $GF(2^m)^*$. Then $D_k := \log_\alpha N_k$ is a difference set with Singer parameters in $(\mathbb{Z}_n, +)$.*

For historical information on this construction and a proof of this theorem, the reader is referred to No, Chung and Yun (1998), Dillon (1999) and Dillon and Dobbertin (2004).

Example 4.18. Let $(m, h) = (5, 2)$ and let α be a generator of $GF(2^m)^*$ with $\alpha^5 + \alpha^2 + 1 = 0$. Then

$$N_k = \{0, 3, 6, 7, 12, 14, 15, 17, 19, 23, 24, 25, 27, 28, 29, 30\}$$

is a $(31, 16, 8)$ difference set in $(\mathbb{Z}_{31}, +)$.

4.11.5 *The Dillon–Dobbertin construction*

The family of cyclic difference sets with Singer parameters in the next theorem were constructed in Dillon and Dobbertin (2004).

Theorem 4.37 (Dillon–Dobbertin). *Let m be a positive integer and let $n = 2^m - 1$. For each h with $1 \le h < m/2$ and $\gcd(h, m) = 1$, define $\Delta_h(x) = (x + 1)^d + x^d + 1$, where $d = 4^h - 2^k + 1$. Then*

$$B_k := \log_\alpha \left[GF(2^m) \backslash \Delta_k(GF(2^m)) \right]$$

is a difference set with Singer parameters in $(\mathbb{Z}_n, +)$, where α is a generator of $GF(2^m)^*$.

Furthermore, for each fixed m, the $\phi(m)/2$ difference sets B_k are pairwise inequivalent, where ϕ is the Euler function.

Example 4.19. Let $(m, h) = (5, 2)$ and let α be a generator of $GF(2^m)^*$ with $\alpha^5 + \alpha^2 + 1 = 0$. Then $B_k = \{0, 1, 2, 4, 5, 7, 8, 9, 10, 14, 16, 18, 19, 20, 25, 28\}$ is a $(31, 16, 8)$ difference set in $(\mathbb{Z}_{31}, +)$.

4.11.6 *The Gordon–Mills–Welch construction*

The following construction of cyclic difference sets is due to Gordon, Mills and Welch (1962).

Theorem 4.38 (Gordon–Mills–Welch). *Let $m_0 > 2$ be a divisor of m and let*

$$R := \{x \in GF(2^m) : \text{Tr}_{2^m/2^{m_0}}(x) = 1\}.$$

If D is any difference set with Singer parameters $(2^{m_0} - 1, 2^{m_0-1}, 2^{m_0-2})$ in $(GF(2^{m_0})^*, \times)$, then $U_D := \log_\alpha[R(D^{(r)})]$ is a DS with Singer parameters in $(\mathbb{Z}_{2^m-1}, +)$, where r is any representative of the 2-cyclotomic coset modulo $(2^{m_0} - 1)$ with $\gcd(r, 2^{m_0} - 1) = 1$, $D^{(r)} := \{y^r : y \in D\}$, and α is a generator of $GF(2^m)^*$.

The Gordon–Mills–Welch (GMW) construction is very powerful and generic. Any difference set with Singer parameters $(2^{m_0} - 1, 2^{m_0-1}, 2^{m_0-2})$ in any subfield

GF(2^{m_0}) can be plugged into it, and may produce new difference set with Singer parameters.

Example 4.20 (Singer Case). Let $m = 6$ and let α be a generator of GF(2^6)* with $\alpha^6 + \alpha^4 + \alpha^3 + \alpha + 1 = 0$. Let D be the Singer difference set in GF(2^3)* and let $r = 1$. Then

$$U_D = \{3, 6, 11, 12, 13, 19, 21, 22, 23, 24, 25, 26, 29, 31, 33, 37, 38,$$

$$41, 42, 43, 44, 46, 47, 48, 50, 52, 53, 55, 58, 59, 61, 62\},$$

which is a $(63, 32, 16)$ difference set in $(\mathbb{Z}_{63}, +)$. Furthermore, $\text{rank}_2(U_D) = 6$.

Example 4.21 (Singer Case). Let $m = 6$ and let α be a generator of GF(2^6)* with $\alpha^6 + \alpha^4 + \alpha^3 + \alpha + 1 = 0$. Let D be the Singer difference set in GF(2^3)* and let $r = 3$. Then

$$U_D = \{1, 2, 4, 5, 7, 8, 10, 14, 15, 16, 17, 20, 21, 28, 30, 31, 32, 34,$$

$$35, 39, 40, 42, 47, 49, 51, 55, 56, 57, 59, 60, 61, 62\},$$

which is a $(63, 32, 16)$ difference set in $(\mathbb{Z}_{63}, +)$. Furthermore, $\text{rank}_2(U_D) = 12$.

The GMW construction is generalized as follows (see Pott (1995) and Jungnickel and Tonchev (1999)).

Theorem 4.39. *Let R be an (m, n, ℓ, h) relative difference set in a group $(A, +)$ relative to a normal subgroup N, and let S be an (n, k, λ) difference set in $(N, +)$. Define*

$$D = S + R = \{s + r : s \in S, \ r \in R\}.$$

If $k^2 h = \ell\lambda$, then D is an $(mn, k\ell, \lambda\ell)$ difference set in $(A, +)$.

To make use of this generalized construction to obtain new difference sets, one needs to find a suitable relative difference set and a difference set in the forbidden subgroup N.

4.11.7 *The No construction*

In this subsection, we introduce a general construction of difference sets with Singer parameters developed in No (2004). This construction contains some earlier constructions as special cases.

A function f from GF(q^m) (respectively, GF(q^m)*) to GF(q) is said to be *d-homogeneous* if

$$f(xy) = y^d f(x)$$

for any $x \in GF(q^m)$ (respectively, $x \in GF(q^m)^*$) and any $y \in GF(q)$. d-homogeneous functions were introduced in Klapper (1995) for the construction of sequences with desirable properties.

A function f from $GF(q^m)$ to $GF(q)$ is called *balanced* if $|f^{-1}(b)| = q^{m-1}$ for each $b \in GF(q)$. A function f from $GF(q^m)^*$ to $GF(q)$ is called *balanced* if $|f^{-1}(b)| = q^{m-1}$ for each $b \in GF(q)^*$ and $|f^{-1}(0)| = q^{m-1} - 1$.

A function f from $GF(q^m)^*$ to $GF(q)$ is called *difference-balanced* if $f(xz) - f(x)$ is balanced for every $z \in GF(q^m)^* \backslash \{1\}$. A function f from $GF(q^m)$ to $GF(q)$ is called *difference-balanced* if $f(xz) - f(x)$ is balanced for every $z \in GF(q^m) \backslash \{0, 1\}$.

We have the following simple but useful conclusion [No (2004)].

Lemma 4.7. *If f from $GF(q^m)^*$ to $GF(q)$ is d-homogeneous and difference-balanced, then f is balanced.*

The general construction of difference sets with Singer parameters developed in No (2004) is the following.

Theorem 4.40 (No). *Let f from $GF(q^m)^*$ to $GF(q)$ be d-homogeneous for some d and difference-balanced. Define*

$$D = \left\{ i : f(\alpha^i) = 0, \ 0 \le i \le n - 1 \right\},$$

where $n = (q^m - 1)/(q - 1)$. Then D is a difference set in $(\mathbb{Z}_n, +)$ with the Singer parameters

$$\left(\frac{q^m - 1}{q - 1}, \frac{q^{m-1} - 1}{q - 1}, \frac{q^{m-2} - 1}{q - 1} \right).$$

It was pointed out in No (2004) that the following functions from $GF(q^m)^*$ to $GF(q)$ are d-homogeneous and difference-balanced for certain d:

(a) $f(x) = \text{Tr}_{q^m/q}(x)$, with $d = 1$.
(b) $f(x) = \text{Tr}_{q^h/q}\left(\left[\text{Tr}_{q^m/q^h}(x) \right]^r \right)$, with $d = r$, where $\gcd(r, q^h - 1) = 1$ and $h \mid m$.
(c) $f(x) = \text{Tr}_{q^l/q}\left[\left(\text{Tr}_{q^h/q^l} \left[\text{Tr}_{q^m/q^h}(x) \right]^r \right)^u \right]$, with $d = ru$, where $\gcd(r, q^h - 1) = 1$, $\gcd(u, q^l - 1) = 1$ and $l \mid h \mid m$.

When the function $f(x)$ in (a) is plugged into Theorem 4.40, we obtain the Singer difference set [Singer (1938)]. When the function $f(x)$ in (b) is plugged into Theorem 4.40, we obtain the GMW difference set [Gordon, Mills and Welch (1962)]. When the function $f(x)$ in (c) is plugged into Theorem 4.40, we obtain the cascaded GMW difference sets [Klapper, Chan and Goresky (1993)].

Theorems 4.33 and 4.34 together with Lemma 4 in No (2004) prove that the following functions from $GF(3^m)^*$ to $GF(q)$ are 1-homogeneous and difference-balanced:

(1) $f(x) = \mathrm{Tr}_{3^m/3}(x + x^d)$, where $m = 3k$ and $d = 3^{2k} - 3^k + 1$.
(2) $f(x) = \mathrm{Tr}_{3^m/3}(x + x^d)$, where $m = 2h + 1$ and $d = 2 \times 3^h + 1$.

Hence the HKM and Lin difference sets are also special cases of the construction in Theorem 4.40.

The following theorem comes from No (2004).

Theorem 4.41. *Let I be a subset of $\mathbb{Z}_{q^h-1}$, where h is a positive integer. Let*

$$B(x) := \sum_{i \in I} b_i \mathrm{Tr}_{q^h/q}(x^i) \in GF(q^h)[x].$$

Assume that $B(x)$ from $GF(q^h)$ to $GF(q)$ is d_1-homogeneous and also difference balanced and $A(x)$ from $GF(q^m)$ to $GF(q^h)$ is d_2-homogeneous and difference balanced, where $\gcd(d_1, q - 1) = 1$, $\gcd(d_2, q^h - 1) = 1$ and $h | m$. Let $1 \le r \le q^h - 1$. Then the function from $GF(q^m)^*$ to $GF(q)$

$$f(x) = \sum_{i \in I} b_i \mathrm{Tr}_{q^m/q}(A(x)^{ir})$$

is both d-homogeneous and difference balanced, where $d = d_1 d_2 r \bmod (q - 1)$. Furthermore, the set

$$D = \left\{ i : f(\alpha^i) = 0, \ 0 \le i \le n - 1 \right\},$$

where $n = (q^m - 1)/(q - 1)$, is a difference set in $(\mathbb{Z}_n, +)$ with the Singer parameters

$$\left(\frac{q^m - 1}{q - 1}, \frac{q^{m-1} - 1}{q - 1}, \frac{q^{m-2} - 1}{q - 1} \right).$$

The difference sets of Theorem 4.41 were called the Generalized GMW difference sets in No (2004).

4.11.8 *Other constructions*

A construction of difference sets with Singer parameters in $GF(3^m)^*/GF(3)^*$ was presented in Arasu and Player (2003), where the 3-rank of these difference sets were also discussed. Some constructions of cyclic difference sets with Singer parameters were documented in Cao (2007).

Codes from Difference Sets

The number of difference sets with Singer parameters in $(\mathbb{Z}_{2^m-1}, +)$ is huge. It may be easy to construct such difference sets. But it is very difficult to settle the equivalence problem due to the GMW and generalized GMW constructions. Below, we introduce a number of conjectured families of cyclic difference sets with Singer parameters.

Conjecture 4.2. *For any* $f \in \mathrm{GF}(2^m)[x]$, *we define* $\mathrm{Im}^*(f) = \{f(x) : x \in \mathrm{GF}(2^m)\}\backslash\{0\}$. *Let* α *be a generator of* $\mathrm{GF}(2^m)^*$. *Then the set* $\log_\alpha(\mathrm{Im}^*(f))$ *is a difference set with Singer parameters in* $(\mathbb{Z}_{2^m-1}, +)$ *for the following polynomials* $f \in \mathrm{GF}(2^m)[x]$:

(a) $f(x) = x + x^{2^m - 2^{(m+4)/2}} + x^{2^m - 4}$, *where* $m \equiv 2$ (mod 4).
(b) $f(x) = x + x^{2^{(m+2)/2}+2} + x^{2^{m-1}+2^{m/2}+2^{(m-2)/2}+2}$, *where* $m \equiv 2$ (mod 4).
(c) $f(x) = x + x^{2^{m-1}+2^{m/2}-1} + x^{2^m - 2^{m/2} - 2^{(m-2)/2}+1}$, *where* $m \equiv 2$ (mod 4).
(d) $f(x) = x + x^{2^{(m+2)/2}+2^{m/2}-1} + x^{2^m - 2^{(m+2)/2}+3}$, *where* $m \equiv 2$ (mod 4).
(e) $f(x) = x + x^{2^{(m+2)/2}+2^{m/2}-1} + x^{2^{(m+4)/2}+2^{m/2}-4}$, *where* $m \equiv 2$ (mod 4).
(f) $f(x) = x + x^{2^{m-1}-2^{(m-4)/2}} + x^{2^m - 2^{m-2} - 2^{(m-2)/2}}$, *where* $m \equiv 2$ (mod 4) *and* $m \geq 6$.
(g) $f(x) = x + x^{2^{m-1}-2^{m/2}} + x^{2^m - 2^{m/2} - 2^{(m-2)/2}}$, *where* $m \equiv 2$ (mod 4).
(h) $f(x) = x + x^{2^{(m+2)/2}-1} + x^{2^m - 2^{(m+2)/2}-4}$, *where* $m \equiv 2$ (mod 4).
(i) $f(x) = x + x^{2^{m/2}-6} + x^{2^{(m+6)/2}-7}$, *where* $m \equiv 2$ (mod 4).
(j) $f(x) = x + x^{2^{m/2}-2} + x^{2^m - 2^{(m+2)/2} - 2^{m/2}+1}$, *where* $m \equiv 2$ (mod 4).
(k) $f(x) = x + x^{2^{m/2}+3} + x^{2^{(m+2)/2}+2}$, *where* $m \equiv 2$ (mod 4).
(l) $f(x) = x + x^{2^{m/2}+3} + x^{2^m - 2^{m/2}+1}$, *where* $m \equiv 2$ (mod 4).
(m) $f(x) = x + x^{2^{m/2}+2} + x^{2^{m-1}+2^{(m-2)/2}}$, *where* $m \equiv 2$ (mod 4).
(n) $f(x) = x + x^{2^{(m-2)/2}+2^{m-1}} + x^{2^{(m-2)/2}+2^{m-1}+1}$, *where* $m \equiv 2$ (mod 4).
(o) $f(x) = x + x^{2^m - 2^{(m+2)/2}} + x^{2^m - 2^{(m+2)/2}+2}$, *where* $m \equiv 0$ (mod 4).
(p) $f(x) = x + x^{2^{m/2}+1} + x^{2^{(m+2)/2}}$, *where* $m \equiv 0$ (mod 2).
(q) $f(x) = x + x^{2^{(m-2)/2}} + x^{2^{m-1}+2^{(m-2)/2}}$, *where* $m \equiv 0$ (mod 2).
(r) $f(x) = x + x^2 + x^{2^{m/2}+1}$, *where* $m \equiv 0$ (mod 2).
(s) $f(x) = x + x^2 + x^{2^m - 2^{m/2}+1}$, *where* $m \equiv 0$ (mod 2).

Conjecture 4.3. *For any* $f \in \mathrm{GF}(2^m)[x]$, *we define* $\mathrm{Im}^*(f) = \{f(x) : x \in \mathrm{GF}(2^m)\}\backslash\{0\}$. *Let* m *be odd and let* α *be a generator of* $\mathrm{GF}(2^m)^*$. *Then the set* $\log_\alpha(\mathrm{Im}^*(f))$ *is a difference set with Singer parameters in* $(\mathbb{Z}_{2^m-1}, +)$ *for the following polynomials* $f \in \mathrm{GF}(2^m)[x]$:

(a) $f(x) = x^{2^m - 17} + x^{(2^m+19)/3} + x$.
(b) $f(x) = x^{2^m - 2^{m-4}-1} + x^{2^m - (2^{m-2}+4)/3} + x$, *where* $m \geq 5$.
(c) $f(x) = x^{2^m - 3} + x^{2^{(m+3)/2}+2^{(m+1)/2}+4} + x$.
(d) $f(x) = x^{2^m - 2^{(m-1)/2}-1} + x^{2^{m-1}-2^{(m-1)/2}} + x$.

(e) $f(x) = x^{2^m-2-(2^{m-1}-2^2)/3} + x^{2^m-2^2-(2^m-2^3)/3} + x$.

(f) $f(x) = x^{2^m-2^{(m+1)/2}+2^{(m-1)/2}} + x^{2^m-2^{(m+1)/2}-1} + x$.

(g) $f(x) = x^{2^m-3(2^{(m+1)/2}-1)} + x^{2^{(m+1)/2}+2^{(m-1)/2}-2} + x$.

(h) $f(x) = x^{2^m-2^{m-2}-1} + x^{2^{m-1}-2} + x$.

(i) $f(x) = x^{2^m-2^{(m+3)/2}-3} + x^{2^{(m+1)/2}+2} + x$.

(j) $f(x) = x^{2^m-3(2^{(m-1)/2}+1)} + x^{2^{m-1}-1} + x$.

(k) $f(x) = x^{2^m-5} + x^6 + x$.

Conjecture 4.4. *For any* $f \in \mathrm{GF}(2^m)[x]$, *we define* $D_f = \{f(x(x+1)) : x \in \mathrm{GF}(2^m)\}\backslash\{0\}$. *Let* α *be a generator of* $\mathrm{GF}(2^m)^*$. *Then the set* $\log_\alpha(D_f)$ *is a difference set with Singer parameters in* $(\mathbb{Z}_{2^m-1}, +)$ *for the following polynomials* $f \in \mathrm{GF}(2^m)[x]$:

(a) $f(x) = x + x^{2^{(m+1)/2}-1} + x^{2^m-2^{(m+1)/2}+1}$, *where m is odd*.

(b) $f(x) = x + x^{(2^m+1)/3} + x^{(2^{m+1}-1)/3}$, *where m is odd*.

(c) $f(x) = x + x^{2^{(m+2)/2}-1} + x^{2^m-2^{m/2}+1}$, *where m is even*.

It is very likely that some of the conjectured difference sets above are obviously equivalent to some existing ones. The reader is invited to work on these conjectures.

4.12 Planar Difference Sets

An (n, k, λ) difference set D is called *planar* if $\lambda = 1$. One fundamental question is whether planar difference sets exist or not.

Example 4.22. When $m = 3$, the Singer difference set of Theorem 4.32 is planar and has parameters $(q^2 + q + 1, q + 1, 1)$, where q is any prime power.

This example shows that planar difference sets with parameters $(\ell^2 + \ell + 1, \ell + 1, 1)$ exist for any prime power ℓ. However, the following conjecture is still open.

Conjecture 4.5 (PPC). *If an abelian planar difference set of order* ℓ *exists, then* ℓ *is a prime power.*

It was verified by Gordon that the Prime Power Conjecture (PPC) is true for order $\ell \leq 2,000,000$ [Gordon (1994)]. There are a number of necessary conditions for the existence of abelian planar difference sets. Below is a summary of them.

Theorem 4.42. *Let D be a planar abelian difference set of order* ℓ. *Then any prime divisor of* ℓ *is a numerical divisor of D.*

Jungnickel and Vedder proved the following two theorems regarding the existence of planar difference sets [Jungnickel and Vedder (1984)].

Theorem 4.43. *If a planar difference set of order $\ell = m^2$ exists in $(A, +)$, then there exists a planar difference set of order m in some subgroup of $(A, +)$.*

Theorem 4.44. *If a planar difference set has even order ℓ, then $\ell = 2, \ell = 4$, or ℓ is a multiple of 8.*

Wilbrink proved the following [Wilbrink (1985)].

Theorem 4.45. *If a planar difference set has order ℓ divisible by 3, then $\ell = 3$ or ℓ is a multiple of 9.*

The next result is due to Lander (1983).

Theorem 4.46. *Let D be a planar abelian difference set of order ℓ in $(A, +)$. If t_1, t_2, t_3, and t_4 are numerical multipliers such that*

$$t_1 - t_2 \equiv t_3 - t_4 \pmod{\exp(A)},$$

then $\exp(A)$ divides the least common multiple $\mathrm{lcm}(t_1 - t_2, t_1 - t_3)$.

Another open problem about planar difference sets is the following.

Conjecture 4.6. *Any abelian planar difference set is cyclic.*

There are much more results regarding the existence of planar difference sets. The reader may consult Jungnickel (1992) and Beth, Jungnickel and Lenz (1999)[Chapter VI] for further information.

Chapter 5

Almost Difference Sets

In this chapter, we introduce almost difference sets and known constructions. We will deal with their applications in linear codes and codebooks later.

5.1 Definitions and Properties

Let $(A, +)$ be an abelian group of order n. A k-subset D of A is an (n, k, λ, t) *almost difference set* of A if the difference function $\mathrm{diff}_D(x)$ takes on λ altogether t times and $\lambda + 1$ altogether $n - 1 - t$ times when x ranges over all the nonzero elements of A, where the difference function $\mathrm{diff}_D(x)$ was defined in (4.1) and is given by

$$\mathrm{diff}_D(w) = |(D + w) \cap D|.$$

If an (n, k, λ, t) almost difference set exits, then

$$k(k - 1) = t\lambda + (n - 1 - t)(\lambda + 1). \tag{5.1}$$

If D is an (n, k, λ, t) almost difference set in an abelian group $(A, +)$, then its *complement*, $D^c = A \setminus D$ is an $(n, n - k, n - 2k + \lambda, t)$ almost difference set in $(A, +)$.

Example 5.1. Let $n = 13$ and let $D = \{1, 3, 4, 9, 10, 12\}$. Then D is a $(13, 6, 2, 6)$ almost difference set in $(\mathbb{Z}_{13}, +)$. Its complement, $D^c = \{0, 2, 5, 6, 7, 8, 11\}$, is a $(7, 7, 3, 6)$ almost difference set in $(\mathbb{Z}_{13}, +)$.

Two different types of almost different sets were introduced in Davis (1992), Ding (1994) and Ding (1995). The almost difference sets defined above are a unified version of the two types [Ding, Helleseth and Martinsen (2001)].

The multipliers, the developments, automorphism groups, and p-ranks of almost difference sets are defined in the same ways as those of difference sets. However, we do not have much general theory about almost difference sets, as the structures of almost difference sets are not as *strong* as those of difference sets.

107

5.2 Characteristic Sequences of Almost Difference Sets in $\mathbb{Z}_n$

Let D be any subset of $(\mathbb{Z}_n, +)$, where $n \geq 2$ and n is a positive integer. The *characteristic sequence* of D, denoted by $s(D)^\infty$, is a binary sequence of period n, where

$$s(D)_i = \begin{cases} 1 & \text{if } i \text{ mod } n \in D, \\ 0 & \text{otherwise.} \end{cases}$$

Almost difference sets in $(\mathbb{Z}_n, +)$ can be characterized with its characteristic sequence $s(D)^\infty$ as follows.

Theorem 5.1. *Let D be any subset of $(\mathbb{Z}_n, +)$ with size k. Then D is an (n, k, λ, t) almost difference set in $(\mathbb{Z}_n, +)$ if and only if*

$$\text{AC}_{s(D)}(w) = \begin{cases} n - 4(k - \lambda), & t \text{ times,} \\ n - 4(k - \lambda - 1), & n - 1 - t \text{ times} \end{cases}$$

when w ranges over all nonzero elements in $\mathbb{Z}_n$, where $\text{AC}_{s(D)}(w)$ is the autocorrelation function of the periodic binary sequence $s(D)^\infty$ defined in Section 1.8.2.

It is straightforward to prove this theorem. Hence, cyclic almost difference sets can be defined with the language of sequences.

The following result follows from Theorem 5.1.

Theorem 5.2. *Let $s(D)^\infty$ denote the characteristic sequence of a subset D of $\mathbb{Z}_n$.*

(1) *Let $n \equiv 1 \pmod{4}$. Then $\text{AC}_{s(D)}(w) \in \{1, -3\}$ for all $w \not\equiv 0 \pmod{n}$ if and only if D is an $(n, k, k - (n + 3)/4, nk - k^2 - (n - 1)^2/4)$ almost difference set in $\mathbb{Z}_n$.*

(2) *Let $n \equiv 2 \pmod{4}$. Then $\text{AC}_{s(D)}(w) \in \{2, -2\}$ for all $w \not\equiv 0 \pmod{n}$ if and only if D is an $(n, k, k - (n + 2)/4, nk - k^2 - (n - 1)(n - 2)/4)$ almost difference set in $\mathbb{Z}_n$.*

(3) *Let $n \equiv 0 \pmod{4}$. Then $\text{AC}_{s(D)}(w) \in \{0, -4\}$ for all $w \not\equiv 0 \pmod{n}$ if and only if D is an $(n, k, k - (n + 4)/4, nk - k^2 - (n - 1)n/4)$ almost difference set in $\mathbb{Z}_n$.*

Theorem 5.2 shows that the characteristic sequence of certain cyclic almost difference sets has optimal autocorrelation. In fact, the original motivation of introducing and studying almost difference sets is the construction of binary sequences with optimal autocorrelation.

5.3 Characteristic Functions of Almost Difference Sets

Let D be any subset of an abelian group $(A, +)$ of order n. The *characteristic function* of D, denoted by ξ_D, is defined by

$$\xi_D(x) = \begin{cases} 1 & \text{if } x \in D, \\ 0 & \text{otherwise.} \end{cases}$$

Almost difference sets in $(A, +)$ can be characterized with its characteristic function ξ_D as follows.

Theorem 5.3. *Let D be any k-subset of an abelian group $(A, +)$ with order n. Then D is an (n, k, λ) difference set in $(A, +)$ if and only if*

$$\mathrm{AC}_{\xi_D}(w) = \sum_{a \in A} (-1)^{\xi_D(a+w) - \xi_D(a)}$$

$$= \begin{cases} n - 4(k - \lambda), & t \text{ times,} \\ n - 4(k - \lambda - 1), & n - 1 - t \text{ times,} \end{cases}$$

when w ranges over all nonzero elements in $\mathbb{Z}_n$.

It is also easy to prove this theorem. This theorem shows that almost difference sets can be defined in terms of the characteristic function ξ_D. In many cases, the characteristic function of almost difference sets has optimum nonlinearity. This is another motivation of studying almost difference sets.

5.4 Cyclotomic Constructions

Cyclotomy is also a powerful tool in constructing almost difference sets. The theorem below summarizes some cyclotomic almost difference sets.

Theorem 5.4. *Below is a list of cyclotomic almost difference sets in $(\mathrm{GF}(r), +)$.*

(1) $C_0^{(2,r)}$ *with parameters* $\left(r, \frac{r-1}{2}, \frac{r-5}{4}, \frac{r-1}{2}\right)$, *where $r \equiv 1$ (mod 4). It is also called the Paley partial difference set.*

(2) $C_0^{(4,r)}$ *with parameters* $\left(r, \frac{r-1}{4}, \frac{r-13}{16}, \frac{r-1}{2}\right)$, *where $r = 25 + 4y^2$ or $r = 9 + 4y^2$ and y^2 is odd [Ding (1997)].*

(3) $C_0^{(4,r)} \cup \{0\}$ *with parameters* $\left(r, \frac{r+3}{4}, \frac{r-5}{16}, \frac{r-1}{2}\right)$, *where $r = 1 + 4y^2$ or $r = 49 + 4y^2$ and y^2 is odd [Ding, Helleseth and Lam (1999)].*

(4) $C_0^{(8,r)}$ *with parameters* $\left(r, \frac{r-1}{8}, \frac{r-41}{64}, \frac{r-1}{2}\right)$, *where $r \equiv 41$ (mod 64) and $r = 19^2 + 4y^2 = 1 + 2b^2$ for some integer y and b or $r \equiv 41$ (mod 64) and $r = 13^2 + 4y^2 = 1 + 2b^2$ for some integer y and b [Ding (1997)].*

(5) $C_0^{(4,r)} \cup C_1^{(4,r)}$ *with parameters* $\left(r, \frac{r-1}{2}, \frac{r-5}{4}, \frac{r-1}{2}\right)$, *where* $r = x^2 + 4$ *and* $x \equiv 1 \pmod 4$ [*Ding, Helleseth and Lam* (1999)].

Example 5.2. Let $r = 29 = 25 + 4 \times 1^2$. Then $C_0^{(4,r)} = \{1, 7, 16, 20, 23, 24, 25\}$, which is a $(29, 7, 1, 14)$ almost difference set in $(\mathbb{Z}_{29}, +)$.

Example 5.3. Let $r = 13 = 9 + 4 \times 1^2$. Then $C_0^{(4,r)} = \{1, 3, 9\}$, which is a $(13, 4, 0, 6)$ almost difference set in $(\mathbb{Z}_{13}, +)$.

Example 5.4. Let $r = 37 = 1 + 4 \times 3^2$. Then

$$C_0^{(4,r)} \cup \{0\} = \{0, 1, 7, 9, 10, 12, 16, 26, 33, 34\},$$

which is a $(37, 10, 2, 18)$ almost difference set in $(\mathbb{Z}_{37}, +)$.

Example 5.5. Let $r = 53 = 49 + 4 \times 1^2$. Then

$$C_0^{(4,r)} \cup \{0\} = \{0, 1, 10, 13, 15, 16, 24, 28, 36, 42, 44, 46, 47, 49\},$$

which is a $(53, 14, 3, 26)$ almost difference set in $(\mathbb{Z}_{53}, +)$.

The existence of a difference set $C_0^{(8,r)}$ depends on the answer to the following open problem whose answer is very likely negative.

Problem 5.1. *Is there a set of integers* r, y *and* b *such that*

- r *is a prime power with* $r \equiv 41 \pmod{64}$, *and*
- $r = 19^2 + 4y^2 = 1 + 2b^2$ *or* $r = 13^2 + 4y^2 = 1 + 2b^2$?

Example 5.6. Let $r = 29 = 5^2 + 4$ and let 2 be the primitive element of GF(r) for defining the cyclotomic classes of order 4. Then

$$C_0^{(4,r)} \cup C_1^{(4,r)} = \{1, 2, 3, 7, 11, 14, 16, 17, 19, 20, 21, 23, 24, 25\},$$

which is a $(29, 14, 6, 14)$ almost difference set in $(\mathbb{Z}_{29}, +)$.

Cyclotomic classes $C_0^{(8,r)}$ of order 8 can be employed to construct almost difference sets in the following way.

Theorem 5.5 ([**Ding, Pott and Wang** (2014)]). *Suppose that* $q = t^2 + 2 \equiv 3 \pmod 8$ *is a prime power, where* t *is an integer. Let* $r = q^2$. *Then the set*

$$D = C_0^{(8,r)} \cup C_1^{(8,r)} \cup C_2^{(8,r)} \cup C_5^{(8,r)}$$

is an almost difference set in (GF(r), +) *with parameters* $(r, (r-1)/2, (r-5)/4, (r-1)/2)$, *provided that the generator* α *of* GF(r)* *employed to define the cyclotomic classes of order 8 is properly chosen.*

Example 5.7. If $q = 3^3 = 5^2 + 2$, then

$$D = C_0^{(8,r)} \cup C_1^{(8,r)} \cup C_2^{(8,r)} \cup C_5^{(8,r)}$$

is a $(729, 364, 181, 364)$ almost difference set in $(GF(3^6), +)$.

Cyclotomic classes of order 12 can also be employed to construct almost difference sets. The next theorem documents a recent construction.

Theorem 5.6 ([Nowak, Olmez and Song (2013)]). *Let r be a prime of the form $r = s^2 + 4 \equiv 1 \pmod{12}$, where $s \equiv 1 \pmod 4$. Then*

$$D = C_0^{(12,r)} \cup C_1^{(12,r)} \cup C_4^{(12,r)} \cup C_5^{(12,r)} \cup C_8^{(12,r)} \cup C_9^{(12,r)}$$

is an almost difference set in $(GF(r), +)$ with parameters $(r, (r-1)/2, (r-5)/4, (r-1)/2$ for a properly chosen primitive element of $GF(r)$ employed to define the cyclotomic classes of order 12.

Example 5.8. Let $r = 13$ and let 2 be the primitive root of 13 that is used to define the cyclotomic classes of order 12. Then

$$D = C_0^{(12,r)} \cup C_1^{(12,r)} \cup C_4^{(12,r)} \cup C_5^{(12,r)} \cup C_8^{(12,r)} \cup C_9^{(12,r)} = \{1, 2, 3, 5, 6, 9\},$$

which is a $(13, 6, 2, 6)$ almost difference set in $(\mathbb{Z}_{13}, +)$.

Example 5.9. Let $r = 229$ and let 5 be the primitive root of 229 that is used to define the cyclotomic classes of order 12. Then

$$D = C_0^{(12,r)} \cup C_1^{(12,r)} \cup C_4^{(12,r)} \cup C_5^{(12,r)} \cup C_8^{(12,r)} \cup C_9^{(12,r)}$$

which is a $(229, 114, 56, 114)$ almost difference set in $(\mathbb{Z}_{229}, +)$.

The following construction of almost difference sets was discovered in Zhang, Lei and Zhang (2006).

Theorem 5.7. *Let r_1 and r_2 are two prime powers. Let $A = (GF(r_1), +) \times (GF(r_2), +)$. Define*

$$D = \{(a, b) \in A : a, b \text{ both squares or nonsquares}\}.$$

Then D is an almost difference set in A if and only if $r_1 - r_2 = \pm 2$ or $r_1 = r_2$.

The following theorem documents a class of almost difference sets constructed in Sidelnikov (1969) and rediscovered in Lempel, Cohn and Eastman (1977).

Theorem 5.8. *Let r be an old prime power. Define $D_r = \log_a(C_1^{(2,r)} - 1)$, where α is a generator of $GF(r)^*$. Then the subset D_r is an almost difference set in*

$(\mathbb{Z}_{r-1}, +)$ *with parameters*

- $\left(r - 1, \frac{r-1}{2}, \frac{r-3}{4}, \frac{3r-5}{4}\right)$ *if* $r \equiv 3$ (mod 4)*, and*
- $\left(r - 1, \frac{r-1}{2}, \frac{r-5}{4}, \frac{r-1}{4}\right)$ *if* $r \equiv 1$ (mod 4)*.*

Example 5.10. Let $r = 3^3$ and let α be a generator of GF$(3^3)^*$ with $\alpha^3 + 2\alpha + 1 = 0$. Then $D = \{0, 1, 2, 3, 5, 6, 8, 9, 15, 18, 19, 20, 24\}$, which is a $(26, 13, 6, 19)$ almost difference set in $(\mathbb{Z}_{26}, +)$. In addition, $\text{rank}_2(D) = 26$.

Example 5.11. Let $r = 3^4$ and let α be a generator of GF$(3^4)^*$ with $\alpha^4 + 2\alpha^3 + 2 = 0$. Then

$$D = \{4, 5, 8, 11, 12, 14, 15, 17, 19, 22, 24, 28, 33, 34, 36, 38, 41, 42, 43, 44, 45,$$

$$46, 49, 51, 52, 53, 55, 56, 57, 58, 59, 66, 67, 68, 71, 72, 73, 76, 77, 79\},$$

which is a $(80, 40, 19, 20)$ almost difference set in $(\mathbb{Z}_{80}, +)$.

After adding one element into the almost difference set of Theorem 5.8, we have the following construction.

Theorem 5.9 ([No, Chung, Song, Yang, Lee and Helleseth (2001)]). *Let* $r \equiv 3$ (mod 4) *be a prime power and let* α *be a generator of* GF$(r)^*$. *Define*

$$C_r = \{(r - 1)/2\} \cup \log_\alpha(D_1^{(2,r)} - 1).$$

Then C_r *is a* $\left(r - 1, \frac{r+1}{2}, \frac{r+1}{4}, \frac{3(r-3)}{4}\right)$ *almost difference set in* $(\mathbb{Z}_{r-1}, +)$.

The next two theorems describe several classes of almost difference sets that are based on cyclotomy.

Theorem 5.10 ([Ding, Helleseth and Martinsen (2001)]). *Let* $r \equiv 5$ (mod 8) *be a prime power. It is known that* $r = s^2 + 4t^2$ *for some s and t with* $s \equiv \pm 1$ (mod 4)*. Set* $n = 2r$.

Let $i, j, l \in \{0, 1, 2, 3\}$ *be three pairwise distinct integers, and define*

$$D = \left[\{0\} \times (C_i^{(4,r)} \cup C_j^{(4,r)})\right] \cup \left[\{1\} \times (C_l^{(4,r)} \cup C_j^{(4,r)})\right].$$

Then D is an $\left(n, \frac{n-2}{2}, \frac{n-6}{4}, \frac{3n-6}{4}\right)$ *almost difference set in* (GF(2) × GF(r), +) *if*

(1) $t = 1$ *and* $(i, j, l) = (0, 1, 3)$ *or* $(0, 2, 1)$; *or*
(2) $s = 1$ *and* $(i, j, l) = (1, 0, 3)$ *or* $(0, 1, 2)$.

Example 5.12. Let $r = 13 = (-3)^2 + 4 \times 1^2$ and let 7 be the generator of $\mathrm{GF}(r)^*$ for defining the cyclotomic classes $C_j^{(4,r)}$ of order 4. Let $(i, j, l) = (0, 2, 1)$. Then

$$\Phi^{-1}(D) = \{4, 7, 10, 11, 12, 14, 16, 17, 21, 22, 23, 25\},$$

which is a $(26, 12, 5, 18)$ almost difference set in $(\mathbb{Z}_{26}, +)$, where $\Phi(x) = (x \bmod 2, x \bmod 13)$ is a one-to-one mapping from $\mathbb{Z}_{26}$ to $\mathbb{Z}_2 \times \mathbb{Z}_{13}$. In addition, $\mathrm{rank}_2(\Phi^{-1}(D)) = 24$.

Example 5.13. Let $r = 37 = 1^2 + 4 \times 3^2$ and let 2 be the generator of $\mathrm{GF}(r)^*$ for defining the cyclotomic classes $C_j^{(4,r)}$ of order 4. Let $(i, j, l) = (1, 0, 3)$. Then

$$\Phi^{-1}(D) = \{1, 2, 5, 7, 9, 10, 12, 13, 14, 16, 17, 18, 19, 20, 23, 24, 26, 32,$$

$$33, 34, 35, 38, 43, 44, 45, 46, 47, 49, 52, 53, 59, 63, 66, 68, 70, 71\}$$

which is a $(74, 36, 17, 54)$ almost difference set in $(\mathbb{Z}_{74}, +)$, where $\Phi(x) = (x \bmod 2, x \bmod 37)$ is a one-to-one mapping from $\mathbb{Z}_{74}$ to $\mathbb{Z}_2 \times \mathbb{Z}_{37}$. In addition, $\mathrm{rank}_2(\Phi^{-1}(D)) = 72$.

Theorem 5.11 ([Ding, Helleseth and Martinsen (2001)]). *Let* $r \equiv 5$ (mod 8) *be a prime power. It is known that* $r = s^2 + 4t^2$ *for some* s *and* t *with* $s \equiv \pm 1$ (mod 4). *Set* $n = 2r$.

Let $i, j, l \in \{0, 1, 2, 3\}$ *be three pairwise distinct integers, and define*

$$D = \left[\{0\} \times \left(C_i^{(4,r)} \cup C_j^{(4,r)} \right) \right] \cup \left[\{1\} \times \left(C_l^{(4,r)} \cup C_j^{(4,r)} \right) \right] \cup \{0, 0\}.$$

Then D *is an* $\left(n, \frac{n}{2}, \frac{n-2}{4}, \frac{3n-2}{4} \right)$ *almost difference set in* $(\mathrm{GF}(2) \times \mathrm{GF}(r), +)$ *if*

(1) $t = 1$ *and* $(i, j, l) \in \{(0, 1, 3), (0, 2, 3), (1, 2, 0), (1, 3, 0)\}$; *or*
(2) $s = 1$ *and* $(i, j, l) \in \{(0, 1, 2), (0, 3, 2), (1, 0, 3), (1, 2, 3)\}$.

Note that the cyclotomic classes $C_i^{(4,r)}$ $(1 \le i \le 3)$ depend on the choice of the generator α of $\mathrm{GF}(r)^*$ that is used to define the cyclotomic classes of order 4. Hence, the correctness of the conclusions of the two theorems above is up to a proper choice of the primitive element α.

At the end of this section, we introduce a construction of almost difference sets with uniform cyclotomy. To this end, we need the following lemma.

Lemma 5.1. *Assume that* p *is a prime,* $N \ge 2$ *is a positive integer,* $r = p^{2j\gamma}$, *where* $N | (p^j + 1)$ *for some* j *and* j *is the smallest such positive integer. Let* S *and* T *be the set of nonzero squares and non-squares of* $\mathrm{GF}(r)$, *respectively. Let* χ *be any nonprincipal additive character* χ.

(1) *If* γ, p, $\frac{p^j+1}{N}$ *are all odd, then*

$$(\chi(S), \chi(T)) = \begin{cases} \left(\frac{\sqrt{r}-1}{2}, \frac{-1-\sqrt{r}}{2}\right) & \text{if } N/2 \text{ is even,} \\ \left(-\frac{1+\sqrt{r}}{2}, -1+\frac{1+\sqrt{r}}{2}\right) & \text{if } N/2 \text{ is odd.} \end{cases}$$

(2) *In all the other cases, we have*

$$(\chi(S), \chi(T)) = \left(\frac{(-1)^\gamma \sqrt{r} - 1}{2}, \frac{-1 - (-1)^\gamma \sqrt{r}}{2}\right).$$

Proof.   The desired conclusions follow from Theorem 1.18 directly.   □

The following result was known in the literature of partial difference sets [Ma (1994)].

Theorem 5.12. *Assume that p is an odd prime, $N \geq 2$ is a positive even integer, $r = p^{2j\gamma}$, where $N|(p^j + 1)$ for some j and j is the smallest such positive integer. If γ, p, $\frac{p^j+1}{N}$ are all odd, then the set*

$$D = \sum_{i \in \mathcal{I}} C_i^{(N,r)}$$

is an $(r, \frac{r-1}{2}, \frac{r-5}{4}; \frac{r-1}{2})$ almost difference set in $(\mathrm{GF}(r), +)$, where these $C_i^{(N,r)}$ are cyclotomic classes of order N in $\mathrm{GF}(r)$, and $\mathcal{I} \subset \{0, 1, \ldots, N - 1\}$ with $|\mathcal{I}| = N/2$.

Proof.   Let S and T be the set of nonzero squares and nonsquares of $\mathrm{GF}(r)$, respectively. We will show that the set D satisfies

$$D - D = \frac{r-1}{2}\{0\} \cup \frac{r-5}{4}X \cup \frac{r-1}{4}(\mathrm{GF}(r)^* \setminus X), \qquad (5.2)$$

where $X = S$ or T depending on r and N, $D - D$ denotes the multiset $\{i - j : i \in D, j \in D\}$, and iX denotes the multiset in which each element of X appears exactly i times.

Equation (5.2) can be proved by distinguishing among the following four cases:

(1) $N/2 \in \mathcal{I}$ and $N/2$ is even; (2) $N/2 \in \mathcal{I}$ and $N/2$ is odd;

(3) $N/2 \notin \mathcal{I}$ and $N/2$ is even; (4) $N/2 \notin \mathcal{I}$ and $N/2$ is odd.

We prove (5.2) only for the first case as the remaining cases can be similarly proved. Let χ be any nonprincipal additive character of $\mathrm{GF}(r)$. In the first case, it follows from Theorem 1.18 that

$$\chi(D)\chi(-D) = \sum_{i,j\in\mathcal{I}} \chi(C_i^{(N,r)})\overline{\chi(C_j^{(N,r)})}$$

$$= \sum_{i,j\in\mathcal{I}} \eta_i^{(N,r)}\overline{\eta_j^{(N,r)}}$$

$$= \eta_{N/2}^{(N,r)}\overline{\eta_{N/2}^{(N,r)}} + \eta_{N/2}^{(N,r)} \sum_{j\in\mathcal{I}\setminus\{N/2\}} \overline{\eta_j^{(N,r)}} + \overline{\eta_{N/2}^{(N,r)}} \sum_{i\in\mathcal{I}\setminus\{N/2\}} \overline{\eta_i^{(N,r)}}$$

$$+ \sum_{i,j\in\mathcal{I}\setminus\{N/2\}} \eta_i^{(N,r)}\overline{\eta_j^{(N,r)}}$$

$$= \left(\sqrt{r} - \frac{\sqrt{r}+1}{N}\right)^2 + 2\left(\sqrt{r} - \frac{\sqrt{r}+1}{N}\right)\left(-\frac{1+\sqrt{r}}{N}\right)$$

$$\cdot\left(\frac{N}{2}-1\right) + \left(-\frac{1+\sqrt{r}}{N}\right)^2\cdot\left(\frac{N}{2}-1\right)^2$$

$$= \frac{(\sqrt{r}-1)^2}{4}.$$

On the other hand, by Lemma 5.1, we have

$$\chi\left[\frac{r-1}{2}\{0\} \cup \frac{r-5}{4}X \cup \frac{r-1}{4}(\mathrm{GF}(r)^*\setminus X)\right] = \frac{r-1}{4} - \chi(S) = \frac{(\sqrt{r}-1)^2}{4}.$$

Then the equality in (5.2) follows. Hence D is an $(r, \frac{r-1}{2}, \frac{r-5}{4}; \frac{r-1}{2})$ almost difference set in $(\mathrm{GF}(r), +)$. $\qquad\square$

Theorem 5.12 produces a lot of almost difference sets in $(\mathrm{GF}(r), +)$, where r is an even power of a prime. Below is an example.

Example 5.14. Let $p = 5$, $r = 5^2$ and let $N = 6$. Then in Theorem 5.12, $j = 1$ and $\gamma = 1$. Hence, γ, j and $(p^j + 1)/N$ are all odd. Let $\mathcal{I} = \{0, 3, 5\}$ and let α be a generator of $\mathrm{GF}(5^2)^*$ with $\alpha^2 + 4\alpha + 2 = 0$. Then

$$D = C_0^{(6,r)} \cup C_3^{(6,r)} \cup C_5^{(6,r)}$$

$$= \{1, \alpha^{17}, 3, \alpha^3, \alpha^{21}, \alpha^5, 2, \alpha^{23}, \alpha^9, \alpha^{11}, 4, \alpha^{15}\},$$

which is a $(25, 12, 5, 12)$ almost difference set in $(\mathrm{GF}(5^2), +)$.

5.5   Generalized Cyclotomic Constructions

Let n_1 and n_2 be two distinct primes, and let $n = n_1 n_2$. Put $N = \gcd(n_1, n_2)$. Recall the generalized cyclotomic classes $W_{2i}^{(N)}$ of order N introduced in Section 1.5. Define

$$D_0^{(2)} = \bigcup_{i=0}^{(N-2)/2} W_{2i}^{(N)} \quad \text{and} \quad D_1^{(2)} = \bigcup_{i=0}^{(N-2)/2} W_{2i+1}^{(N)}.$$

Clearly $D_0^{(2)}$ is a subgroup of $\mathbb{Z}_n^*$ and $D_1^{(2)} = \varrho D_0^{(2)}$, where ϱ was defined in Section 1.5.

The following result was proved in Ding (1998), Mertens and Bessenrodt (1998), and Brandstatter and Winterhof (2005).

Theorem 5.13. *Let*

$$D = D_1^{(2)} \cup \{n_1, 2n_1, \ldots, (n_2 - 1)n_1\}. \tag{5.3}$$

If $n_2 - n_1 = 4$, then D is an almost difference set in $(\mathbb{Z}_{n_1(n_1+4)}, +)$ with parameters

$$\left(n_1(n_1 + 4), \frac{(n_1 + 3)(n_1 + 1)}{2}, \frac{(n_1 + 3)(n_1 + 1)}{4}, \frac{(n_1 - 1)(n_1 + 5)}{2} \right).$$

Example 5.15. Let $(n_1, n_2) = (3, 7)$. Then

$$D = D_1^{(2)} \cup \{n_1, 2n_1, \ldots, (n_2 - 1)n_1\}$$
$$= \{2, 3, 6, 8, 9, 10, 11, 12, 13, 15, 18, 19\},$$

which is a $(21, 12, 6, 8)$ almost difference set in $(\mathbb{Z}_{21}, +)$.

5.6   Constructions with Difference Sets

Certain types of difference sets can be employed to construct almost difference sets. The following two theorems document two generic constructions.

Theorem 5.14 ([Arasu, Ding, Helleseth, Kumar and Martinsen (2001)]). *Let $\mathbb{C}$ be an $\left(\ell, \frac{\ell-1}{2}, \frac{\ell-3}{4}\right)$ (respectively, $\left(\ell, \frac{\ell+1}{2}, \frac{\ell+1}{4}\right)$) difference set in $(\mathbb{Z}_\ell, +)$, where $\ell \equiv 3 \pmod 4$. Define a subset of $\mathbb{Z}_{4\ell}$ by*

$$D = [(\ell + 1)\mathbb{C} \bmod 4\ell] \cup [(\ell + 1)(\mathbb{C} - \delta)^c + 3\ell \bmod 4\ell] \cup$$
$$[(\ell + 1)\mathbb{C}^c + 2\ell \bmod 4\ell] \cup [(\ell + 1)(\mathbb{C} - \delta)^c + 3\ell \bmod 4\ell], \tag{5.4}$$

where $\mathbb{C}^c$ *and* $(\mathbb{C} - \delta)^c$ *denote the complement of* $\mathbb{C}$ *and* $\mathbb{C} - \delta$ *in* $\mathbb{Z}_\ell$ *respectively. Then* D *is a* $(4\ell, 2\ell - 1, \ell - 2, \ell - 1)$ *(respectively,* $(4\ell, 2\ell + 1, \ell, \ell - 1)$*) almost difference set in* $(\mathbb{Z}_{4\ell}, +)$.

Lemma 5.2 ([Jungnickel (1982)]). *Let* D_1 *be a* (v, a, λ) *difference set in a group* A, *and let* D_2 *be a difference set with parameters* $(4u^2, 2u^2 - u, u^2 - u)$ *in a group* B. *Then* $D := (D_2 \times D_1^c) \cup (D_2^c \times D_1)$ *is a divisible difference set in* $B \times A$ *relative to* $\{1\} \times A$, *with parameters* $(4u^2, v, 2u^2v + 2au - uv, \lambda_1, \lambda_2)$, *where*

$$\lambda_1 = (2u^2 - u)(v - 2a) + 4u^2\lambda, \ \lambda_2 = u^2v - uv + 2au,$$

and D_2^c *denotes the complement of* D_2.

As a consequence of Lemma 5.2, we have the following.

Theorem 5.15 ([Arasu, Ding, Helleseth, Kumar and Martinsen (2001)]). *Let* D_1 *be an* $\left(\ell, \frac{\ell-1}{2}, \frac{\ell-3}{4}\right)$ *(respectively,* $\left(\ell, \frac{\ell+1}{2}, \frac{\ell+1}{4}\right)$*) difference set in* $(\mathbb{Z}_\ell, +)$, *let* D_2 *be a difference set in* $(\mathbb{Z}_4, +)$ *with parameters* $(4, 1, 0)$. *Then*

$$D := (D_2 \times D_1^c) \cup (D_2^c \times D_1)$$

is a $(4\ell, 2\ell - 1, \ell - 2, \ell - 1)$ *(respectively,* $(4\ell, 2\ell + 1, \ell, \ell - 1)$*) almost difference set in* $(\mathbb{Z}_4 \times \mathbb{Z}_\ell, +)$.

The following construction generalizes those of Theorems 5.14 and 5.15.

Theorem 5.16 ([Tang and Ding (2010)]). *Suppose that* A *and* B *are, respectively, difference sets with parameters* $(\ell, (\ell+1)/2, (\ell+1)/4)$ *or* $(\ell, (\ell-1)/2, (\ell - 3)/4)$ *in* $(G, +)$. *Define*

$$D := (\{0, 2\} \times A) \cup (\{1\} \times B) \cup (\{3\} \times B^c),$$

where B^c *is the complement of* B, *then* D *is a* $(4\ell, 2\ell + 1, \ell, \ell - 1)$ *or* $(4\ell, 2\ell - 1, \ell - 2, \ell - 1)$ *almost difference set in* $(\mathbb{Z}_4, +) \times (G, +)$.

The following is another construction of almost difference sets.

Theorem 5.17 ([Zhang, Lei and Zhang (2006)]). *Let* $r \equiv 3 \pmod 4$ *be a prime power. Then the following set*

$$D := (\{0\} \times C_0^{(2,r)}) \cup (\{1, 2, 3\} \times C_0^{(2,r)}) \cup \{(0, 0), (1, 0), (3, 0)\}$$

is a $(4r, 2r + 1, r, r - 1)$ *almost difference set in* $\mathbb{Z}_4 \times \mathrm{GF}(r)$.

The following is a construction of almost difference sets using skew Hadamard difference sets, which is analogous to the approach in Ding, Wang and Xiang (2007).

Theorem 5.18 ([Ding, Pott and Wang (2014)]). *Let r and $r + 4$ be two prime powers with $r \equiv 3 \pmod 4$. Let E and F be two skew Hadamard difference sets in two abelian groups $(\mathrm{GF}(r), +)$ and $(\mathrm{GF}(r + 4), +)$, respectively. Then the set*

$$D := \{(x, y) : (x, y) \in [(E \times F) \cup ((-E) \times (-F))]\} \cup \{(x, 0) : x \in \mathrm{GF}(r)\}$$

is a $(r^2 + 4r, (r^2 + 4r - 3)/2, (r^2 + 4r - 9)/4, (r^2 + 4r - 5)/2)$ almost difference set in $(\mathrm{GF}(r), +) \times (\mathrm{GF}(r + 4), +)$.

Proof. We now consider the difference function of D defined in (4.1), and have

$$\begin{aligned}
\mathrm{diff}_D(d_1, d_2) &= \left|[D + (d_1, d_2)] \cap D\right| \\
&= \left|[E \times F \cap (E + d_1) \times (F + d_2)] \cup [E \times F \cap (-E + d_1) \times (-F + d_2)]\right. \\
&\quad \cup [E \times F \cap \mathrm{GF}(r) \times \{d_2\}] \cup [(-E) \times (-F) \cap (E + d_1) \times (F + d_2)] \\
&\quad \cup [(-E) \times (-F) \cap (-E + d_1) \times (-F + d_2)] \\
&\quad \cup [\mathrm{GF}(r) \times \{0\} \cap \mathrm{GF}(r) \times \{d_2\}] \\
&\quad \cup [(-E) \times (-F) \cap \mathrm{GF}(r) \times \{d_2\}] \cup [\mathrm{GF}(r) \times \{0\} \cap (E + d_1) \times (F + d_2)] \\
&\quad \left. \cup [\mathrm{GF}(r) \times \{0\} \cap (-E + d_1) \times (-F + d_2)]\right|.
\end{aligned} \tag{5.5}$$

We discuss the possible values of $\mathrm{diff}_D(d_1, d_2)$ in the following cases.

- Suppose that $d_1 \neq 0$ and $d_2 = 0$. By (5.5), we have

$$\begin{aligned}
\mathrm{diff}_D(d_1, 0) &= \left|[E \cap (E + d_1)] \times F\right| + \left|[(-E) \cap (-E + d_1)] \times (-F)\right| + r \\
&= \frac{r - 3}{4} \times \frac{r + 3}{2} + \frac{r - 3}{4} \times \frac{r + 3}{2} + r \\
&= \frac{r^2 + 4r - 9}{4}.
\end{aligned}$$

- Suppose that $d_1 = 0$ and $d_2 \in F$. It then follows from (5.5) that

$$\begin{aligned}
\mathrm{diff}_D(0, d_2) &= \left|E \times [F \cap (F + d_2)]\right| + |E| \\
&\quad + \left|(-E) \times [(-F) \cap (-F + d_2)]\right| + |-E|
\end{aligned}$$

$$= \frac{r-1}{2} \times \frac{r+1}{4} + \frac{r-1}{2} + \frac{r-1}{2} \times \frac{r+1}{4} + \frac{r-1}{2}$$

$$= \frac{r^2 + 4r - 5}{4}.$$

Similarly, for $d_2 \in -F$, we also have $\text{diff}_D(0, d_2 \in -F) = \frac{r^2 + 4r - 5}{4}$.

• Suppose that $d_1 \in E$ and $d_2 \in F$. By (5.5), we have

$$\text{diff}_D(d_1, d_2)$$
$$= \left| [E \cap (E + d_1)] \times [F \cap (F + d_2)] \right|$$
$$+ \left| [E \cap (-E + d_1)] \times [F \cap (-F + d_2)] \right|$$
$$+ |E| + \left| [(-E) \cap (E + d_1)] \times [(-F) \cap (F + d_2)] \right|$$
$$+ \left| [(-E) \cap (-E + d_1)] \times [(-F) \cap (-F + d_2)] \right| + |-E|.$$

Note that $d_1 \in E$, then we have

$$\left| E \cap (-E + d_1) \right| = |-E| - |(-E) \cap (-E + d_1)| - 1.$$

Thus, in this case, we get

$$\text{diff}_D(d_1, d_2)$$
$$= \frac{r-3}{4} \frac{r+1}{4} + \left(\frac{r-1}{2} - \frac{r-3}{4} - 1 \right) \left(\frac{r+3}{2} - \frac{r+1}{4} - 1 \right)$$
$$+ \frac{r-1}{2} + \left(\frac{r-1}{2} - \frac{r-3}{4} \right) \left(\frac{r+3}{2} - \frac{r+1}{4} \right) + \frac{r-3}{4} \frac{r+1}{4} + \frac{r-1}{2}$$
$$= \frac{r^2 + 4r - 5}{4}.$$

• Suppose that $d_1 \in E$ and $d_2 \in -F$. In this case, we can similarly prove that

$$\text{diff}_D(d_1, d_2) = \frac{r^2 + 4r - 9}{4}.$$

• Suppose that $d_1 \in -E$ and $d_2 \in F$. In this case, we can similarly prove that

$$\text{diff}_D(d_1, d_2) = \frac{r^2 + 4r - 9}{4}.$$

- Suppose that $d_1 \in -E$ and $d_2 \in -F$. In this case, we can similarly prove that

$$\text{diff}_D(d_1, d_2) = \frac{r^2 + 4r - 5}{4}.$$

Summarizing the conclusions in all the cases above proves this theorem. □

Plugging all the families of skew Hadamard difference sets presented in Section 4.9 into Theorem 5.18 will produce a lot of almost difference sets. Information on the inequivalence of these almost difference sets can be found in Ding, Pott and Wang (2014).

5.7 Generic Constructions with Planar Functions

Planar functions can be used to construct almost difference sets. Below is such a construction.

Theorem 5.19 ([Arasu, Ding, Helleseth, Kumar and Martinsen (2001)]). *Let f be a function from an abelian group $(A, +)$ of order n to another abelian group $(B, +)$ of order n with perfect nonlinearity.*
Define $C_b = \{x \in A \mid f(x) = b\}$ and

$$D = \bigcup_{b \in B} \{b\} \times C_b \subseteq B \times A.$$

Then D is an $(n^2, n, 0, n - 1)$ almost difference set in $B \times A$.

All the known planar functions in Section 1.7 can be plugged into this generic construction to obtain almost difference sets.

5.8 Planar Almost Difference Sets

Almost difference sets with parameters $(n, k, 0, t)$ are called *planar*. Section 5.7 presents a generic construction of planar almost difference sets with planar functions. In this section, we present two classes of planar almost difference sets that are cyclic.

Theorem 5.20. *Let q be a power of a prime, and let α be a generator of $\text{GF}(q^2)^*$. Define*

$$D_q = \left\{ 0 \leq i \leq n - 1 : \text{Tr}(\alpha^i) = 1 \right\} \subset \mathbb{Z}_n,$$

where $n = q^2 - 1$ and $\text{Tr}(x) = x + x^q$ is the trace function from $\text{GF}(q^2)$ to $\text{GF}(q)$.
If $q \geq 3$, then D_q is a $[q^2 - 1, q, 0, q - 2]$ almost difference set in $\mathbb{Z}_n$.

Proof. For any integer ℓ with $0 < \ell \leq n - 1$, we now prove that

$$\left| D_q \cap (D_q + \ell) \right| = \begin{cases} 0 & \text{if } \ell \in \{q + 1, 2(q + 1), \ldots, (q - 2)(q + 1)\}, \\ 1 & \text{otherwise,} \end{cases} \quad (5.6)$$

where $D_q + \ell = \{x + \ell : x \in D_q\} \subset \mathbb{Z}_n$.

Let $h \in D_q$ and $h - \ell \in D_q$. Then $\text{Tr}(\alpha^h) = 1$ and $\text{Tr}(\alpha^{h-\ell}) = 1$. Hence

$$\begin{cases} \alpha^h + \alpha^{hq} = 1, \\ \alpha^{h-\ell} + \alpha^{(h-\ell)q} = 1. \end{cases} \quad (5.7)$$

Define $y = \alpha^h$. Then (5.7) becomes

$$\begin{cases} y + y^q = 1, \\ \alpha^{-\ell} y + \alpha^{-\ell q} y^q = 1. \end{cases} \quad (5.8)$$

It then follows from (5.8) that

$$\left(\alpha^{-\ell} - \alpha^{-\ell q} \right) y = 1 - \alpha^{-\ell q}. \quad (5.9)$$

It is easily seen that $\alpha^{-\ell} - \alpha^{-\ell q} = 0$ with $0 < \ell < n$ if and only if

$$\ell \in \{q + 1, 2(q + 1), \ldots, (q - 2)(q + 1)\}.$$

On the other hand, for each $\ell \in \{q + 1, 2(q + 1), \ldots, (q - 2)(q + 1)\}$ we have $1 - \alpha^{-\ell q} \neq 0$. Then desired conclusion in (5.6) follows from (5.9). This completes the proof of this theorem. $\qquad \square$

Example 5.16. Let $q = 7$, and let α be the generator of $\text{GF}(q^2)^*$ with minimal polynomial $\alpha^2 + 6\alpha + 3 = 0$ over $\text{GF}(7)$. Then

$$D_q = \{1, 7, 27, 32, 34, 45, 46\},$$

which is a $[48, 7, 0, 5]$ almost difference set in $(\mathbb{Z}_{48}, +)$.

The following theorem introduces another construction of planar cyclic almost difference sets in $(\mathbb{Z}_n, +)$.

Theorem 5.21. *Let α be a generator of $\text{GF}(q^2)^*$ and let $n = q^2 - 1$. Define*

$$U_q = \log_\alpha \left(C_1^{(q+1,q^2)} - 1 \right) \subset \mathbb{Z}_n,$$

where $C_1^{(q+1,q^2)}$ is the cyclotomic class of order $q + 1$ in $\text{GF}(q^2)$.

Let $D_q = U_q \cup \{\mu\}$, where $\mu = 0$ if q is even and $\mu = (q^2 - 1)/2$ if q is odd. Then D_q is a $[q^2 - 1, q, 0, q - 2]$ almost difference set in $(\mathbb{Z}_n, +)$.

Proof.   Let $r = q^2$. By the definition of μ, we have $\alpha^\mu = -1$. It is proved in Baumert, Mills and Ward (1982) that

$$(0, 0)^{(q+1,r)} = q - 2,$$
$$(0, i)^{(q+1,r)} = (i, 0)^{(q+1,r)} = (i, i)^{(q+1,r)} = 0 \text{ for all } 0 < i \leq q,$$
$$(i, j)^{(q+1,r)} = 1 \text{ for all } 0 \neq i \neq j \neq 0.$$

We shall make use of these cyclotomic numbers of order $q + 1$ below.

For any $1 \leq \ell \leq n - 1 = r - 2$, we have

$$\left| (U_q \cup \{\mu\}) \cap (U_q \cup \{\mu\} + \ell) \right|$$

$$= \left| U_q \cap (U_q + \ell) \right| + \left| U_q \cap \{\mu - \ell, \mu + \ell\} \right|$$

$$= \left| \log_\alpha \left(C_1^{(q+1,r)} - 1 \right) \cap \left(\log_\alpha \left(C_1^{(q+1,r)} - 1 \right) + \ell \right) \right| + \left| U_q \cap \{\mu - \ell, \mu + \ell\} \right|$$

$$= \left| \left(C_1^{(q+1,r)} - 1 \right) \cap \alpha^\ell \left(C_1^{(q+1,r)} - 1 \right) \right| + \left| U_q \cap \{\mu - \ell, \mu + \ell\} \right|$$

$$= \left| C_1^{(q+1,r)} \cap \left(\alpha^\ell C_1^{(q+1,r)} + 1 - \alpha^\ell \right) \right| + \left| U_q \cap \{\mu - \ell, \mu + \ell\} \right|$$

$$= \left| \frac{1}{1 - \alpha^\ell} C_1^{(q+1,r)} \cap \left(\frac{\alpha^\ell}{1 - \alpha^\ell} C_1^{(q+1,r)} + 1 \right) \right| + \left| U_q \cap \{\mu - \ell, \mu + \ell\} \right|$$

$$= \left| a C_1^{(q+1,r)} \cap \left(b C_1^{(q+1,r)} + 1 \right) \right| + \left| \{\alpha^{\mu-\ell}, \alpha^{\mu+\ell}\} \cap \left(C_1^{(q+1,r)} - 1 \right) \right|$$

$$= \left| a C_1^{(q+1,r)} \cap \left(b C_1^{(q+1,r)} + 1 \right) \right| + \left| \{-\alpha^{-\ell}, -\alpha^\ell\} \cap \left(C_1^{(q+1,r)} - 1 \right) \right|,$$

where

$$a = \frac{1}{1 - \alpha^\ell} \quad \text{and} \quad b = a\alpha^\ell.$$

We first consider the case that $\ell \equiv 0 \pmod{q + 1}$. In this case, we have that $\alpha^\ell \in \mathrm{GF}(q)^* = C_0^{(q+1,r)}$ and thus $a = (1 - \alpha^\ell)^{-1} \notin C_q^{(q+1,r)}$. Since $b = a\alpha^\ell$ and $\ell \equiv 0 \pmod{q + 1}$, a and b are in the same cyclotomic class. Therefore, there is an integer $1 \leq i \leq q$ such that

$$\left| a C_1^{(q+1,r)} \cap \left(b C_1^{(q+1,r)} + 1 \right) \right| = (i, i)_{q+1} = 0.$$

In this case, since $1 - \alpha^\ell \in \mathrm{GF}(q)$ and $1 - \alpha^{-\ell} \in \mathrm{GF}(q)$, we have

$$1 - \alpha^\ell \notin C_1^{(q+1,r)} \quad \text{and} \quad 1 - \alpha^{-\ell} \notin C_1^{(q+1,r)}.$$

It then follows that

$$\left| \{-\alpha^{-\ell}, -\alpha^{\ell}\} \cap \left(C_1^{(q+1,r)} - 1 \right) \right| = 0.$$

Hence, in this case, we have

$$\left| (U_q \cup \{\mu\}) \cap (U_q \cup \{\mu\} + \ell) \right| = 0.$$

We now deal with the second case, i.e., $\ell \not\equiv 0 \pmod{q+1}$. In this case a and b belong to two distinct cyclotomic classes of order $q+1$. Note that $1 - \alpha^{-\ell} = -\alpha^{-\ell}(1 - \alpha^{\ell})$ and $\ell \not\equiv 0 \pmod{q+1}$. We know that $1 - \alpha^{-\ell}$ and $1 - \alpha^{\ell}$ belong to different cyclotomic classes of order $q + 1$. Hence

$$\left| \{-\alpha^{-\ell}, -\alpha^{\ell}\} \cap \left(C_1^{(q+1,r)} - 1 \right) \right| \le 1.$$

When $\left| \{-\alpha^{-\ell}, -\alpha^{\ell}\} \cap \left(C_1^{(q+1,r)} - 1 \right) \right| = 0$, we have

$$\left| (U_q \cup \{\mu\}) \cap (U_q \cup \{\mu\} + \ell) \right| = \left| a C_1^{(q+1,r)} \cap \left(b C_1^{(q+1,r)} + 1 \right) \right| = (i, j)^{(q+1,r)}$$

for some i and j. In this subcase, it is clear that $i \ne j$, as a and b belong to distinct cyclotomic classes. We now prove that both i and j are nonzero. Notice that $-1 \in C_0^{(q+1,r)}$ for both even and odd q. Hence for any nonzero element $c \in GF(r)$, c and $-c$ are in the same cyclotomic class. On the other hand, it is obvious that $c \in C_1^{(q+1,r)}$ if and only if $c^{-1} \in C_q^{(q+1,r)}$. Since $a^{-1} = 1 - \alpha^{\ell} \notin C_1^{(q+1,r)}$, we have $a \notin C_q^{(q+1,r)}$ and thus $i \ne 0$. Similarly, we have $j \ne 0$ as $1 - \alpha^{-\ell} \notin C_1^{(q+1,r)}$. Hence in this subcase we have always $(i, j)^{(q+1,r)} = 1$.

When $\left| \{-\alpha^{-\ell}, -\alpha^{\ell}\} \cap \left(C_1^{(q+1,r)} - 1 \right) \right| = 1$, due to symmetry we assume that $1 - \alpha^{\ell} \in C_1^{(q+1,r)}$. Then we have $a = (1 - \alpha^{\ell})^{-1} \in C_q^{(q+1,r)}$. It then follows that

$$\left| (U_q \cup \{\mu\}) \cap (U_q \cup \{\mu\} + \ell) \right|$$

$$= \left| a C_1^{(q+1,r)} \cap \left(b C_1^{(q+1,r)} + 1 \right) \right| + \left| \{-\alpha^{-\ell}, -\alpha^{\ell}\} \cap \left(C_1^{(q+1,r)} - 1 \right) \right|$$

$$= \left| a C_1^{(q+1,r)} \cap \left(b C_1^{(q+1,r)} + 1 \right) \right| + 1$$

$$= (0, j)^{(q+1,r)} + 1$$

$$= 1,$$

where $j \not\equiv 0 \pmod{q+1}$.

In summary, in the second case we have

$$\left| (U_q \cup \{\mu\}) \cap (U_q \cup \{\mu\} + \ell) \right| = 1$$

for all $1 \le \ell \le n - 1$ with $\ell \not\equiv 0 \pmod{q + 1}$.

Obviously, $|D_q| = q$. The desired conclusion then follows. $\square$

Example 5.17. Let $q = 2^3$, and let α be the generator of $GF(q^2)^*$ with minimal polynomial $\alpha^6 + \alpha^4 + \alpha^3 + \alpha + 1 = 0$ over $GF(2)$. Then

$$U_q \cup \{0\} = \{0, 22, 32, 43, 48, 56, 60, 62\},$$

which is a $[65, 8, 0, 6]$ almost difference set in $(\mathbb{Z}_{63}, +)$.

Example 5.18. Let $q = 7$, and let α be the generator of $GF(q^2)^*$ with minimal polynomial $\alpha^2 + 6\alpha + 3 = 0$ over $GF(7)$. Then

$$U_q \cup \{(q^2 - 1)/2\} = \{4, 24, 29, 31, 42, 43, 46\},$$

which is a $[48, 7, 0, 5]$ almost difference set in $(\mathbb{Z}_{48}, +)$.

It is very likely that the difference sets documented in Theorems 5.20 and 5.21 are equivalent. In fact, we have the following question.

Problem 5.2. *For any prime power q, are all planar almost difference sets with parameters $(q^2, q, 0, q - 2)$ in $(\mathbb{Z}_{q^2-1}, +)$ equivalent?*

An (n, k) *modular Golomb ruler* is a set of k distinct elements, $\{a_0, a_1, \ldots, a_{k-1}\} \subset \mathbb{Z}_n$, such that the multiset $\{(a_i - a_j) \bmod n : 0 \le i \ne j \le k-1\}$ contains each element of $\mathbb{Z}_n$ at most once. In design theory, modular Golomb rulers are called a planar cyclic difference packing modulo n [Swanson (2000)]. It is showed in Drakakis (2009) that the almost difference set of Theorem 5.21 is a cyclic shift of the Bose–Chowla construction of modular Golomb rulers [Bose (1942); Bose and Chowla (1962)].

The almost difference set in Theorem 5.21 is equivalent to the one of Yu, Feng and Zhang (2014)[Theorem 1] and is further equivalent to the support of a decimated N-ary Sidelnikov sequence.

5.9   A Generic Construction of Almost Difference Sets

For almost difference sets, we have a GMW-like construction as follows.

Theorem 5.22 ([Cai and Ding (2009)]). *Let R_2 be any $(2^{m/2} - 1, 2^{(m-2)/2} - 1, 2^{(m-4)/2} - 1)$ difference set in $GF(2^{m/2})^*$. Define*

$$R_1 = \{x \in GF(2^m) : Tr_{2^m/2^{m/2}}(x) = 1\}, \quad R = \{r_1 r_2 : r_1 \in R_1, r_2 \in R_2\}.$$

Then R is a $(2^m - 1, 2^{m-1} - 2^{m/2}, 2^{m-2} - 2^{m/2}, 2^{m/2} - 2)$ *almost difference set in* $\mathrm{GF}(2^m)^*$. *Furthermore, the characteristic sequence of the set* $\log_\alpha R$ *has only the out-of-phase autocorrelation values* $\{-1, 3\}$, *where* α *is any generator of* $\mathrm{GF}(2^m)^*$.

Proof. For the convenience of description, define $G = \mathrm{GF}(2^m)^*$ and $H = \mathrm{GF}(2^{m/2})^*$. We first prove that R_1 is a relative difference set with parameters $(2^{m/2} + 1, 2^{m/2} - 1, 2^{m/2}, 1)$ in G relative to H. This is to prove that

$$R_1 R_1^{(-1)} = k_1 + \lambda_1 (G \setminus H),$$

where $R_1^{(-1)} := \{r^{-1} : r \in R_1\}$, $k_1 = 2^{m/2}$ and $\lambda_1 = 1$. Clearly, $|R_1| = k_1$. Note that $\mathrm{Tr}_{2^m/2^{m/2}}(x) = x + x^\gamma$, where $\gamma := 2^{m/2}$. We need to compute the number of solutions $(x, y) \in \mathrm{GF}(2^m)^* \times \mathrm{GF}(2^m)^*$ of the following set of equations

$$x + x^\gamma = 1, \quad y + y^\gamma = 1, \quad xy^{-1} = a, \tag{5.10}$$

where $a \in \mathrm{GF}(2^m)$. It is easily seen that the number of solutions $(x, y) \in \mathrm{GF}(2^m)^* \times \mathrm{GF}(2^m)^*$ of (5.10) is the same as the number of solutions $y \in \mathrm{GF}(2^m)^*$ of the following set of equations

$$y + y^\gamma = 1, \quad (a - a^\gamma)y^q = a - 1. \tag{5.11}$$

If $a \in \mathrm{GF}(2^{m/2})^*$, $a^\gamma - a = 0$. So (5.11) has no solution. If $a \in \mathrm{GF}(2^m) \setminus \mathrm{GF}(2^{m/2})$, then $a^\gamma - a \neq 0$ and (5.11) has the unique solution $y = (a^\gamma - 1)/(a^\gamma - a)$. This proves the difference set property of R_1.

Since R_2 is a $(2^{m/2} - 1, 2^{(m-2)/2} - 1, 2^{(m-4)/2} - 1)$ difference set in $\mathrm{GF}(2^m)^*$, we have

$$R_2 R_2^{(-1)} = k_2 + \lambda_2 (H \setminus \{1\}),$$

where $k_2 = 2^{(m-2)/2} - 1$ and $\lambda_2 = 2^{(m-4)/2} - 1$.

Now, we have

$$(R_1 R_2)(R_1 R_2)^{-1}$$

$$= (k_1 + \lambda_1(G \setminus H))((k_2 - \lambda_2) + \lambda_2 H)$$

$$= k_1(k_2 - \lambda_2) + \lambda_1(k_2 - \lambda_2)(G \setminus H) + k_1\lambda_2 H + \lambda_1\lambda_2(G \setminus H)H$$

$$= k_1(k_2 - \lambda_2) + \lambda_1(k_2 - \lambda_2)G + \lambda_1\lambda_2|H|G$$

$$\quad - \lambda_1(k_2 - \lambda_2)H + k_1\lambda_2 H - \lambda_1\lambda_2|H|H$$

$$= k_1(k_2 - \lambda_2) + (\lambda_1(k_2 - \lambda_2) + \lambda_1\lambda_2|H|)G$$

$$\quad + (k_1\lambda_2 - \lambda_1(k_2 - \lambda_2) - \lambda_1\lambda_2|H|)H.$$

Note that

$$\lambda_1(k_2 - \lambda_2) + \lambda_1\lambda_2|H| = 2^{m-2} - 2^{m/2} + 1$$

and

$$k_1\lambda_2 - \lambda_1(k_2 - \lambda_2) - \lambda_1\lambda_2|H| = -1.$$

We obtain

$$(R_1R_2)(R_1R_2)^{-1}$$
$$= (2^{m-2} - 2^{m/2} + 1)(G \setminus H) + (2^{m-2} - 2^{m/2})(H \setminus \{1\}) + 2^{m-1} - 2^{m/2}.$$

This proves the almost difference set property of R. It follows from Theorem 5.1 that the characteristic sequence of $\log_\alpha R$ has only the out-of-phase autocorrelation values $\{-1, 3\}$. $\square$

This construction is generic in the sense that the difference sets with Singer parameters described in Section 4.11 can be plugged in to obtain many classes of cyclic almost difference sets.

As seen before, the Gordon–Mills–Welch construction of Section 4.11 is generic and powerful in constructing difference sets with Singer parameters. The construction idea of Theorem 5.22 is similar to that of the Gordon–Mills–Welch construction. However, the objective here is to construct almost difference sets.

We point out here that Theorem 5.22 can be generalized as follows.

Theorem 5.23. *Let R_2 be any $(2^{m/2} - 1, 2^{(m-2)/2} - 1, 2^{(m-4)/2} - 1)$ difference set in $GF(2^{m/2})^*$ and let h be any positive integer with $\gcd(h, 2^{m/2} - 1) = 1$. Define*

$$R_1 = \{x \in GF(2^m) : \mathrm{Tr}_{2^m/2^{m/2}}(x) = 1\}, \quad R = \{r_2^h r_1 : r_1 \in R_1, r_2 \in R_2\}.$$

Then R is a $(2^m - 1, 2^{m-1} - 2^{m/2}, 2^{m-2} - 2^{m/2}, 2^{m/2} - 2)$ almost difference set in $GF(2^m)^*$. Furthermore, the characteristic sequence of the set $\log_\alpha R$ has only the out-of-phase autocorrelation values $\{-1, 3\}$, where α is any generator of $GF(2^m)$.

Proof. Note that $R_2^h := \{r_2^h : r_2 \in R_2\}$ is also a $(2^{m/2} - 1, 2^{(m-2)/2} - 1, 2^{(m-4)/2} - 1)$ difference set in $GF(2^{m/2})^*$. The desired conclusion then follows from Theorem 5.22. $\square$

Example 5.19. Let $m = 6$ and let α be a generator of $GF(2^m)$ with $\alpha^6 + \alpha^4 + \alpha^3 + \alpha + 1 = 0$. Let R_2 be the complement of the Segre difference set with parameters

$(2^{m/2} - 1, 2^{(m-2)/2}, 2^{(m-4)/2})$ in GF$(2^{m/2})^*$ and let $h = 3$. Then

$$\log_\alpha(R) = \{1, 2, 4, 5, 7, 8, 10, 14, 15, 16, 17, 20, 28,$$

$$30, 32, 34, 35, 39, 40, 49, 51, 56, 57, 60\},$$

which is a $(63, 24, 8, 6)$ almost difference set in $(\mathbb{Z}_{63}, +)$.

Example 5.20. Let $m = 6$ and let α be a generator of GF(2^m) with $\alpha^6 + \alpha^4 + \alpha^3 + \alpha + 1 = 0$. Let R_2 be the complement of the Segre difference set with parameters $(2^{m/2} - 1, 2^{(m-2)/2}, 2^{(m-4)/2})$ in GF$(2^{m/2})^*$ and let $h = 1$. Then

$$\log_\alpha(R) = \{3, 6, 11, 12, 13, 19, 22, 23, 24, 25, 26, 29, 33,$$

$$37, 38, 41, 43, 44, 46, 48, 50, 52, 53, 58\},$$

which is a $(63, 24, 8, 6)$ almost difference set in $(\mathbb{Z}_{63}, +)$.

5.10   Comments and Open Problems

Some cyclotomic almost difference sets were described in Wang and Wang (2011), but rediscovered earlier ones. A connection between partitioned difference families and almost difference sets was demonstrated in Wang and Wang (2011).

The equivalence problem of almost difference sets is a very important problem. In the literature, there are partial results about the inequivalence of the almost difference sets presented in this chapter. However, many of the equivalence problems remain open.

It would be interesting if progress on the following problems can be made.

Problem 5.3.

- *Construct new almost difference sets.*
- *Study the equivalence among the almost difference sets presented in this chapter.*
- *Does a $\left(n, \frac{n-1}{2}, \lambda, t\right)$ almost difference set exist for all odd n?*

Chapter 6

Linear Codes of Difference Sets

The study of difference-set cyclic codes dates back at least to Weldon (1966). In this chapter, we study linear codes from some difference sets and almost difference sets. Our focus is the cyclic case.

6.1 Basic Theory of Linear Codes of Difference Sets

In Section 3.2, we studied t-designs, (n, k, λ) designs and their linear codes. In this section, we deal with (n, k, λ) designs from difference sets and their linear codes. We first introduce an incidence structure from difference sets.

Let D be an (n, k, λ) difference set in an abelian group $(A, +)$. Recall the development $\mathbb{D} = (\mathcal{P}, \mathcal{B}, \mathcal{I})$ of D, where $\mathcal{P}$ is the set of the elements in A, $\mathcal{B} = \{a + D : a \in A\}$, and the incidence relation $\mathcal{I}$ is the membership of sets. Each block $a + D = \{a + x : x \in D\}$ is called a translate of D.

The development $\mathbb{D}$ of a difference set, as an (n, k, λ) design, has its linear code $\mathcal{C}_{\mathrm{GF}(q)}(\mathbb{D})$ over $\mathrm{GF}(q)$, which was defined in Sections 3.1.4 and treated in Section 3.2. For simplicity, we call $\mathcal{C}_{\mathrm{GF}(q)}(\mathbb{D})$ the *linear code* of the difference set D and write $\mathcal{C}_{\mathrm{GF}(q)}(D)$ for $\mathcal{C}_{\mathrm{GF}(q)}(\mathbb{D})$. Similarly, by the *incidence matrix* of D we mean that of its development $\mathbb{D}$, and we write M_D for $M_{\mathbb{D}}$. We also call the p-rank of the development $\mathbb{D}$ the *p-rank of the difference set D*.

If D is a cyclic difference set in $(\mathbb{Z}_n, +)$, its incidence matrix M_D must be circulant, i.e., every row of the incident matrix is a cyclic shift of the first row. In this case, the first row in the incidence matrix is the first periodic segment of the characteristic sequence of D, and the code $\mathcal{C}_{\mathrm{GF}(q)}(D)$ must be a cyclic code. Hence, we have the following basic theorem.

Theorem 6.1. *Let D be a difference set of $(\mathbb{Z}_n, +)$ and let q be a power of a prime p. Then $\mathcal{C}_{\mathrm{GF}(q)}(D)$ is an $[n, \kappa]$ cyclic code over $\mathrm{GF}(q)$ with generator polynomial*

$g(x) = \gcd(x^n - 1, D(x))$, *where the Hall polynomial $D(x)$ of D is given by*

$$D(x) = \sum_{i \in D} x^i \in GF(q)[x]$$

and κ is the p-rank of the difference set D. In addition, $\kappa = n - \mathcal{L}_s$, where $\mathcal{L}_s$ denotes the linear span of the characteristic sequence of the difference set D.

The complement code $\mathcal{C}_{GF(q)}(D)^c$ has parameters $[n, n - \kappa]$ and generator polynomial $h(x) = (x^n - 1)/g(x)$, and the dual code $\mathcal{C}_{GF(q)}(D)^\perp$ has parameters $[n, n - \kappa]$ and generator polynomial $h^*(x)/h(0)$, where $h^*(x)$ is the reciprocal of $h(x)$.

Since (n, k, λ) difference sets are symmetric 2-(n, k, λ) designs with order $k - \lambda$, the following is a corollary of Theorem 3.5.

Corollary 6.1. *Let D be an (n, k, λ) design with order $k - \lambda$. Let p be a prime and let F be a field of characteristic p. Then, we have the following.*

(a) *If p divides $(k - \lambda)$ and p divides k, then $\mathcal{C}_F(\mathbb{D})$ is self-orthogonal and has dimension at most $n/2$.*
(b) *If p does not divide $(k - \lambda)$ but p divides k, then $\mathcal{C}_F(\mathbb{D})$ has dimension $n - 1$.*
(c) *If p does not divide $(k - \lambda)$ and p does not divide k, then $\mathcal{C}_F(\mathbb{D})$ has dimension n.*

Corollary 6.1 shows that the linear code $\mathcal{C}_F(\mathbb{D})$ of a difference set is nontrivial and interesting only when the characteristic p of F divides the order $k - \lambda$ of the difference set D. In this monograph, we consider the linear code $\mathcal{C}_F(\mathbb{D})$ only for the case that p divides $(k - \lambda)$.

Similarly, the following is a corollary of Theorem 3.7 and is very useful in the sequel.

Corollary 6.2. *Let D be an (n, k, λ) difference set. Let p be the characteristic of a finite field F. Assume that p divides $k - \lambda$, but p^2 does not divide $k - \lambda$. Then n is odd, and we have the following.*

(a) *If p divides k, then $\mathcal{C}_F(\mathbb{D})$ is self-orthogonal and has dimension $(n - 1)/2$.*
(b) *If p does not divide k, then $\mathcal{C}_F(\mathbb{D})^\perp \subset \mathcal{C}_F(\mathbb{D})$ and $\mathcal{C}_F(\mathbb{D})$ has dimension $(n + 1)/2$.*

The code $\mathcal{C}_F(\mathbb{D})$ in the cases covered in Corollary 6.2 is interesting, as it has dimension either $(n - 1)/2$ or $(n + 1)/2$. The most interesting case is when p^2 divides the order $k - \lambda$, as the dimension of the code $\mathcal{C}_F(\mathbb{D})$ varies from case to case.

6.2 Linear Codes from Hadamard Difference Sets

6.2.1 *Bent functions and Hadamard difference sets*

Let f be a Boolean function from $\mathrm{GF}(2)^m$ to $\mathrm{GF}(2)$. The Walsh transform of f is defined by

$$\hat{f}(w) = \sum_{x \in \mathrm{GF}(2)^m} (-1)^{f(x)+w \cdot x}, \qquad (6.1)$$

where $w \cdot x = w_1 x_1 + w_2 x_2 + \cdots + w_m x_m$, $w = (w_1, w_2, \cdots, w_m) \in \mathrm{GF}(2)^m$ and $x = (x_1, x_2, \ldots, x_m) \in \mathrm{GF}(2)^m$.

Let f be a function from $\mathrm{GF}(2^m)$ to $\mathrm{GF}(2)$. The Walsh transform of f is defined by

$$\hat{f}(w) = \sum_{x \in \mathrm{GF}(2^m)} (-1)^{f(x)+\mathrm{Tr}_{2^m/2}(wx)}, \qquad (6.2)$$

where $w \in \mathrm{GF}(2^m)$.

A function from $\mathrm{GF}(2)^m$ $\big($respectively, $\mathrm{GF}(2^m)\big)$ to $\mathrm{GF}(2)$ is *bent* if $|\hat{f}(w)| = 2^{m/2}$ for every $w \in \mathrm{GF}(2)^m$ $\big($respectively, $w \in \mathrm{GF}(2^m)\big)$. Bent functions exist only for even m, and were coined by Rothaus (1976). The following result is well-known.

Theorem 6.2. *A function from* $\mathrm{GF}(2)^m$ $\big($*respectively,* $\mathrm{GF}(2^m)\big)$ *to* $\mathrm{GF}(2)$ *is bent if and only if it is perfect nonlinear, i.e.,*

$$|\{x \in A : f(x + a) - f(x) = b\}| = 2^{m-1} \qquad (6.3)$$

for each nonzero element a *of* $\mathrm{GF}(2)^m$ $\big($*respectively,* $\mathrm{GF}(2^m)\big)$ *and every* $b \in \mathrm{GF}(2)$.

The *support* of a function from a group A to $\mathrm{GF}(2)$ is defined to be

$$\mathrm{suppt}(f) = \{x \in A : f(x) = 1\} \subseteq A.$$

Let f be bent. Then by definition $\hat{f}(0) = \pm 2^{m/2}$. It then follows that

$$|\mathrm{suppt}(f)| = 2^{m-1} \pm 2^{(m-2)/2}. \qquad (6.4)$$

It is known that the algebraic degree of any bent function on $\mathrm{GF}(2)^m$ is at most $m/2$. Two bent functions f and g on $\mathrm{GF}(2)^m$ are *equivalent* if $g(x) = f(\sigma(x) + b) + \ell(x)$ for an automorphism $\sigma \in \mathrm{Aut}((\mathrm{GF}(2)^m, +))$, an element $b \in \mathrm{GF}(2)^m$ and an affine function ℓ from $\mathrm{GF}(2)^m$ to $\mathrm{GF}(2)$.

A relation between bent functions and difference sets is demonstrated in the following theorem.

Theorem 6.3. *A function f from* $\mathrm{GF}(2)^m$ *(respectively,* $\mathrm{GF}(2^m)$*) to* $\mathrm{GF}(2)$ *is bent if and only if* $\mathrm{suppt}(f)$ *is a difference set in* $(\mathrm{GF}(2)^m, +)$ *(respectively,* $(\mathrm{GF}(2^m), +)$*) with parameters*

$$(2^m, 2^{m-1} \pm 2^{(m-2)/2}, 2^{m-2} \pm 2^{(m-2)/2}). \tag{6.5}$$

Proof.   We have

$$|\{x \in A : f(x + a) - f(x) = 1\}|$$

$$= |(\mathrm{suppt}(f) + a) \cap (A \backslash \mathrm{suppt}(f))| + |((A \backslash \mathrm{suppt}(f)) + a) \cap \mathrm{suppt}(f)|$$

$$= k - |(\mathrm{suppt}(f) + a) \cap \mathrm{suppt}(f)| + k - |(\mathrm{suppt}(f) + a) \cap \mathrm{suppt}(f)|$$

$$= 2k - 2\mathrm{diff}_{\mathrm{suppt}(f)}(a).$$

Similarly, we have

$$|\{x \in A : f(x + a) - f(x) = 0\}| = 2^m - (2k - 2\mathrm{diff}_{\mathrm{suppt}(f)}(a)).$$

The desired conclusion then follows from Theorem 6.2 and 6.4. $\square$

Recall that difference sets with parameters of (6.5) are called Hadamard difference sets.

6.2.2   Constructions of Hadamard difference sets

From Theorem 6.3, we know that bent functions on $\mathrm{GF}(2)^m$ (respectively, $\mathrm{GF}(2^m)$) and Hadamard difference sets in $(\mathrm{GF}(2)^m, +)$ (respectively, $(\mathrm{GF}(2^m), +)$) are the same. Below, we summarize some constructions of bent functions.

Quadratic Boolean functions on $\mathrm{GF}(2)^m$ are those of the form:

$$f(x) = \sum_{1 \le i < j \le m} a_{i,j} x_i x_j + \sum_{1 \le i \le m} a_i x_i + a,$$

where all $a_{i,j}$, a_i and a are elements of $\mathrm{GF}(2)$.

Clearly, affine functions cannot be bent. The following lemma characterizes quadratic bent functions.

Lemma 6.1. *Let* $\ell(x)$ *be any affine Boolean function on* $\mathrm{GF}(2)^m$, *where m is even. A quadratic function*

$$f(x) = \sum_{1 \le i \le m} a_{i,j} x_i x_j + \ell(x)$$

is bent if and only if $|\mathrm{suppt}(f)| = 2^{m-1} \pm 2^{(m-2)/2}$.

Up to equivalence, there are only two quadratic bent functions, which are

$$f(x_1, x_2, \ldots, x_m) = x_1 x_2 + x_3 x_4 + \cdots + x_{m-1} x_m + b, \qquad (6.6)$$

where $b \in \mathrm{GF}(2)$.

The following is the Maiorana–McFarland construction (see Dillon (1974) and McFarland (1973)).

Theorem 6.4 (Maiorana–McFarland). *Let m be any even positive integer and let π be a bijective mapping from $\mathrm{GF}(2)^{m/2}$ to $\mathrm{GF}(2)^{m/2}$. We denote its coordinate functions by $\pi_1, \ldots, \pi_{m/2}$. Let g be a function from $\mathrm{GF}(2)^{m/2}$ to $\mathrm{GF}(2)$. Then*

$$f(x_1, x_2, \ldots, x_m) = x_1 \pi_1(x_{m/2+1}, \ldots, x_m) + x_2 \pi_2(x_{m/2+1}, \ldots, x_m)$$

$$+ \cdots + x_{m/2} \pi_{m/2}(x_{m/2+1}, \ldots, x_m) + g(x_{m/2+1}, \ldots, x_m)$$

is a bent function.

The finite field version of the Maiorana–McFarland construction is stated as follows.

Theorem 6.5 (Maiorana–McFarland). *Let m be any even positive integer and let π be a permutation of $\mathrm{GF}(2^{m/2})$. The function from $\mathrm{GF}(2^{m/2}) \times \mathrm{GF}(2^{m/2})$ to $\mathrm{GF}(2)$*

$$f(x, y) = \mathrm{Tr}_{2^{m/2}/2}(x \pi(y)) + g(y)$$

is bent, where g is any function from $\mathrm{GF}(2^{m/2})$ to $\mathrm{GF}(2)$.

The Maiorana–McFarland construction is generic and yields a huge number of bent functions and hence Hadamard difference sets. This family of bent functions have various algebraic degrees, including quadratic and cubic bent functions. The degree of some bent functions in this family could be high. Plugging any permutation on $\mathrm{GF}(2^{m/2})$ into this construction gives a bent function. Hence, every permutation in the following theorem produces a bent function on $\mathrm{GF}(2^{m/2}) \times \mathrm{GF}(2^{m/2})$.

Theorem 6.6. *The following is a list of permutations on $\mathrm{GF}(2^t)$.*

(a) *Dickson permutation polynomials $D_h(x, a)$, where $\gcd(h, 2^{2t} - 1) = 1$ (see Section 1.3.3.2).*

(b) *$x + x^3 + x^{2^{(t+1)/2}+1}$, where t is odd [Dobbertin (1999a)]. (This gives a cubic bent function with the Maiorana–McFarland construction.)*

(c) *$x^{2^k} + (ax)^{2^k+1} + ax^2$, where $t = 3k$ and $a^{(2^t-1)/(2^k-1)} \neq 1$ [Blokhuis, Coulter, Henderson and O'Keefe (2001)].*

(d) $x^{3 \times 2^{(t+1)/2}+4} + x^{2^{(t+1)/2}+2} + x^{2^{(t+1)/2}}$, *where t is odd* (*see Cherowitzo* (1998) *and Dobbertin* (2002) [*Theorem* 4]).

(e) $x + x^{(2^t+1)/3} + x^{(2^{t+1}-1)/3}$, *where t is odd* [*Ding, Qu, Wang and Yuan* (2014)].

(f) $x + x^{2^{(t+1)/2}+1} + x^{2^{(t+3)/2}+3}$, *where t is odd* [*Ding, Qu, Wang and Yuan* (2014)].

(g) $x^{2^{t-2}-1} + x^{2^{t-1}-1} + x^{2^t-2^{t-2}-1}$, *where t is odd* [*Ding, Qu, Wang and Yuan* (2014)].

(h) $x + x^{2^{(t+1)/2}-1} + x^{2^t-2^{(t+1)/2}+1}$, *where t is odd* [*Ding, Qu, Wang and Yuan* (2014)].

(i) $x^{2^{(t-1)/2}-1} + x^{2^t-2^{(t-1)/2}-2} + x^{2^t-2^{(t-1)/2}-1}$, *where t is odd* [*Ding, Qu, Wang and Yuan* (2014)].

(j) $x + x^3 + x^{2^t-2^{(t+3)/2}+2}$, *where t is odd* [*Ding, Qu, Wang and Yuan* (2014)].

(k) $x^{2^{t-1}-1} + x^{2^{t-2}+2^{(t-1)/2}-1} + x^{2^t-2^{t-2}-1}$, *where t is odd* [*Ding, Qu, Wang and Yuan* (2014)].

(l) $x + x^{2^d} + x^{2^d+1}$, *where* $t \geq 3$ *is odd and* $1 \leq d \leq m-1$ [*Ding, Qu, Wang and Yuan* (2014)].

(m) $x^{2^{(t-1)/2}-1} + x^{2^{t-1}-1} + x^{2^{t-1}-2^{(t-1)/2}}$, *where t is odd* [*Ding, Qu, Wang and Yuan* (2014)].

(n) $x^{2^{t-2}-2^{(t-3)/2}} + x^{2^{(t-1)/2}-1} + x^{2^{t-2}-1+2^{(t-3)/2}}$, *where t is odd* [*Ding, Qu, Wang and Yuan* (2014)].

(o) $x + x^{2^{(t+2)/2}-1} + x^{2^t-2^{t/2}+1}$, *where t is even* [*Ding, Qu, Wang and Yuan* (2014)].

(p) $x + x^{2^{t/2+1}-1} + x^{2^t-2^{t/2+1}+2}$, *where* $t \equiv 4 \pmod 8$ [*Ding, Qu, Wang and Yuan* (2014)].

(q) $x + x^{2^{t/2}} + x^{2^t-2^{t/2}+1}$, *where* $t \equiv 4 \pmod 8$ [*Ding, Qu, Wang and Yuan* (2014)].

A permutation polynomial f on GF(2^m) is called an *o-polynomial* if $f(0) = 0$, $f(1) = 1$, and for each $s \in$ GF(2^m),

$$f_s(x) = (f(x+s) + f(s))x^{2^m-2} \tag{6.7}$$

is a permutation polynomial. Such a polynomial is called an o-polynomial as it can be used to construct hyperovals [Xiang (2005)]. o-polynomials automatically give bent functions with the Maiorana–McFarland construction. Below is a list of o-polynomials over GF(2^m):

(1) $f(x) = x^{2^i}$, where $\gcd(i, m) = 1$.

(2) $f(x) = x^6$, where m is odd [Segre (1962)]).

(3) $f(x) = x^{3 \times 2^{(m+1)/2}+4}$, where m is odd [Glynn (1983)].

(4) $f(x) = x^{2^{(m+1)/2}+2^{(3m+1)/4}}$, where $m \equiv 1 \pmod 4$ [Glynn (1983)].

(5) $f(x) = x^{2^{(m+1)/2}+2^{(m+1)/4}}$, where $m \equiv 3 \pmod 4$ [Glynn (1983)].

(6) $f(x) = x^{\frac{1}{6}} + x^{\frac{3}{6}} + x^{\frac{5}{6}}$, where m is odd [Payne (1985)].

(7) $f(x) = x^{2^{(m+1)/2}} + x^{2^{(m+1)/2}+2} + x^{3 \times 2^{(m+1)/2}+4}$, where m is an odd integer [Cherowitzo (1998)].

More o-polynomials can be found in Cherowitzo, Penttila, Pinneri and Royle (1996) and Cherowitzo, OKeefe and Penttila (2003). Other connections between bent functions and o-polynomials are given in Carlet and Mesnager (2011). A generalization of the Maiorana–McFarland construction can be found in Carlet (2004a).

By the p-rank of a bent function, we mean the p-rank of its difference set. Since the order of any Maiorana–McFarland difference set is $2^{(m-2)/2}$, we need to consider only the 2-rank of bent functions for our coding applications.

The following is an upper bound on the 2-rank of the Maiorana–McFarland bent functions.

Theorem 6.7 ([Weng, Feng and Qiu (2007)]). *Let $m \geq 4$ be even, and let f be any Maiorana–McFarland bent function with m variables. We have*

$$\text{rank}_2(f) \leq 2^{(m+2)/2} - 2. \tag{6.8}$$

The following result demonstrates that the Weng–Feng–Qiu upper bound is tight [Weng, Feng and Qiu (2007)].

Theorem 6.8 ([Weng, Feng and Qiu (2007)]). *Let $\sigma(x) = x^{2^{m/2}-2}$ be the inverse permutation on $GF(2^{m/2})$, and let g be any function on $GF(2^{m/2})$. Then the following Maiorana–McFarland bent function on $GF(2^{m/2}) \times GF(2^{m/2})$*

$$f(x, y) = \text{Tr}_{2^{m/2}/2}(x\sigma(y)) + g(y) \tag{6.9}$$

achieves the rank upper bound of (6.8).

The following theorem documents the partial spread construction.

Theorem 6.9 ([Dillon (1974)]). *Let g be any balanced function from $GF(2^{m/2})$ to $GF(2)$. Then the following function f from $GF(2^{m/2}) \times GF(2^{m/2})$ to $GF(2)$*

$$f(x, y) = g(xy^{2^{m/2}-2}), \quad x, y \in GF(2^{m/2})$$

is bent.

The Dillon construction above is also generic, as there are a large number of balanced functions from g from $GF(2^{m/2})$ to $GF(2)$.

The following is a lower bound on the 2-rank of the Dillon bent functions [Weng, Feng and Qiu (2007)].

Theorem 6.10 ([Weng, Feng and Qiu (2007)]). *For the bent functions in Theorem 6.9, we have*

$$\text{rank}_2(f) \geq 2^{(m+2)/2} - 2.$$

For other constructions of bent functions, the reader is referred to Carlet (2010) and Canteaut, Charpin and Kyureghyan (2008).

The classification of bent functions with eight variables were attacked in Hou (1998). Up to equivalence, there are only four cubic bent functions with eight variables, which are the following:

$$x_7(x_1x_2 + x_8) + x_1x_3 + x_2x_4 + x_5x_6,$$

$$x_7(x_1x_2 + x_3x_4 + x_8) + x_1x_4 + x_2x_5 + x_3x_6,$$

$$x_7(x_1x_2 + x_3x_4 + x_8) + x_3(x_4 + x_2) + x_5x_6,$$

$$x_7(x_1x_2 + x_3x_4 + x_5x_6 + x_8) + x_5(x_4 + x_2) + x_1x_4 + x_3x_6.$$

6.2.3 *Linear codes from Hadamard difference sets*

In this subsection, we investigate the linear code $\mathcal{C}_{\text{GF}(q)}(D)$ of Hadamard difference sets D with the parameters of (6.5) in $(\text{GF}(2)^m, +)$ and $(\text{GF}(2^m), +)$. Note that the order of these difference sets is 2^{m-2}. Due to Theorem 6.1, we consider the code only for the case $q = 2$.

The next result was stated in Assmus and Key (1992b) without proof, and was proved in Jungnickel and Tonchev (1992).

Theorem 6.11. *Let D be a difference set in $(\text{GF}(2^m), +)$ or in $(\text{GF}(2)^m, +)$ with parameters*

$$(2^m, 2^{m-1} - 2^{(m-2)/2}, 2^{m-2} - 2^{(m-2)/2}) \tag{6.10}$$

then $\text{rank}_2(D) \geq m + 2$.

A symmetric design is said to have the *symmetric difference property* if the symmetric difference of any three blocks is either a block or the complement of a block [Kantor (1975)]. The following lemma is proved in Kantor (1975).

Lemma 6.2. *Let $\mathbb{D}$ be a symmetric design with the symmetric difference property. Then $\mathbb{D}$ has the parameters of (6.10).*

The following theorem shows that symmetric designs with the symmetric difference property has the minimum rank.

Theorem 6.12 ([Dillon and Shatz (1987)]). *Let $\mathbb{D}$ be a symmetric design with the parameters of* (6.10). *Then* $\mathrm{rank}_2(D) = m + 2$ *and the all-one vector* $\bar{1} \in \mathcal{C}_{\mathrm{GF}(2)}(\mathbb{D})$ *if and only if $\mathbb{D}$ has the symmetric difference property. In this case, $\mathcal{C}_{\mathrm{GF}(2)}(\mathbb{D})$ consists of the all-zero vector $\bar{0}$, the all-one vector $\bar{1}$, 2^m codewords of weight $2^{m-1} - 2^{(m-2)/2}$ corresponding to the blocks, 2^m codewords of weight $2^{m-1} + 2^{(m-2)/2}$ corresponding to the complements of the blocks and $2^{m+1} - 2$ codewords of weight 2^{m-1}.*

It is open whether or not the binary code of a design with the parameters of (6.10) contains $\bar{1}$, even if one assumes that $\mathrm{rank}_2(D) = m + 2$. If $\mathbb{D}$ is the development of a difference set with the parameters of (6.10), it is then known that $\bar{1}$ is contained in the binary code of the design [Assmus and Key (1992a)][p. 282].

The following is proved in Assmus and Key (1992a) [Theorem 7.10.5].

Theorem 6.13. *Let D be a difference set in* $(\mathrm{GF}(2)^m, +)$ *or* $(\mathrm{GF}(2^m), +)$ *with the parameters of* (6.10). *Then*

$$\mathrm{rank}_2(D) \leq 2^{m-1} + 1 - \frac{1}{2}\binom{m}{m/2}.$$

This upper bound on the 2-rank of Hadamard difference sets may not be tight for $m \geq 4$. It would be very interesting to develop a tight upper bound.

A Boolean function f on $\mathrm{GF}(2)^m$ (respectively, $\mathrm{GF}(2^m)$) is said to be *difference-linear* if $f(x + a) + f(x) + f(a) + f(0)$ is a linear function for every $a \in \mathrm{GF}(2)^m$ (respectively, $a \in \mathrm{GF}(2^m)$). All quadratic bent functions are difference-linear.

We now prove the following result.

Theorem 6.14. *Let D be a difference set in* $(\mathrm{GF}(2)^m, +)$ *or* $(\mathrm{GF}(2^m), +)$ *with the parameters of* (6.10) *and let ξ_D be characteristic function of D. If ξ_D is difference-linear, then $\mathcal{C}_{\mathrm{GF}(2)}(\mathbb{D})$ has parameters $[2^m, m + 2]$, and consists of the all-zero vector $\bar{0}$, the all-one vector $\bar{1}$, 2^m codewords of weight $2^{m-1} - 2^{(m-2)/2}$ corresponding to the blocks, 2^m codewords of weight $2^{m-1} + 2^{(m-2)/2}$ corresponding to the complements of the blocks and $2^{m+1} - 2$ codewords of weight 2^{m-1}.*

In addition, $\mathcal{C}_{\mathrm{GF}(2)}(\mathbb{D})$ contains the first-order Reed–Muller code $\mathcal{R}_2(1, m)$ as a subcode.

Proof.   To simplify our proof, we use A to denote $GF(2^m)$ and $GF(2)^m$. Let $x_1, x_2, \ldots, x_{2^m}$ denote the elements in A. By definition, the incidence matrix M_D of the difference set D is given by

$$M_D = \begin{bmatrix} (\xi_D(x_i + x_1))_{i=1}^{2^m} \\ (\xi_D(x_i + x_2))_{i=1}^{2^m} \\ \vdots \\ (\xi_D(x_i + x_{2^m-1}))_{i=1}^{2^m} \\ (\xi_D(x_i + x_{2^m}))_{i=1}^{2^m} \end{bmatrix},$$

which generates the same linear code as the following matrix:

$$\tilde{M}_D = \begin{bmatrix} (\xi_D(x_i + x_1))_{i=1}^{2^m} \\ (\xi_D(x_i + x_2) + \xi_D(x_i + x_1))_{i=1}^{2^m} \\ \vdots \\ (\xi_D(x_i + x_{2^m-1}) + \xi_D(x_i + x_1))_{i=1}^{2^m} \\ (\xi_D(x_i + x_{2^m}) + \xi_D(x_i + x_1))_{i=1}^{2^m} \end{bmatrix}.$$

Let $\mathring{M}_D$ demote the matrix by deleting the first row of the matrix $\tilde{M}_D$.

Since ξ_D is difference-linear, each function $\xi_D(x + x_j) + \xi_D(x + x_1)$ is affine, i.e., it is a linear function plus a constant in $GF(2)$. For any pair of distinct i and j in $\{2, 3, \ldots, 2^m\}$, we have

$$(\xi_D(x + x_i) + \xi_D(x + x_1)) - (\xi_D(x + x_j) + \xi_D(x + x_1))$$

$$= \xi_D(x + x_i) - \xi_D(x + x_j)$$

which is a balanced function by Theorem 6.2. Hence, all the $2^m - 1$ nonzero affine functions

$$\xi_D(x + x_j) + \xi_D(x + x_1), \quad j = 2, 3, \ldots, 2^m$$

are pairwise distinct. It then follows that the 2-rank of the matrix $\mathring{M}_D$ is at most $m + 1$.

On the other hand, Theorem 6.11 shows that the 2-rank of the matrix $\mathring{M}_D$ is at least $m + 1$. Hence $\text{rank}_2(\mathring{M}_D) = m + 1$, and the binary code with generator matrix $\mathring{M}_D$ is the first-order Reed–Muller code $\mathcal{R}_2(1, m)$, which is a subcode of $\mathcal{C}_{GF(2)}(D)$.

Note that $\xi_D(x + x_1) + \ell(x)$ is also bent for any affine function $\ell(x)$ on A. The weight of the codeword $(\xi_D(x + x_1) + \ell(x))_{x \in A}$ is either $2^{m-1} + 2^{(m-2)/2}$

or $2^{m-1} - 2^{(m-2)/2}$, as the size of the support of any bent function is equal to $2^{m-1} \pm 2^{(m-2)/2}$. The rest of the desired conclusions then follow easily. $\quad\square$

Only a few designs with the symmetric difference property are known [Parker, Spence and Tonchev (1994)]. Thus, the following problem is worthy of investigation.

Problem 6.1. *Construct new symmetric designs with the parameters of* (6.10) *and the symmetric difference property.*

The following two problems are open and interesting.

Problem 6.2. *Find out the parameters and the weight distribution of the code* $\mathcal{C}_{\mathrm{GF}(2)}(D)$ *for the Maiorana–McFarland difference sets with the parameters of* (6.5).

Problem 6.3. *Find out the parameters and the weight distribution of the code* $\mathcal{C}_{\mathrm{GF}(2)}(D)$ *for the Dillon difference sets with the parameters of* (6.5).

6.3 Cyclic Codes of Paley (Almost) Difference Sets

Throughout this section, let n be an odd prime, and let γ be a primitive element of $\mathrm{GF}(n)$. Recall that the cyclotomic classes of order two in $\mathrm{GF}(n)$ are defined by

$$C_i^{(2,n)} = \gamma^i \langle \gamma^2 \rangle,$$

where $i \in \{0, 1\}$. The elements in $C_0^{(2,n)}$ and $C_1^{(2,n)}$ are quadratic residues and quadratic nonresidues modulo n, respectively.

If $n \equiv 3 \pmod 4$, the sets $C_i^{(2,n)}$ and $C_i^{(2,n)} \cup \{0\}$ are Paley difference sets in $(\mathbb{Z}_n, +)$. If $n \equiv 1 \pmod 4$, they are Paley almost difference sets. In this section, we study the p-ranks and the cyclic codes of these difference sets and almost difference sets in $(\mathbb{Z}_n, +)$.

We now need to recall quadratic residue codes of length n over $\mathrm{GF}(q)$, where q is a power of p and $\gcd(q, n) = 1$. Put $m = \mathrm{ord}_n(q)$. Let α be a generator of $\mathrm{GF}(q^m)^*$ and let $\beta = \alpha^{(q^m-1)/n}$. Then β is a primitive nth root of unity in $\mathrm{GF}(q^m)$. Define

$$g_0(x) = \prod_{i \in C_0^{(2,n)}} (x - \beta^i) \quad \text{and} \quad g_1(x) = \prod_{i \in C_1^{(2,n)}} (x - \beta^i).$$

It is known that $g_i(x) \in \mathrm{GF}(q)[x]$ for all $i \in \{0, 1\}$ if and only if $q \in C_0^{(2,n)}$.

When $q \in C_0^{(2,n)}$, we have

$$x^n - 1 = (x - 1)g_0(x)g_1(x).$$

In this case, recall that $\text{QRC}_i^{(n,q)}$ and $\overline{\text{QRC}}_i^{(n,q)}$ denote the cyclic code over $\text{GF}(q)$ of length n with generator polynomial $g_i(x)$ and $(x-1)g_i(x)$, respectively, for each $i \in \{0, 1\}$. The four codes $\text{QRC}_i^{(n,q)}$ and $\overline{\text{QRC}}_i^{(n,q)}$ are quadratic residue codes.

We are now ready to study the codes of the Paley difference sets and almost difference sets. Let D denote $C_0^{(2,n)}$ or $C_0^{(2,n)} \cup \{0\}$. Then the Hall polynomial over $\text{GF}(q)$ of D is

$$D(x) = \rho + \sum_{i \in C_0^{(2,n)}} x^i \in \text{GF}(q)[x],$$

where $\rho = 0$ if $D = C_0^{(2,n)}$ and $\rho = 1$ if $D = C_0^{(2,n)} \cup \{0\}$.

For the code $\mathcal{C}_{\text{GF}(q)}(D)$ of the (almost) difference set D, we have the following result.

Theorem 6.15. *The cyclic code $\mathcal{C}_{\text{GF}(q)}(D)$ of the (almost) difference set D has parameters $[n, \text{rank}_p(D)]$ and parity check polynomial $h(x)$, where $\text{rank}_p(D)$ and $h(x)$ are given below.*

(1) *When $q \in C_0^{(2,n)}$, $n \equiv 1 \pmod 4$ and $\frac{n-1}{4} \equiv 0 \pmod p$, we have $D(\beta) \in \{0, 2\rho - 1\}$,*

$$\text{rank}_p(D) = \begin{cases} \dfrac{n-1}{2} & \text{if } \rho = 0, \\[2mm] \dfrac{n+1}{2} & \text{if } \rho = 1 \end{cases}$$

and

$$h(x) = \begin{cases} g_1(x) & \text{if } D(\beta) = 0 \text{ and } \rho = 0, \\ g_0(x) & \text{if } D(\beta) = 2\rho - 1 \text{ and } \rho = 0, \\ (1-x)g_1(x) & \text{if } D(\beta) = 0 \text{ and } \rho = 1, \\ (1-x)g_0(x) & \text{if } D(\beta) = 2\rho - 1 \text{ and } \rho = 1. \end{cases}$$

(2) When $q \in C_0^{(2,n)}$, $n \equiv 1 \pmod 4$ and $\frac{n-1}{4} \not\equiv 0 \pmod p$, we have

$$\operatorname{rank}_p(D) = \begin{cases} n - 1 & \text{if } \dfrac{n-1+2\rho}{2} \equiv 0 \pmod p, \\[2ex] n & \text{if } \dfrac{n-1+2\rho}{2} \not\equiv 0 \pmod p \end{cases}$$

and

$$h(x) = \begin{cases} \dfrac{x^n - 1}{x - 1} & \text{if } \dfrac{n-1+2\rho}{2} \equiv 0 \pmod p, \\[2ex] x^n - 1 & \text{if } \dfrac{n-1+2\rho}{2} \not\equiv 0 \pmod p. \end{cases}$$

(3) When $q \in C_0^{(2,n)}$, $n \equiv 3 \pmod 4$ and $\frac{n+1}{4} \equiv 0 \pmod p$, we have $D(\beta) \in \{0, 2\rho - 1\}$,

$$\operatorname{rank}_p(D) = \begin{cases} \dfrac{n-1}{2} & \text{if } \rho = 1, \\[2ex] \dfrac{n+1}{2} & \text{if } \rho = 0 \end{cases}$$

and

$$h(x) = \begin{cases} g_1(x) & \text{if } D(\beta) = 0 \text{ and } \rho = 1, \\ g_0(x) & \text{if } D(\beta) = 2\rho - 1 \text{ and } \rho = 1, \\ (1-x)g_1(x) & \text{if } D(\beta) = 0 \text{ and } \rho = 0, \\ (1-x)g_0(x) & \text{if } D(\beta) = 2\rho - 1 \text{ and } \rho = 0. \end{cases}$$

(4) When $q \in C_0^{(2,n)}$, $n \equiv 3 \pmod 4$ and $\frac{n+1}{4} \not\equiv 0 \pmod p$, we have

$$\operatorname{rank}_p(D) = \begin{cases} n - 1 & \text{if } \dfrac{n-1+2\rho}{2} \equiv 0 \pmod p, \\[2ex] n & \text{if } \dfrac{n-1+2\rho}{2} \not\equiv 0 \pmod p \end{cases}$$

and

$$h(x) = \begin{cases} \dfrac{x^n - 1}{x - 1} & \text{if } \dfrac{n-1+2\rho}{2} \equiv 0 \pmod p, \\[2ex] x^n - 1 & \text{if } \dfrac{n-1+2\rho}{2} \not\equiv 0 \pmod p. \end{cases}$$

(5) *When $q \in C_1^{(2,n)}$ and $\frac{n-1+2\rho}{2} \equiv 0$ (mod p), we have* $\mathrm{rank}_p(D) = n - 1$ *and* $h(x) = (x^n - 1)/(x - 1)$.

(6) *When $q \in C_1^{(2,n)}$ and $\frac{n-1+2\rho}{2} \not\equiv 0$ (mod p), we have* $\mathrm{rank}_p(D) = n$ *and* $h(x) = x^n - 1$.

Proof. Note that the generator and parity check polynomials of the cyclic code $\mathcal{C}_{\mathrm{GF}(q)}(D)$ are given by $g(x) = \gcd(x^n - 1, D(x))$ and $h(x) = (x^n - 1)/g(x)$, where $\gcd(x^n - 1, D(x))$ is computed over $\mathrm{GF}(q)$. Notice also that $\mathrm{rank}_p(D) = \deg(h(x))$.

The following three statements are easy to prove:

B1 $C_0^{(2,n)}$ is a subgroup of $\mathbb{Z}_n^*$ with $|C_0^{(2,n)}| = (n-1)/2$ and $uC_1^{(2,n)} = C_1^{(2,n)}$ for any $u \in C_0^{(2,n)}$.

B2 $D(\beta^u) = D(\beta)$ for any $u \in C_0^{(2,n)}$ and $D(\beta^v) = 2\rho - 1 - D(\beta)$ for any $v \in C_1^{(2,n)}$.

B3 $D(\beta) \in \mathrm{GF}(q)$ iff $(D(\beta))^q = D(\beta)$ iff $q \in C_0^{(2,n)}$.

Suppose that $n \equiv 1$ (mod 4). We now prove that

$$D(\beta)(D(\beta) - (2\rho - 1)) = \frac{n-1}{4}. \tag{6.11}$$

In this case, $-1 \in C_0^{(2,n)}$. By Lemma 1.5, we have

$$D(\beta)^2 = \left(\rho + \sum_{i \in C_0^{(2,n)}} \beta^i \right) \left(\rho + \sum_{j \in C_0^{(2,n)}} \beta^j \right)$$

$$= \rho^2 + 2\rho \left(\sum_{i \in C_0^{(2,n)}} \beta^i \right) + \left(\sum_{i \in C_0^{(2,n)}} \beta^i \right) \left(\sum_{j \in C_0^{(2,n)}} \beta^{-j} \right)$$

$$= -\rho^2 + 2\rho D(\beta) + \left(\sum_{i \in C_0^{(2,n)}} \beta^i \right) \left(\sum_{j \in C_0^{(2,n)}} \beta^{-j} \right)$$

$$= -\rho^2 + 2\rho D(\beta) + \frac{n-1}{2} + \sum_{i,j \in C_0^{(2,n)}, i \neq j} \beta^{i-j}$$

$$= -\rho^2 + 2\rho D(\beta) + \frac{n-1}{2} + \frac{n-5}{4}\left(\sum_{i \in C_0^{(2,n)}} \beta^i\right) + \frac{n-1}{4}\left(\sum_{i \in C_1^{(2,n)}} \beta^i\right)$$

$$= \frac{n-1}{4} + (2\rho - 1)D(\beta).$$

Hence, $D(\beta) \in \{0, 2\rho - 1\}$ if and only if $(n-1)/4 \equiv 0 \pmod{p}$ in the case that $n \equiv 1 \pmod 4$.

Suppose that $n \equiv 3 \pmod 4$. We now prove that

$$D(\beta)(D(\beta) - (2\rho - 1)) = -\frac{n+1}{4}. \tag{6.12}$$

In this case, $-1 \in C_1^{(2,n)}$. By Lemma 1.5, we have

$$D(\beta)^2 = \left(\rho + \sum_{i \in C_0^{(2,n)}} \beta^i\right)\left(\rho + \sum_{j \in C_0^{(2,n)}} \beta^j\right)$$

$$= \rho^2 + 2\rho\left(\sum_{i \in C_0^{(2,n)}} \beta^i\right) + \left(\sum_{i \in C_0^{(2,n)}} \beta^i\right)\left(\sum_{j \in C_1^{(2,n)}} \beta^{-j}\right)$$

$$= -\rho^2 + 2\rho D(\beta) + \sum_{i \in C_0^{(2,n)}, j \in C_1^{(2,n)}} \beta^{i-j}$$

$$= -\rho^2 + 2\rho D(\beta) + \frac{n-3}{4}\left(\sum_{i \in C_0^{(2,n)}} \beta^i\right) + \frac{n+1}{4}\left(\sum_{i \in C_1^{(2,n)}} \beta^i\right)$$

$$= -\frac{n+1}{4} + (2\rho - 1)D(\beta).$$

Hence, $D(\beta) \in \{0, 2\rho - 1\}$ if and only if $(n+1)/4 \equiv 0 \pmod{p}$ in the case that $n \equiv 3 \pmod 4$.

We are now ready to compute the p-rank of D and the parity check polynomial of the cyclic code. We have

$$\mathrm{rank}_p(D) = \deg[(x^n - 1)/\gcd(x^n - 1, D(x))]$$

$$= n - |\{j : D(\beta^j) = 0, 0 \leq j \leq n - 1\}|. \tag{6.13}$$

We consider the following two cases depending on whether q is a quadratic residue modulo n or not.

The case that $q \in C_0^{(2,n)}$

It follows from basic fact B3 that in this case we have $D(\beta) \in \mathrm{GF}(q)$.

When $n \equiv 1 \pmod{4}$ and $\frac{n-1}{4} \equiv 0 \pmod{p}$, we have $D(\beta) \in \{0, 2\rho - 1\}$. If $D(\beta) = 0$, we have $D(\beta^u) = 0$ for all $u \in C_0^{(2,n)}$ and $D(\beta^v) \neq 0$ for all $v \in C_1^{(2,n)}$. In this case, we have $\gcd(x^n - 1, D(x)) = g_0(x)(x - 1)$ or $\gcd(x^n - 1, D(x)) = g_0(x)$. If $D(\beta) = 2\rho - 1$, we have $D(\beta^u) \neq 0$ for all $u \in C_0^{(2,n)}$ and $D(\beta^v) = 0$ for all $v \in C_1^{(2,n)}$. In this case, we have $\gcd(x^n - 1, D(x)) = g_1(x)(x - 1)$ or $\gcd(x^n - 1, D(x)) = g_1(x)$.

When $n \equiv 1 \pmod{4}$ and $\frac{n-1}{4} \not\equiv 0 \pmod{p}$, we have $D(\beta) \notin \{0, 2\rho - 1\}$. Hence $D(\beta^i) \neq 0$ for all i with $1 \leq i \leq n - 1$.

Note that

$$D(1) \equiv \frac{n - 1 + 2\rho}{2} \pmod{p}.$$

The desired conclusions on the parity check polynomial and its degree follow.

The conclusions for the case that $n \equiv 3 \pmod{4}$ and $q \in C_0^{(2,n)}$ can be similarly proved. The details are left to the reader.

The case that $q \in C_1^{(2,n)}$

It follows from basic fact B3 that in this case we have $D(\beta) \notin \mathrm{GF}(q)$. Hence $D(\beta^j) \neq 0$ for all $1 \leq j \leq n - 1$. It follows that

$$\mathrm{rank}_p(D) = n - |\{j : D(\beta^j) = 0, 0 \leq j \leq n - 1\}| = n - 1$$

$$h(x) = \frac{x^n - 1}{x - 1}$$

if $\frac{n-1+2\rho}{2} \bmod p = 0$, and

$$\mathrm{rank}_p(D) = n - |\{j : D(\beta^j) = 0, 0 \leq j \leq n - 1\}| = n$$

$$h(x) = x^n - 1$$

if $\frac{n-1+2\rho}{2} \bmod p \neq 0$.

The desired conclusions on the code then follow in this case. $\qquad \square$

Theorem 6.15 tells us that the code $\mathcal{C}_{\mathrm{GF}(q)}(D)$ is either a quadratic residue code or a trivial cyclic code with dimension 1 or 0.

The 2-rank of the Paley difference sets and the minimal polynomial of their characteristic sequences were determined in Turyn (1964), and were rediscovered in Ding, Helleseth and Shan (1998). The p-rank of the Paley difference set was also obtained for the case that $\frac{n+1}{4} \equiv 0 \pmod 4$ in MacWilliams and Mann (1968).

The following lemma follows from the Law of Quadratic Reciprocity and is useful when we consider the code $\mathcal{C}_{\mathrm{GF}(q)}(D)$ over specific finite fields $\mathrm{GF}(q)$.

Lemma 6.3. *Let n be an odd prime. We have the following conclusions:*

- *2 is a quadratic residue modulo n if $n \equiv \pm 1 \pmod 8$ and a nonresidue if $n \equiv \pm 3 \pmod 8$.*
- *3 is a quadratic residue modulo n if $n \equiv \pm 1 \pmod{12}$ and a nonresidue if $n \equiv \pm 5 \pmod{12}$.*
- *5 is a quadratic residue modulo n if $n \equiv \pm 1 \pmod 5$ and a nonresidue if $n \equiv \pm 3 \pmod 5$.*
- *7 is a quadratic residue modulo n if $n \equiv \pm 1, \pm 3, \pm 9 \pmod{28}$ and a nonresidue if $n \equiv \pm 5, \pm 11, \pm 13 \pmod{28}$.*

In this section, we study the code $\mathcal{C}_{\mathrm{GF}(q)}(D)$ when n is a prime. When $n = \ell^s$, where ℓ is a prime and $s \geq 2$, the code $\mathcal{C}_{\mathrm{GF}(q)}(D)$ is not cyclic. The following problem is worth of investigating.

Problem 6.4. *Let $n = \ell^s$, where ℓ is a prime and $s \geq 2$. Determine the parameters of the code $\mathcal{C}_{\mathrm{GF}(q)}(D)$.*

6.4 Linear Codes of Hall Difference Sets

Let n be a prime in the form $n = 4t^2 + 27$ for some t with $\gcd(3, t) = 1$ throughout this section. Recall that the Hall difference set is defined by

$$D = C_0^{(6,n)} \cup C_1^{(6,n)} \cup C_3^{(6,n)},$$

where these $C_i^{(6,n)}$ are cyclotomic classes of order 6 defined by a properly chosen primitive element α of $\mathrm{GF}(n)$.

Since n is a prime, the code $\mathcal{C}_{\mathrm{GF}(q)}$ of the Hall difference set D is cyclic. In this section, we consider the case $q = 2$. The general case can also be treated, but will involve Gaussian periods of order 6 and is much more technical.

Since we assume that $n = 4t^2 + 27$, either $n \equiv 7 \pmod 8$ or $n \equiv 3 \pmod 8$. One can prove that $2 \in C_0^{(6,n)}$ if and only if $n \equiv 7 \pmod 8$. Let $m = \mathrm{ord}_n(2)$, and let γ be a generator of $\mathrm{GF}(2^m)^*$. Put $\beta = \gamma^{(2^m - 1)/n}$. Then β is primitive nth root of unity in $GF(2^m)$. Define

$$g_0(x) = \sum_{i \in C_0^{(6,n)}} (x - \beta^i).$$

When $n \equiv 7 \pmod 8$, we have that $2 \in C_0^{(6,n)}$. In this case, $g_0(x) \in GF(2)[x]$.
The following theorem follows from Kim and Song (2001).

Theorem 6.16. *The binary cyclic code $\mathcal{C}_{GF(2)}(D)$ of the Hall difference set D has parameters $[n, \mathrm{rank}_2(D)]$ and parity check polynomial $h(x)$, where $\mathrm{rank}_2(D)$ and $h(x)$ are given by*

$$\mathrm{rank}_2(D) = \begin{cases} \dfrac{n+5}{6} & \text{if } n \equiv 7 \pmod 8, \\[2mm] n & \text{if } n \equiv 3 \pmod 8 \end{cases}$$

and

$$h(x) = \begin{cases} (x-1)g_0(x) & \text{if } n \equiv 7 \pmod 8, \\ x^n - 1 & \text{if } n \equiv 3 \pmod 8. \end{cases}$$

Example 6.1. Let $n = 31$ and let the primitive root α of n be 3. The cyclic code $\mathcal{C}_{GF(2)}(D)$ of the Hall difference set has parameters $[31, 6, 15]$ and parity check polynomial

$$h(x) = x^6 + x^5 + x^3 + x^2 + x + 1.$$

This is an optimal linear code. Its dual has parameters $[31, 25, 4]$ and is also optimal.

Example 6.2. Let $n = 127$ and let the primitive root α of n be 3. The cyclic code $\mathcal{C}_{GF(2)}(D)$ of the Hall difference set has parameters $[127, 22, 47]$ and parity check polynomial

$$h(x) = x^{22} + x^{21} + x^{20} + x^{16} + x^{14} + x^8 + x^7 + x^6 + x + 1.$$

This code has the same minimum weight as the best binary linear code of length 127 and dimension 22 in the Database.

The next problem is interesting and may be challenging.

Problem 6.5. *Develop a tight lower bound on the minimum weight of the code $\mathcal{C}_{GF(2)}(D)$ of the Hall difference set for the case $n \equiv 7 \pmod 8$.*

The following problem is also open.

Problem 6.6. *Let $n = \ell^s$, where ℓ is a prime of the form $\ell = 4t^2 + 27$ and $s \geq 2$. Determine the parameters of the linear code $\mathcal{C}_{\mathrm{GF}(q)}(D)$ of the Hall difference set D in the noncyclic group* $(\mathrm{GF}(n), +)$.

6.5 Cyclic Codes of Cyclotomic (Almost) Difference Sets of Order 4

6.5.1 *Basic notations and results*

Throughout this section, let n be an odd prime such that $n \equiv 1 \pmod 4$. It is well-known that n can be expressed as $n = u^2 + 4v^2$, where u is an integer with $u \equiv 1 \pmod 4$ and the sign of v is undetermined. As usual, q is a power of a prime p and satisfies $\gcd(n, q) = 1$. Let $\mathrm{ord}_n(q)$ denote the multiplicative order of q modulo n. Let η be an nth primitive root of unity over $\mathrm{GF}(q^{\mathrm{ord}_n(q)})$. Define for each i with $0 \leq i \leq 3$

$$\Omega_i^{(4,n)}(x) = \prod_{j \in C_i^{(4,n)}} (x - \eta^j),$$

where $C_i^{(4,n)}$ denote the cyclotomic classes of order 4 in $\mathrm{GF}(n)$. We have

$$x^n - 1 = (x - 1) \prod_{i=0}^{3} \Omega_i^{(4,n)}(x).$$

It is straightforward to prove that $\Omega_i^{(4,n)}(x) \in \mathrm{GF}(q)[x]$ if $q \in C_0^{(4,n)}$.

Note that the cyclotomic classes $C_0^{(4,n)}$ and $C_2^{(4,n)}$ do not depend on the choice of the generator of $\mathrm{GF}(n)^*$ employed to define the cyclotomic classes. However, different choices of the generator may lead to a swapping of $C_1^{(4,n)}$ and $C_3^{(4,n)}$. So, we have the same conclusions for the four polynomials $\Omega_i^{(4,n)}(x)$.

Note that the cyclotomic classes of order 2 are given by

$$C_0^{(2,n)} = C_0^{(4,n)} \cup C_2^{(4,n)}, \quad C_1^{(2,n)} = C_1^{(4,n)} \cup C_3^{(4,n)}.$$

Define $\theta_0^{(2,n)} = \sum_{i \in C_0^{(2,n)}} \eta^i$. We now prove that

$$\theta_0^{(2,n)}(\theta_0^{(2,n)} + 1) = \frac{n - 1}{4}. \tag{6.14}$$

In this case, $-1 \in C_0^{(2,n)}$. By Lemma 1.5, we have

$$
\begin{aligned}
(\theta_0^{(2,n)})^2 &= \left(\sum_{i \in C_0^{(2,n)}} \eta^i \right) \left(\sum_{j \in C_0^{(2,n)}} \eta^j \right) \\
&= \left(\sum_{i \in C_0^{(2,n)}} \eta^i \right) \left(\sum_{j \in C_0^{(2,n)}} \eta^{-j} \right) \\
&= \frac{n-1}{2} + \sum_{\substack{i,j \in C_0^{(2,n)} \\ i \neq j}} \eta^{i-j} \\
&= \frac{n-1}{2} + \frac{n-5}{4} \left(\sum_{i \in C_0^{(2,n)}} \eta^i \right) + \frac{n-1}{4} \left(\sum_{i \in C_1^{(2,n)}} \eta^i \right) \\
&= \frac{n-1}{4} - \theta_0^{(2,n)}.
\end{aligned}
$$

Hence, $\theta_0^{(2,n)} \in \{0, -1\}$ if and only if $(n-1)/4 \equiv 0 \pmod{p}$.

The following lemma will be useful later and can be proved with the Law of Biquadratic Reciprocity.

Lemma 6.4. *We have the following conclusions:*

- *2 is a biquadratic residue modulo $n \equiv 1 \pmod 4$ if and only if $n = a^2 + 64b^2$ for some integers a and b.*
- *5 is a biquadratic residue modulo $n = a^2 + b^2$, where b is even, if and only if $b \equiv 0 \pmod 5$.*

Lemma 6.5 ([Ding (2013)]). *Let $\mathrm{ord}_n(q) = (n-1)/4$ and $q - 1 < n$. Assume that $q \in C_0^{(4,n)}$. Then the cyclic code over $\mathrm{GF}(q)$ with parity check polynomial $\Omega_i^{(4,n)}(x)$ has parameters $[n, (n-1)/4, d_i]$, and every nonzero weight w in this code satisfies*

$$
\left\lceil \frac{q^{\frac{n-1}{4}} - \left\lfloor (N_1 - 1)\sqrt{q^{\frac{n-1}{4}}} \right\rfloor}{qN} \right\rceil \leq \frac{w}{q-1} \leq \left\lfloor \frac{q^{\frac{n-1}{4}} + \left\lfloor (N_1 - 1)\sqrt{q^{\frac{n-1}{4}}} \right\rfloor}{qN} \right\rfloor,
$$

where

$$N = \frac{q^{\frac{n-1}{4}} - 1}{n} \quad \text{and} \quad N_1 = \frac{N}{q-1}.$$

Proof. Since $\operatorname{ord}_n(q) = (n-1)/4$ and $q \in C_0^{(4,n)}$, the four polynomials $\Omega_i^{(4,n)}(x)$ are irreducible over $GF(q)$. Hence the code with parity check polynomial $\Omega_i^{(4,n)}(x)$ is an irreducible cyclic code with dimension $(n-1)/4$.

Note that $q - 1 < n$ and n is prime. We have then

$$\gcd\left(\frac{q^{\frac{n-1}{4}} - 1}{q - 1}, N\right) = \frac{N}{q-1}.$$

The desired bounds on the nonzero weights follow from Theorem 2.52. $\square$

Example 6.3. Let $q = 3$ and $n = 13$. We have then the canonical factorization

$$x^{13} - 1 = (x + 2)(x^3 + 2x + 2)(x^3 + 2x^2 + 2)(x^3 + x^2 + x + 2)$$
$$\times (x^3 + 2x^2 + 2x + 2).$$

The cyclic code with parity check polynomial $x^3 + 2x + 2$ has parameters $[13, 3, 9]$ and weight enumerator $1 + 26x^9$.

In this case, $N = 2$ and $N_1 = 1$. The lower and upper bounds in Lemma 6.5 are equal to 9.

In general, the bounds are tight only when N_1 is small.

Lemma 6.6 ([Ding (2013)]). *Let* $\operatorname{ord}_n(q) = (n-1)/4$ *and* $q - 1 < n$. *Assume that* $q \in C_0^{(4,n)}$. *Then the cyclic code over* $GF(q)$ *with parity check polynomial* $(x - 1)\Omega_i^{(4,n)}(x)$ *has parameters* $[n, (n+3)/4, d_i]$, *where*

$$d_i \geq \frac{(q - 1)(q^{\frac{n-1}{4}} - 1) - 1}{qN} - \frac{q-1}{q}\left\lfloor \frac{(N - 1)\sqrt{q^{\frac{n-1}{4}}}}{N}\right\rfloor$$

and $N = (q^{\frac{n-1}{4}} - 1)/n$.

Proof. Since $\operatorname{ord}_n(q) = (n-1)/4$ and $q \in C_0^{(4,n)}$, the four polynomials $\Omega_0^{(4,n)}(x)$ are irreducible and over $GF(q)$. Hence the code with parity check polynomial $(x - 1)\Omega_i^{(4,n)}(x)$ has dimension $(n + 3)/4$, and is the same as the code of (2.15).

Note that $q - 1 < n$ and n is prime. We have then

$$\gcd\left(\frac{q^{\frac{n-1}{4}} - 1}{q - 1}, N\right) = \frac{N}{q - 1}.$$

The desired lower bound on the minimum weight follows from Theorem 2.53.

$\square$

Example 6.4. Let $q = 3$ and $n = 13$. We have then the canonical factorization

$$x^{13} - 1 = (x + 2)(x^3 + 2x + 2)(x^3 + 2x^2 + 2)(x^3 + x^2 + x + 2)$$
$$\times (x^3 + 2x^2 + 2x + 2).$$

The cyclic code with parity check polynomial $(x^3 + 2x + 2)(x - 1)$ has parameters $[13, 4, 7]$.

In this case $N = 2$ and $N_1 = 1$. The lower and upper bounds in Lemma 6.5 are equal to 7.

In general, the lower bound is tight only when N is small.

6.5.2 *The cyclic codes of the (almost) difference sets*

Throughout this section, let n be a prime such that $n \equiv 1 \pmod 4$. Let D denote $C_1^{(4,n)} \cup C_2^{(4,n)} \cup C_3^{(4,n)} \cup \{0\}$ or $C_1^{(4,n)} \cup C_2^{(4,n)} \cup C_3^{(4,n)}$. The Hall polynomial of D is given by

$$D(x) = \rho + \sum_{i \in C_1^{(4,n)} \cup C_2^{(4,n)} \cup C_3^{(4,n)}} x^i \in \mathrm{GF}(q)[x],$$

where $\rho \in \{0, 1\}$.

The cyclic code $\mathcal{C}_{\mathrm{GF}(q)}(D)$ of the set D has generator and parity check polynomials

$$g(x) = \gcd(x^n - 1, D(x)) \quad \text{and} \quad h(x) = \frac{x^n - 1}{g(x)},$$

where $\gcd(x^n - 1, D(x))$ is computed over $\mathrm{GF}(q)$. In this subsection, we treat the cyclic codes $\mathcal{C}_{\mathrm{GF}(q)}(D)$, and always assume that $q \in C_0^{(4,n)}$. This ensures that the polynomials $\Omega_i^{(4,n)}(x)$ defined in Section 6.5.1 are over $\mathrm{GF}(q)$. In this subsection, we also assume that $\frac{n-1}{4} \bmod p = 0$.

Let η be an nth primitive root of unity over $\mathrm{GF}(q^{\mathrm{ord}_n(q)})$. We define $\eta_i = \sum_{\ell \in C_i^{(4,n)}} \eta^\ell$ for each $i \in \{0, 1, 2, 3\}$. Because of the assumption that $\frac{n-1}{4} \bmod$

$p = 0$, by (6.14) we have

$$\eta_0 + \eta_2 = \sum_{i \in C_0^{(4,n)} \cup C_2^{(4,n)}} \eta^i = \sum_{i \in C_0^{(2,n)}} \eta^i \in \{0, -1\}. \tag{6.15}$$

The value $\eta_0 + \eta_2$ depends on the choice of η. Throughout this section, we fix an η such that $\eta_0 + \eta_2 = 0$. Notice that $\eta_0 + \eta_1 + \eta_2 + \eta_3 = -1$. We have then $\eta_1 + \eta_3 = -1$. It is easily seen that $D(\eta) = \rho - 1 - \sum_{i \in C_0^{(4,n)}} \eta^i$ and

$$D(\eta^j) = \rho - 1 - \eta_i \tag{6.16}$$

if $j \in C_i^{(4,n)}$.

Due to the assumption that $\frac{n-1}{4} \bmod p = 0$,

$$D(\eta^0) = D(1) = \rho. \tag{6.17}$$

When $n \equiv 1 \pmod 8$, the dimension and generator polynomial of the code $\mathcal{C}_{\mathrm{GF}(q)}(D)$ are given in the following theorem, where the polynomials $\Omega_i^{(4,n)}(x)$ were defined in Section 6.5.1.

Theorem 6.17 ([Ding (2013)]). *Let $\frac{n-1}{4} \equiv 0 \pmod p$ and $q \in C_0^{(4,n)}$, and let $n \equiv 1 \pmod 8$. As before, let $n = u^2 + 4v^2$ with $u \equiv 1 \pmod 4$. The cyclic code $\mathcal{C}_{\mathrm{GF}(q)}(D)$ has generator polynomial $g(x)$ and parameters $[n, n - \mathrm{rank}_p(D), d]$, where $\mathrm{rank}_p(D), h(x)$ and information on d are given below.*

(a) *When $\frac{n+1-2u}{16} \equiv 0 \pmod p$ and $\frac{n-3+2u}{16} \equiv 0 \pmod p$,*

$$g(x) = \begin{cases} (x-1)\Omega_3^{(4,n)}(x) & \text{if } \eta_1 = 0 \text{ and } \rho = 0, \\ (x-1)\Omega_1^{(4,n)}(x) & \text{if } \eta_1 = -1 \text{ and } \rho = 0, \\ \Omega_1^{(4,n)}(x)\Omega_0^{(4,n)}(x)\Omega_2^{(4,n)}(x) & \text{if } \eta_1 = 0 \text{ and } \rho = 1, \\ \Omega_3^{(4,n)}(x)\Omega_0^{(4,n)}(x)\Omega_2^{(4,n)}(x) & \text{if } \eta_1 = -1 \text{ and } \rho = 1 \end{cases}$$

and

$$\mathrm{rank}_p(D) = \begin{cases} n - \dfrac{n+3}{4} & \text{if } \eta_1 = 0 \text{ and } \rho = 0, \\ n - \dfrac{n+3}{4} & \text{if } \eta_1 = -1 \text{ and } \rho = 0, \\ n - \dfrac{3n-3}{4} & \text{if } \eta_1 = 0 \text{ and } \rho = 1, \\ n - \dfrac{3n-3}{4} & \text{if } \eta_1 = -1 \text{ and } \rho = 1. \end{cases}$$

In addition, if $\eta_1 = 0$ and $\rho = 0$ or $\eta_1 = -1$ and $\rho = 0$, the minimum distance d of the code has the lower bound of Lemma 6.6, provided that $\mathrm{ord}_n(q) = (n-1)/4$.

(b) *When $\frac{n+1-2u}{16} \equiv 0 \pmod{p}$ and $\frac{n-3+2u}{16} \not\equiv 0 \pmod{p}$,*

$$g(x) = \begin{cases} x - 1 & \text{if } \rho = 0, \\ \Omega_0^{(4,n)}(x)\Omega_2^{(4,n)}(x) & \text{if } \rho = 1 \end{cases}$$

and

$$\mathrm{rank}_p(D) = \begin{cases} n - 1 & \text{if } \rho = 0, \\ n - \dfrac{n-1}{2} & \text{if } \rho = 1. \end{cases}$$

Furthermore,

$$\begin{cases} d = n & \text{if } \rho = 0, \\ d \geq \sqrt{n} & \text{if } \rho = 1. \end{cases}$$

(c) *When $\frac{n+1-2u}{16} \equiv 1 \pmod{p}$ and $\frac{n-3+2u}{16} \equiv 0 \pmod{p}$, we distinguish between the two subcases: p odd and $p = 2$.*
If p is odd, we have

$$g(x) = \begin{cases} (x-1)\Omega_3^{(4,n)}(x)\Omega_2^{(4,n)}(x) & \text{if } \eta_0 = 1, \eta_1 = 0, \rho = 0, \\ (x-1)\Omega_1^{(4,n)}(x)\Omega_2^{(4,n)}(x) & \text{if } \eta_0 = 1, \eta_1 = -1, \rho = 0, \\ (x-1)\Omega_3^{(4,n)}(x)\Omega_0^{(4,n)}(x) & \text{if } \eta_0 = -1, \eta_1 = 0, \rho = 0, \\ (x-1)\Omega_1^{(4,n)}(x)\Omega_0^{(4,n)}(x) & \text{if } \eta_0 = -1, \eta_1 = -1, \rho = 0, \\ \Omega_1^{(4,n)}(x) & \text{if } \eta_1 = 0, \rho = 1, \\ \Omega_3^{(4,n)}(x) & \text{if } \eta_1 = -1, \rho = 1 \end{cases}$$

and

$$
\text{rank}_p(D) =
\begin{cases}
n - \dfrac{n+1}{2} & \text{if } \eta_0 = 1, \eta_1 = 0, \rho = 0, \\[2mm]
n - \dfrac{n+1}{2} & \text{if } \eta_0 = 1, \eta_1 = -1, \rho = 0, \\[2mm]
n - \dfrac{n+1}{2} & \text{if } \eta_0 = -1, \eta_1 = 0, \rho = 0, \\[2mm]
n - \dfrac{n+1}{2} & \text{if } \eta_0 = -1, \eta_1 = -1, \rho = 0, \\[2mm]
n - \dfrac{n-1}{4} & \text{if } \eta_1 = 0, \rho = 1, \\[2mm]
n - \dfrac{n-1}{4} & \text{if } \eta_1 = -1, \rho = 1.
\end{cases}
$$

In addition, if $\eta_1 = 0$ and $\rho = 1$ or $\eta_1 = -1$ and $\rho = 1$, the minimum distance d of the code has the lower bound of Lemma 6.5, *provided that* $\text{ord}_n(q) = (n-1)/4$. *In the rest four cases, the code is a duadic code and the minimum odd-like weigh $d_{\text{odd}} \geq \sqrt{n}$.*

If $p = 2$, we have

$$
g(x) =
\begin{cases}
(x - 1)\Omega_3^{(4,n)}(x)\Omega_0^{(4,n)}(x)\Omega_2^{(4,n)}(x) & \text{if } \eta_1 = 0, \rho = 0, \\[2mm]
(x - 1)\Omega_1^{(4,n)}(x)\Omega_0^{(4,n)}(x)\Omega_2^{(4,n)}(x) & \text{if } \eta_1 = -1, \rho = 0, \\[2mm]
\Omega_1^{(4,n)}(x) & \text{if } \eta_1 = 0, \rho = 1, \\[2mm]
\Omega_3^{(4,n)}(x) & \text{if } \eta_1 = -1, \rho = 1
\end{cases}
$$

and

$$
\text{rank}_p(D) =
\begin{cases}
n - \dfrac{3n+1}{4} & \text{if } \eta_1 = 0, \rho = 0, \\[2mm]
n - \dfrac{3n+1}{4} & \text{if } \eta_1 = -1, \rho = 0, \\[2mm]
n - \dfrac{n-1}{4} & \text{if } \eta_1 = 0, \rho = 1, \\[2mm]
n - \dfrac{n-1}{4} & \text{if } \eta_1 = -1, \rho = 1.
\end{cases}
$$

In addition, if $\eta_1 = 0$ and $\rho = 1$ or $\eta_1 = -1$ and $\rho = 1$, the minimum distance d of the code has the lower bound of Lemma 6.5, provided that $\mathrm{ord}_n(q) = (n-1)/4$.

(d) *When $\frac{n+1-2u}{16} \equiv 1 \pmod{p}$ and $\frac{n-3+2u}{16} \not\equiv 0 \pmod{p}$, we distinguish between the two cases: p odd and $p = 2$.*
If p is odd,

$$g(x) = \begin{cases} (x-1)\Omega_2^{(4,n)}(x) & \text{if } \eta_0 = 1, \rho = 0, \\ (x-1)\Omega_0^{(4,n)}(x) & \text{if } \eta_0 = -1, \rho = 0, \\ 1 & \text{if } \rho = 1 \end{cases}$$

and

$$\mathrm{rank}_p(D) = \begin{cases} n - \dfrac{n+3}{4} & \text{if } \eta_0 = 1, \rho = 0, \\ n - \dfrac{n+3}{4} & \text{if } \eta_0 = -1, \rho = 0. \\ n & \text{if } \rho = 1. \end{cases}$$

In addition, if $\eta_0 = 1$ and $\rho = 0$ or $\eta_0 = -1$ and $\rho = 0$, the minimum distance d of the code has the lower bound of Lemma 6.6, provided that $\mathrm{ord}_n(q) = (n-1)/4$.
If $p = 2$,

$$g(x) = \begin{cases} (x-1)\Omega_0^{(4,n)}(x)\Omega_2^{(4,n)}(x) & \text{if } \rho = 0, \\ 1 & \text{if } \rho = 1 \end{cases}$$

and

$$\mathrm{rank}_p(D) = \begin{cases} n - \dfrac{n+1}{2} & \text{if } \rho = 0, \\ n & \text{if } \rho = 1. \end{cases}$$

Furthermore, the code is a quadratic residue code and hence $d \geq \sqrt{n}$ if $\rho = 0$.
(e) *When $\frac{n+1-2u}{16} \not\equiv 0, 1 \pmod{p}$ and $\frac{n-3+2u}{16} \equiv 0 \pmod{p}$,*

$$g(x) = \begin{cases} (x-1)\Omega_3^{(4,n)}(x) & \text{if } \eta_1 = 0, \rho = 0, \\ (x-1)\Omega_1^{(4,n)}(x) & \text{if } \eta_1 = -1, \rho = 0, \\ \Omega_1^{(4,n)}(x) & \text{if } \eta_1 = 0, \rho = 1, \\ \Omega_3^{(4,n)}(x) & \text{if } \eta_1 = -1, \rho = 1 \end{cases}$$

and

$$\text{rank}_p(D) = \begin{cases} n - \dfrac{n+3}{4} & \text{if } \eta_1 = 0, \rho = 0, \\[2ex] n - \dfrac{n+3}{4} & \text{if } \eta_1 = -1, \rho = 0, \\[2ex] n - \dfrac{n-1}{4} & \text{if } \eta_1 = 0, \rho = 1, \\[2ex] n - \dfrac{n-1}{4} & \text{if } \eta_1 = -1, \rho = 1. \end{cases}$$

In addition, if $\eta_1 = 0$ and $\rho = 0$ or $\eta_1 = -1$ and $\rho = 0$, the minimum distance d of the code has the lower bound of Lemma 6.6, provided that $\text{ord}_n(q) = (n-1)/4$. If $\eta_1 = 0$ and $\rho = 1$ or $\eta_1 = -1$ and $\rho = 1$, the minimum distance d of the code has the lower bound of Lemma 6.5, provided that $\text{ord}_n(q) = (n-1)/4$.

(f) *When $\frac{n+1-2u}{16} \not\equiv 0, 1 \pmod{p}$ and $\frac{n-3+2u}{16} \not\equiv 0 \pmod{p}$,*

$$g(x) = \begin{cases} x - 1 & \text{if } \rho = 0, \\ 1 & \text{if } \rho = 1 \end{cases}$$

and

$$\text{rank}_p(D) = \begin{cases} n - 1 & \text{if } \rho = 0, \\ n & \text{if } \rho = 1. \end{cases}$$

In addition, $d = n$ if $\rho = 0$.

Proof. We prove the conclusions for Case (a) only. The conclusions of other cases can be similarly proved.

Since $n \equiv 1 \pmod 8$, $-1 \in C_0^{(4,n)}$. By the definition of cyclotomic numbers, we have

$$\eta_\ell^2 = \left(\sum_{i \in C_\ell^{(4,n)}} \eta^i \right)^2$$

$$= (\ell, \ell)_4 \eta_0 + (\ell+3, \ell+3)_4 \eta_1 + (\ell+2, \ell+2)_4 \eta_2$$

$$+ (\ell+1, \ell+1)_4 \eta_3 + \frac{n-1}{4},$$

where and whereafter $(i, j)_4$ means the cyclotomic number $(i, j)^{(4,n)}$.

It then follows from Table 1.3 and the cyclotomic numbers of order 4 for the case $n \equiv 1 \pmod 8$ that

$$\eta_0^2 = \frac{3n - 1 - 2u}{16} - \frac{u+1}{2}\eta_0 + \frac{v}{2}\eta_1 - \frac{v}{2}\eta_3,$$

$$\eta_1^2 = \frac{3n - 1 - 2u}{16} - \frac{u+1}{2}\eta_1 + \frac{v}{2}\eta_2 - \frac{v}{2}\eta_0,$$

$$\eta_2^2 = \frac{3n - 1 - 2u}{16} - \frac{u+1}{2}\eta_2 + \frac{v}{2}\eta_3 - \frac{v}{2}\eta_1,$$

$$\eta_3^2 = \frac{3n - 1 - 2u}{16} - \frac{u+1}{2}\eta_3 + \frac{v}{2}\eta_0 - \frac{v}{2}\eta_2.$$

Whence,

$$\begin{cases} \eta_0^2 + \eta_2^2 = \dfrac{3n - 1 - 2u}{8} - \dfrac{u+1}{2}(\eta_0 + \eta_2), \\[2mm] \eta_1^2 + \eta_3^2 = \dfrac{3n - 1 - 2u}{8} - \dfrac{u+1}{2}(\eta_1 + \eta_3). \end{cases} \tag{6.18}$$

Since $n \equiv 1 \pmod 8$, $-1 \in C_0^{(4,n)}$. By the definition of cyclotomic numbers, we have

$$\eta_\ell \eta_{\ell+2} = \sum_{i \in C_\ell^{(4,n)}} \sum_{j \in C_{\ell+2}^{(4,n)}} \eta^{i-j}$$

$$= (\ell + 2, \ell)_4 \eta_0 + (\ell + 1, \ell + 3)_4 \eta_1 + (\ell, \ell + 2)_4 \eta_2$$

$$+ (\ell + 3, \ell + 1)_4 \eta_3.$$

It then follows from the cyclotomic numbers of order 4 that

$$\begin{cases} \eta_0 \eta_2 = -\dfrac{n + 1 - 2u}{16} + \dfrac{u-1}{4}(\eta_0 + \eta_2), \\[2mm] \eta_1 \eta_3 = -\dfrac{n + 1 - 2u}{16} + \dfrac{u-1}{4}(\eta_1 + \eta_3). \end{cases} \tag{6.19}$$

Since $\frac{n-1}{4} \bmod p = 0$,

$$D(1) = \rho. \tag{6.20}$$

Recall that $\eta_0 + \eta_2 = 0$ and $\eta_1 + \eta_3 = -1$. In Case (a), by (6.18) and (6.19), we have

$$\eta_0 = \eta_2 = 0, \quad \eta_1(\eta_1 + 1) = \eta_3(\eta_3 + 1) = 0.$$

It then follows from (6.16) and (6.20) that

$$
\gcd(D(x), x^n - 1) = \begin{cases} (x-1)\Omega_3^{(4,n)}(x) & \text{if } \eta_1 = 0 \text{ and } \rho = 0, \\[2mm] (x-1)\Omega_1^{(4,n)}(x) & \text{if } \eta_1 = -1 \text{ and } \rho = 0, \\[2mm] \Omega_1^{(4,n)}(x)\Omega_0^{(4,n)}(x)\Omega_2^{(4,n)}(x) & \text{if } \eta_1 = 0 \text{ and } \rho = 1, \\[2mm] \Omega_3^{(4,n)}(x)\Omega_0^{(4,n)}(x)\Omega_2^{(4,n)}(x) & \text{if } \eta_1 = -1 \text{ and } \rho = 1. \end{cases}
$$

The desired conclusions on the dimension and the generator polynomial of the code $\mathcal{C}_{\mathrm{GF}(q)}(D)$ then follow.

The conclusion on the minimum weight for each case follows from Lemma (6.5) or (6.6), or the square-root bound on the minimum weight in quadratic residue codes, or the square-root bound on the minimum odd-like weight in duadic codes.

$\square$

Example 6.5. Let $(q, n) = (2, 113)$. Then $q \in C_0^{(4,n)}$ and $n = u^2 + 4v^2 = (-7)^2 + 4 \times 4^2$. Then

$$
\frac{n+1-2u}{16} \bmod p = 0 \quad \text{and} \quad \frac{n-3+2u}{16} \bmod p = 0.
$$

So this is Case (a). Let $\rho = 0$. Then $\mathcal{C}_{\mathrm{GF}(2)}(D)$ is a $[113, 84, 8]$ cyclic code with generator polynomial $x^{29} + x^{27} + x^{26} + x^{22} + x^{21} + x^{18} + x^{16} + x^{13} + x^{11} + x^8 + x^7 + x^3 + x^2 + 1$. The best binary linear code of length 113 and dimension 84 known has minimum distance 10. The code of this example is the best binary cyclic code of length 113 and dimension 84 according to Table A.68.

Example 6.6. Let $(q, n) = (2, 113)$. Then $q \in C_0^{(4,n)}$ and $n = u^2 + 4v^2 = (-7)^2 + 4 \times 4^2$. Then

$$
\frac{n+1-2u}{16} \bmod p = 0 \quad \text{and} \quad \frac{n-3+2u}{16} \bmod p = 0.
$$

So this is Case (a). Let $\rho = 1$. Then $\mathcal{C}_{\mathrm{GF}(2)}(D)$ is a $[113, 29, 28]$ cyclic code with generator polynomial

$$
x^{84} + x^{82} + x^{81} + x^{80} + x^{76} + x^{75} + x^{74} + x^{73} + x^{72} + x^{70} + x^{68} + x^{66}
$$

$$
+ x^{65} + x^{64} + x^{63} + x^{62} + x^{60} + x^{59} + x^{58} + x^{57} + x^{56} + x^{55} + x^{53}
$$

$$
+ x^{47} + x^{46} + x^{43} + x^{42} + x^{41} + x^{38} + x^{37} + x^{31} + x^{29} + x^{28} + x^{27}
$$

$$
+ x^{26} + x^{25} + x^{24} + x^{22} + x^{21} + x^{20} + x^{19} + x^{18} + x^{16} + x^{14} + x^{12}
$$

$$
+ x^{11} + x^{10} + x^9 + x^8 + x^4 + x^3 + x^2 + 1.
$$

The best binary linear code known of length 113 and dimension 29 has minimum distance 32. The code of this example is the best cyclic code of length 113 and dimension 29 according to Table A.68.

Example 6.7. Let $(q, n) = (4, 41)$. Then $q \in C_0^{(4,n)}$ and $n = u^2 + 4v^2 = 5^2 + 4 \times 2^2$. Then

$$\frac{n + 1 - 2u}{16} \bmod p = 0 \quad \text{and} \quad \frac{n - 3 + 2u}{16} \bmod p = 1.$$

So this is Case (b). Let $\rho = 1$. Then $\mathcal{C}_{\mathrm{GF}(4)}(D)$ is a $[41, 21, 9]$ cyclic code with generator polynomial $x^{20} + x^{18} + x^{17} + x^{16} + x^{15} + x^{14} + x^{11} + x^{10} + x^9 + x^6 + x^5 + x^4 + x^3 + x^2 + 1$. The best linear code over GF(4) with length 41 and dimension 21 has minimum weight 12.

Example 6.8. Let $(q, n) = (2, 73)$. Then $q \in C_0^{(4,n)}$ and $n = u^2 + 4v^2 = (-3)^2 + 4 \times 4^2$. Then

$$\frac{n + 1 - 2u}{16} \bmod p = 1 \quad \text{and} \quad \frac{n - 3 + 2u}{16} \bmod p = 0.$$

So this is Case (c). Let $\rho = 1$. Then $\mathcal{C}_{\mathrm{GF}(2)}(D)$ is a $[73, 55, 6]$ cyclic code with generator polynomial

$$x^{18} + x^{16} + x^{15} + x^{14} + x^{11} + x^{10} + x^9 + x^8 + x^7 + x^4 + x^3 + x^2 + 1.$$

This is the best binary cyclic code with length 73 and dimension 55 according to Table A.38.

Example 6.9. Let $(q, n) = (2, 73)$. Then $q \in C_0^{(4,n)}$ and $n = u^2 + 4v^2 = (-3)^2 + 4 \times 4^2$. Then

$$\frac{n + 1 - 2u}{16} \bmod p = 1 \quad \text{and} \quad \frac{n - 3 + 2u}{16} \bmod p = 0.$$

So this is Case (c). Let $\rho = 0$. Then $\mathcal{C}_{\mathrm{GF}(2)}(D)$ is a $[73, 18, 24]$ cyclic code. This is the best binary cyclic code with length 73 and dimension 18 according to Table A.38.

Example 6.10. Let $(q, n) = (2, 89)$. Then $q \in C_0^{(4,n)}$ and $n = u^2 + 4v^2 = 5^2 + 4 \times 4^2$. Then

$$\frac{n + 1 - 2u}{16} \bmod p = 1 \quad \text{and} \quad \frac{n - 3 + 2u}{16} \bmod p = 0.$$

So this is Case (c). Let $\rho = 0$. Then $\mathcal{C}_{GF(2)}(D)$ is a [89, 22, 28] cyclic code. This is the best cyclic code of length 89 and dimension 22 according to Table A.48. The best binary linear code with length 89 and dimension 22 known has minimum distance 28.

Example 6.11. Let $(q, n) = (4, 17)$. Then $q \in C_0^{(4,n)}$ and $n = u^2 + 4v^2 = 1^2 + 4 \times 2^2$. Then

$$\frac{n + 1 - 2u}{16} \mod p = 1 \quad \text{and} \quad \frac{n - 3 + 2u}{16} \mod p = 1.$$

So this is Case (d). Let $\rho = 0$. Then $\mathcal{C}_{GF(4)}(D)$ is a [17, 8, 6] cyclic code with generator polynomial $x^9 + x^8 + x^6 + x^3 + x + 1$. The optimal linear code GF(4) with length 17 and dimension 8 has minimum distance 8.

Example 6.12. Let $(q, n) = (7, 29)$. Then $q \in C_0^{(4,n)}$ and $n = u^2 + 4v^2 = 5^2 + 4 \times 1^2$. Then

$$\frac{n + 1 - 2u}{16} \mod p = 1 \quad \text{and} \quad \frac{n - 3 + 2u}{16} \mod p = 2.$$

So this is Case (d). Let $\rho = 1$. Then $\mathcal{C}_{GF(7)}(D)$ is a [29, 29, 1] cyclic code. In this example D is an almost difference set.

Example 6.13. Let $(q, n) = (9, 61)$. Then $q \in C_0^{(4,n)}$ and $n = u^2 + 4v^2 = 5^2 + 4 \times 3^2$. Then

$$\frac{n + 1 - 2u}{16} \mod p = 0 \quad \text{and} \quad \frac{n - 3 + 2u}{16} \mod p = 1.$$

So this is Case (b). Let $\rho = 1$. Then $\mathcal{C}_{GF(3)}(D)$ is a [61, 31, 11] cyclic code. The best linear code GF(9) with length 61 and dimension 31 has minimum distance 19. In this example D is an almost difference set.

Example 6.14. Let $(q, n) = (3, 13)$. Then $q \in C_0^{(4,n)}$ and $n = u^2 + 4v^2 = (-3)^2 + 4 \times 1^2$. Then

$$\frac{n + 1 - 2u}{16} \mod p = 1 \quad \text{and} \quad \frac{n - 3 + 2u}{16} \mod p = 0.$$

So this is Case (c). Let $\rho = 1$. Then $\mathcal{C}_{GF(3)}(D)$ is a [13, 10, 3] cyclic code with generator polynomial $x^3 + 2x + 2$. The best linear code GF(3) with length 13 and dimension 10 has minimum distance 10. In this example D is an almost difference set.

Example 6.15. Let $(q, n) = (3, 109)$. Then $q \in C_0^{(4,n)}$ and $n = u^2 + 4v^2 = (-3)^2 + 4 \times 5^2$. Then

$$\frac{n + 1 - 2u}{16} \bmod p = 1 \quad \text{and} \quad \frac{n - 3 + 2u}{16} \bmod p = 0.$$

So this is Case (c). Let $\rho = 1$. Then $\mathcal{C}_{\mathrm{GF}(3)}(D)$ is a $[109, 82, 10]$ cyclic code with generator polynomial

$$x^{27} + x^{25} + x^{24} + x^{22} + x^{21} + x^{19} + x^{18} + 2x^{13} + x^{12}$$
$$+ 2x^{11} + 2x^9 + x^8 + 2x^7 + x^5 + x^4 + 2x^2 + 2x + 2.$$

The best linear code GF(3) with length 189 and dimension 82 has minimum distance 10. In this example D is an almost difference set.

Remark 6.1. Note that $D = \{0\} \cup C_1^{(4,n)} \cup C_2^{(4,n)} \cup C_3^{(4,n)}$ is an almost difference set in $(\mathrm{GF}(n), +)$ when $n = 5^2 + 4v^2$ for odd v or $n = (-3)^2 + 4v^2$ for odd v. So, Theorem 6.17 covers the cyclic codes of these almost difference sets under the restriction that $\frac{n-1}{4} \pmod{p} = 0$.

Examples 6.12–6.15 show that the cyclic code of such almost difference sets could be optimal.

When $n \equiv 5 \pmod{8}$, the dimension and generator polynomial of the code $\mathcal{C}_{\mathrm{GF}(q)}(D)$ are given in the following theorem, where the polynomials $\Omega_i^{(4,n)}(x)$ were defined in Section 6.5.1.

Theorem 6.18 ([Ding (2013)]). *Let* $\frac{n-1}{4} \equiv 0 \pmod{p}$ *and* $q \in C_0^{(4,n)}$, *and let* $n \equiv 5 \pmod{8}$. *As before, let* $n = u^2 + 4v^2$ *with* $u \equiv 1 \pmod{4}$. *The cyclic code* $\mathcal{C}_{\mathrm{GF}(q)}(D)$ *has generator polynomial* $g(x)$ *and parameters* $[n, n - \mathrm{rank}_p(D), d]$, *where* $\mathrm{rank}_p(D)$, $g(x)$ *and information on* d *are given below.*

(a) *When* $\frac{3n-1+2u}{16} \equiv 0 \pmod{p}$ *and* $\frac{3n+3-2u}{16} \equiv 0 \pmod{p}$,

$$g(x) = \begin{cases} (x-1)\Omega_3^{(4,n)}(x) & \text{if } \eta_1 = 0 \text{ and } \rho = 0, \\ (x-1)\Omega_1^{(4,n)}(x) & \text{if } \eta_1 = -1 \text{ and } \rho = 0, \\ \Omega_1^{(4,n)}(x)\Omega_0^{(4,n)}(x)\Omega_2^{(4,n)}(x) & \text{if } \eta_1 = 0 \text{ and } \rho = 1, \\ \Omega_3^{(4,n)}(x)\Omega_0^{(4,n)}(x)\Omega_2^{(4,n)}(x) & \text{if } \eta_1 = -1 \text{ and } \rho = 1 \end{cases}$$

and

$$
\mathrm{rank}_p(D) = \begin{cases}
n - \dfrac{n+3}{4} & \text{if } \eta_1 = 0 \text{ and } \rho = 0, \\[2mm]
n - \dfrac{n+3}{4} & \text{if } \eta_1 = -1 \text{ and } \rho = 0, \\[2mm]
n - \dfrac{3n-3}{4} & \text{if } \eta_1 = 0 \text{ and } \rho = 1, \\[2mm]
n - \dfrac{3n-3}{4} & \text{if } \eta_1 = -1 \text{ and } \rho = 1.
\end{cases}
$$

In addition, if $\eta_1 = 0$ and $\rho = 0$ or $\eta_1 = -1$ and $\rho = 0$, the minimum distance
d of the code has the lower bound of Lemma 6.6, provided that $\mathrm{ord}_n(q) = (n-1)/4$.

(b) *When* $\frac{3n-1+2u}{16} \equiv 0 \pmod p$ *and* $\frac{3n+3-2u}{16} \not\equiv 0 \pmod p$,

$$
g(x) = \begin{cases}
x - 1 & \text{if } \rho = 0, \\
\Omega_0^{(4,n)}(x)\Omega_2^{(4,n)}(x) & \text{if } \rho = 1
\end{cases}
$$

and

$$
\mathrm{rank}_p(D) = \begin{cases}
n - 1 & \text{if } \rho = 0, \\[2mm]
n - \dfrac{n-1}{2} & \text{if } \rho = 1.
\end{cases}
$$

In addition,

$$
\begin{cases}
d = n & \text{if } \rho = 0, \\
d \geq \sqrt{n} & \text{if } \rho = 1.
\end{cases}
$$

(c) *When* $\frac{3n-1+2u}{16} \equiv p - 1 \pmod p$ *and* $\frac{3n+3-2u}{16} \equiv 0 \pmod p$,

$$
g(x) = \begin{cases}
(x-1)\Omega_3^{(4,n)}(x)\Omega_2^{(4,n)}(x) & \text{if } \eta_0 = 1, \eta_1 = 0, \rho = 0, \\
(x-1)\Omega_1^{(4,n)}(x)\Omega_2^{(4,n)}(x) & \text{if } \eta_0 = 1, \eta_1 = -1, \rho = 0, \\
(x-1)\Omega_3^{(4,n)}(x)\Omega_0^{(4,n)}(x) & \text{if } \eta_0 = -1, \eta_1 = 0, \rho = 0, \\
(x-1)\Omega_1^{(4,n)}(x)\Omega_0^{(4,n)}(x) & \text{if } \eta_0 = -1, \eta_1 = -1, \rho = 0, \\
\Omega_1^{(4,n)}(x) & \text{if } \eta_1 = 0, \rho = 1, \\
\Omega_3^{(4,n)}(x) & \text{if } \eta_1 = -1, \rho = 1
\end{cases}
$$

and

$$\text{rank}_p(D) = \begin{cases} n - \dfrac{n+1}{2} & \textit{if } \eta_0 = 1, \eta_1 = 0, \rho = 0, \\[2ex] n - \dfrac{n+1}{2} & \textit{if } \eta_0 = 1, \eta_1 = -1, \rho = 0, \\[2ex] n - \dfrac{n+1}{2} & \textit{if } \eta_0 = -1, \eta_1 = 0, \rho = 0, \\[2ex] n - \dfrac{n+1}{2} & \textit{if } \eta_0 = -1, \eta_1 = -1, \rho = 0, \\[2ex] n - \dfrac{n-1}{4} & \textit{if } \eta_1 = 0, \rho = 1, \\[2ex] n - \dfrac{n-1}{4} & \textit{if } \eta_1 = -1, \rho = 1. \end{cases}$$

In addition, if $\eta_1 = 0$ and $\rho = 1$ or $\eta_1 = -1$ and $\rho = 1$, the minimum distance d of the code has the lower bound of Lemma 6.5, provided that $\text{ord}_n(q) = (n-1)/4$. In the rest four cases, the code is a duadic code and hence the minimum odd-like weigh $d_{\text{odd}} \geq \sqrt{n}$.

(d) *When $\frac{3n-1+2u}{16} \equiv p - 1 \pmod{p}$ and $\frac{3n+3-2u}{16} \not\equiv 0 \pmod{p}$,*

$$g(x) = \begin{cases} (x-1)\Omega_2^{(4,n)}(x) & \textit{if } \eta_0 = 1, \rho = 0, \\[1ex] (x-1)\Omega_0^{(4,n)}(x) & \textit{if } \eta_0 = -1, \rho = 0, \\[1ex] x^n - 1 & \textit{if } \rho = 1 \end{cases}$$

and

$$\text{rank}_p(D) = \begin{cases} n - \dfrac{n+3}{4} & \textit{if } \eta_0 = 1, \rho = 0, \\[2ex] n - \dfrac{n+3}{4} & \textit{if } \eta_0 = -1, \rho = 0, \\[2ex] n & \textit{if } \rho = 1. \end{cases}$$

In addition, if $\eta_0 = 1$ and $\rho = 0$ or $\eta_0 = -1$ and $\rho = 0$, the minimum distance d of the code has the lower bound of Lemma 6.6, provided that $\text{ord}_n(q) = (n-1)/4$.

(e) *When* $\frac{3n-1+2u}{16} \not\equiv 0, p-1 \pmod{p}$ *and* $\frac{3n+3-2u}{16} \equiv 0 \pmod{p}$,

$$g(x) = \begin{cases} (x-1)\Omega_3^{(4,n)}(x) & \text{if } \eta_1 = 0, \rho = 0, \\ (x-1)\Omega_1^{(4,n)}(x) & \text{if } \eta_1 = -1, \rho = 0, \\ \Omega_1^{(4,n)}(x) & \text{if } \eta_1 = 0, \rho = 1, \\ \Omega_3^{(4,n)}(x) & \text{if } \eta_1 = -1, \rho = 1 \end{cases}$$

and

$$\text{rank}_p(D) = \begin{cases} n - \dfrac{n+3}{4} & \text{if } \eta_1 = 0, \rho = 0, \\ n - \dfrac{n+3}{4} & \text{if } \eta_1 = -1, \rho = 0, \\ n - \dfrac{n-1}{4} & \text{if } \eta_1 = 0, \rho = 1, \\ n - \dfrac{n-1}{4} & \text{if } \eta_1 = -1, \rho = 1. \end{cases}$$

Furthermore, if $\eta_1 = 0$ and $\rho = 0$ or $\eta_1 = -1$ and $\rho = 0$, the minimum distance d of the code has the lower bound of Lemma 6.6, provided that $\text{ord}_n(q) = (n-1)/4$. If $\eta_1 = 0$ and $\rho = 1$ or $\eta_1 = -1$ and $\rho = 1$, the minimum distance d of the code has the lower bound of Lemma 6.5, provided that $\text{ord}_n(q) = (n-1)/4$.

(f) *When* $\frac{3n-1+2u}{16} \not\equiv 0, p-1 \pmod{p}$ *and* $\frac{3n+3-2u}{16} \not\equiv 0 \pmod{p}$,

$$g(x) = \begin{cases} x-1 & \text{if } \rho = 0, \\ 1 & \text{if } \rho = 1 \end{cases}$$

and

$$\text{rank}_p(D) = \begin{cases} n-1 & \text{if } \rho = 0, \\ n & \text{if } \rho = 1. \end{cases}$$

Furthermore, $d = n$ if $\rho = 0$.

Proof. We prove only the conclusion of Case (c). The conclusions of other cases can be similarly proved.

Since $n \equiv 5 \pmod 8$, $-1 \in C_2^{(4,n)}$. Note that p must be odd, as $n \equiv 5$ (mod 8) and $\frac{n-1}{4}$ mod $p = 0$. By the definition of cyclotomic numbers, we have

$$
\begin{aligned}
\eta_\ell^2 &= \left(\sum_{i \in C_\ell^{(4,n)}} \eta^i \right)^2 \\
&= \sum_{i \in C_\ell^{(4,n)}} \sum_{j \in C_{\ell+2}^{(4,n)}} \eta^{i-j} \\
&= (\ell+2, \ell)_4 \eta_0 + (\ell+1, \ell+3)_4 \eta_1 + (\ell, \ell+2)_4 \eta_2 + (\ell+3, \ell+1)_4 \eta_3.
\end{aligned}
$$

It then follows from Table 1.2 and the cyclotomic numbers of order 4 for the case $n \equiv 1 \pmod 8$ that

$$
\begin{aligned}
\eta_0^2 &= -\frac{n+1-6u}{16} + \frac{u-1}{2}\eta_0 + \frac{u-v}{2}\eta_1 - \frac{u+v}{2}\eta_3, \\
\eta_1^2 &= -\frac{n+1-6u}{16} + \frac{u-1}{2}\eta_1 + \frac{u-v}{2}\eta_2 - \frac{u+v}{2}\eta_0, \\
\eta_2^2 &= -\frac{n+1-6u}{16} + \frac{u-1}{2}\eta_2 + \frac{u-v}{2}\eta_3 - \frac{u+v}{2}\eta_1, \\
\eta_3^2 &= -\frac{n+1-6u}{16} + \frac{u-1}{2}\eta_3 + \frac{u-v}{2}\eta_0 - \frac{u+v}{2}\eta_2.
\end{aligned}
$$

Whence,

$$
\begin{cases}
\eta_0^2 + \eta_2^2 = -\dfrac{n+1+2u}{8} - \dfrac{u+1}{2}(\eta_0 + \eta_2), \\[3mm]
\eta_1^2 + \eta_3^2 = -\dfrac{n+1+2u}{8} - \dfrac{u+1}{2}(\eta_1 + \eta_3).
\end{cases}
\tag{6.21}
$$

Since $n \equiv 1 \pmod 8$, $-1 \in C_2^{(4,n)}$. By the definition of cyclotomic numbers, we have

$$
\begin{aligned}
\eta_\ell \eta_{\ell+2} &= \sum_{i \in C_\ell^{(4,n)}} \sum_{j \in C_\ell^{(4,n)}} \eta^{i-j} \\
&= (\ell, \ell)_4 \eta_0 + (\ell+3, \ell+3)_4 \eta_1 + (\ell+2, \ell+2)_4 \eta_2 \\
&\quad + (\ell+1, \ell+1)_4 \eta_3 + \frac{n-1}{4}.
\end{aligned}
$$

It then follows from the cyclotomic numbers of order 4 that

$$
\begin{cases}
\eta_0 \eta_2 = \dfrac{3n - 1 + 2u}{16} + \dfrac{u - 1}{4}(\eta_0 + \eta_2), \\[3mm]
\eta_1 \eta_3 = \dfrac{3n - 1 + 2u}{16} + \dfrac{u - 1}{4}(\eta_1 + \eta_3).
\end{cases}
\tag{6.22}
$$

Since $\frac{n-1}{4} \bmod p = 0$,

$$
D(1) = \rho. \tag{6.23}
$$

Recall that $\eta_0 + \eta_2 = 0$ and $\eta_1 + \eta_3 = -1$. In Case 3, by (6.21) and (6.22), we have

$$
\eta_0 = \eta_2 = 1, \quad \eta_1(\eta_1 + 1) = \eta_3(\eta_3 + 1) = 0.
$$

It then follows from (6.16) and (6.23) that

$$
\begin{aligned}
&\gcd\left(D(x), x^n - 1\right) \\[2mm]
&= \begin{cases}
(x - 1)\Omega_3^{(4,n)}(x)\Omega_2^{(4,n)}(x) & \text{if } \eta_0 = 1, \eta_1 = 0, \rho = 0, \\
(x - 1)\Omega_1^{(4,n)}(x)\Omega_2^{(4,n)}(x) & \text{if } \eta_0 = 1, \eta_1 = -1, \rho = 0, \\
(x - 1)\Omega_3^{(4,n)}(x)\Omega_0^{(4,n)}(x) & \text{if } \eta_0 = -1, \eta_1 = 0, \rho = 0, \\
(x - 1)\Omega_1^{(4,n)}(x)\Omega_0^{(4,n)}(x) & \text{if } \eta_0 = -1, \eta_1 = -1, \rho = 0, \\
\Omega_1^{(4,n)}(x) & \text{if } \eta_1 = 0, \rho = 1, \\
\Omega_3^{(4,n)}(x) & \text{if } \eta_1 = -1, \rho = 1.
\end{cases}
\end{aligned}
$$

The desired conclusions on the dimension and the generator polynomial of the code $\mathcal{C}_{\mathrm{GF}(q)}(D)$ then follow. The conclusion on the minimum weight for each case follows from Lemma 6.5 or 6.6, or the square-root bound on the minimum weight in quadratic residue codes, or the square-root bound on the minimum odd-like weight in duadic codes. □

Example 6.16. Let $(q, n) = (9, 37)$. Then $q \in C_0^{(4,n)}$ and $n = u^2 + 4v^2 = 1^2 + 4 \times 3^2$. Then

$$
\frac{3n - 1 + 2u}{16} \bmod p = 1 \quad \text{and} \quad \frac{3n + 3 - 2u}{16} \bmod p = 1.
$$

So this is Case (f). Let $\rho = 0$. Then $\mathcal{C}_{\mathrm{GF}(9)}(D)$ is a $[37, 36, 2]$ cyclic code and optimal. In this example, D is an almost difference set.

Example 6.17. Let $(q, n) = (9, 37)$. Then $q \in C_0^{(4,n)}$ and $n = u^2 + 4v^2 = 1^2 + 4 \times 3^2$. Then

$$\frac{3n - 1 + 2u}{16} \bmod p = 1 \quad \text{and} \quad \frac{3n + 3 - 2u}{16} \bmod p = 1.$$

So this is Case (f). Let $\rho = 1$. Then $C_{\text{GF}(9)}(D)$ is a $[37, 37, 1]$ cyclic code and optimal.

Example 6.18. Let $(q, n) = (9, 61)$. Then $q \in C_0^{(4,n)}$ and $n = u^2 + 4v^2 = 5^2 + 4 \times 3^2$. Then

$$\frac{3n - 1 + 2u}{16} \bmod p = 0 \quad \text{and} \quad \frac{3n + 3 - 2u}{16} \bmod p = 2.$$

So this is Case (b). Let $\rho = 0$. Then $C_{\text{GF}(9)}(D)$ is a $[61, 60, 2]$ cyclic code and optimal.

Example 6.19. Let $(q, n) = (9, 61)$. Then $q \in C_0^{(4,n)}$ and $n = u^2 + 4v^2 = 5^2 + 4 \times 3^2$. Then

$$\frac{3n - 1 + 2u}{16} \bmod p = 1 \quad \text{and} \quad \frac{3n + 3 - 2u}{16} \bmod p = 2.$$

So this is Case (b). Let $\rho = 0$. Then $C_{\text{GF}(9)}(D)$ is a $[61, 31, 11]$ cyclic code. The best linear code over GF(9) with length 61 and dimension 30 has minimum distance 20.

Example 6.20. Let $(q, n) = (3, 13)$. Then $q \in C_0^{(4,n)}$ and $n = u^2 + 4v^2 = (-3)^2 + 4 \times 1^2$. Then

$$\frac{3n - 1 + 2u}{16} \bmod p = 2 \quad \text{and} \quad \frac{3n + 3 - 2u}{16} \bmod p = 0.$$

So this is Case (c). Let $\rho = 0$. Then $C_{\text{GF}(3)}(D)$ is a $[13, 6, 6]$ cyclic code. The best ternary cyclic code of length 13 and dimension 6 has minimum weight 6 according to Table A.83. The optimal linear code over GF(3) with length 13 and dimension 6 has minimum distance 6.

Example 6.21. Let $(q, n) = (3, 13)$. Then $q \in C_0^{(4,n)}$ and $n = u^2 + 4v^2 = (-3)^2 + 4 \times 1^2$. Then

$$\frac{3n - 1 + 2u}{16} \bmod p = 2 \quad \text{and} \quad \frac{3n + 3 - 2u}{16} \bmod p = 0.$$

So this is Case (c). Let $\rho = 1$. Then $C_{\text{GF}(q)}(D)$ is a $[13, 10, 3]$ cyclic code and is optimal, as the known optimal linear code over GF(3) with length 13 and dimension 10 has minimum distance 3.

Example 6.22. Let $(q, n) = (3, 109)$. Then $q \in C_0^{(4,n)}$ and $n = u^2 + 4v^2 = (-3)^2 + 4 \times 5^2$. Then

$$\frac{3n - 1 + 2u}{16} \bmod p = 2 \quad \text{and} \quad \frac{3n + 3 - 2u}{16} \bmod p = 0.$$

So this is Case (c). Let $\rho = 1$. Then $\mathcal{C}_{\mathrm{GF}(3)}(D)$ is a $[109, 82, 10]$ cyclic code. It has the same parameters as the best known ternary linear code with length 109 and dimension 82 which is also cyclic.

Example 6.23. Let $(q, n) = (3, 109)$. Then $q \in C_0^{(4,n)}$ and $n = u^2 + 4v^2 = (-3)^2 + 4 \times 5^2$. Then

$$\frac{3n - 1 + 2u}{16} \bmod p = 2 \quad \text{and} \quad \frac{3n + 3 - 2u}{16} \bmod p = 0.$$

So this is Case (c). Let $\rho = 0$. Then $\mathcal{C}_{\mathrm{GF}(3)}(D)$ is a $[109, 54]$ cyclic code.

Example 6.24. Let $(q, n) = (7, 29)$. Then $q \in C_0^{(4,n)}$ and $n = u^2 + 4v^2 = 5^2 + 4 \times 1^2$. Then

$$\frac{3n - 1 + 2u}{16} \bmod p = 6 \quad \text{and} \quad \frac{3n + 3 - 2u}{16} \bmod p = 5.$$

So this is Case (d). Let $\rho = 0$. Then $\mathcal{C}_{\mathrm{GF}(7)}(D)$ is a $[29, 8, 15]$ cyclic code. The known optimal linear code over GF(7) with length 29 and dimension 8 has minimum distance 17.

Example 6.25. Let $(q, n) = (5, 101)$. Then $q \in C_0^{(4,n)}$ and $n = u^2 + 4v^2 = 1^2 + 4 \times 5^2$. Then

$$\frac{3n - 1 + 2u}{16} \bmod p = p - 1 \quad \text{and} \quad \frac{3n + 3 - 2u}{16} \bmod p = p - 1.$$

So this is Case (d). Let $\rho = 0$. Then $\mathcal{C}_{\mathrm{GF}(5)}(D)$ is a $[101, 75]$ cyclic code. In this example, D is an almost difference set.

Example 6.26. Let $(q, n) = (5, 101)$. Then $q \in C_0^{(4,n)}$ and $n = u^2 + 4v^2 = 1^2 + 4 \times 5^2$. Then

$$\frac{3n - 1 + 2u}{16} \bmod p = p - 1 \quad \text{and} \quad \frac{3n + 3 - 2u}{16} \bmod p = p - 1.$$

So this is Case (d). Let $\rho = 1$. Then $\mathcal{C}_{\mathrm{GF}(5)}(D)$ is a $[101, 101, 1]$ cyclic code.

Example 6.27. Let $(q, n) = (13, 53)$. Then $q \in C_0^{(4,n)}$ and $n = u^2 + 4v^2 = (-7)^2 + 4 \times 1^2$. Then

$$\frac{3n - 1 + 2u}{16} \bmod p = 9 \quad \text{and} \quad \frac{3n + 3 - 2u}{16} \bmod p = 11.$$

So this is Case (f). Let $\rho = 0$. Then $\mathcal{C}_{GF(13)}(D)$ is a $[53, 52, 2]$ cyclic code and is optimal. In this example, D is an almost difference set.

Example 6.28. Let $(q, n) = (13, 53)$. Then $q \in C_0^{(4,n)}$ and $n = u^2 + 4v^2 = (-7)^2 + 4 \times 1^2$. Then

$$\frac{3n - 1 + 2u}{16} \bmod p = 9 \quad \text{and} \quad \frac{3n + 3 - 2u}{16} \bmod p = 11.$$

So this is Case (f). Let $\rho = 1$. Then $\mathcal{C}_{GF(13)}(D)$ is a $[53, 53, 1]$ cyclic code and is optimal.

Example 6.29. Let $(q, n) = (9, 373)$. Then $q \in C_0^{(4,n)}$ and $n = u^2 + 4v^2 = (-7)^2 + 4 \times 9^2$. Then

$$\frac{3n - 1 + 2u}{16} \bmod p = 0 \quad \text{and} \quad \frac{3n + 3 - 2u}{16} \bmod p = 2.$$

So this is Case (b). Let $\rho = 0$. Then $\mathcal{C}_{GF(9)}(D)$ is a $[373, 372, 2]$ cyclic code. In this example, D is an almost difference set.

Example 6.30. Let $(q, n) = (9, 373)$. Then $q \in C_0^{(4,n)}$ and $n = u^2 + 4v^2 = (-7)^2 + 4 \times 9^2$. Then

$$\frac{3n - 1 + 2u}{16} \bmod p = 0 \quad \text{and} \quad \frac{3n + 3 - 2u}{16} \bmod p = 2.$$

So this is Case (b). Let $\rho = 1$. Then $\mathcal{C}_{GF(9)}(D)$ is a $[373, 187]$ cyclic code.

Remark 6.2. Note that $D = C_1^{(4,n)} \cup C_2^{(4,n)} \cup C_3^{(4,n)}$ is an almost difference set in $(GF(n), +)$ when $n = 1^2 + 4v^2$ for odd v or $n = (-7)^2 + 4v^2$ for odd v. In this case, we have $n \equiv 5 \pmod 8$. So, Theorem 6.18 covers the cyclic code of these almost difference sets under the restriction that $\frac{n-1}{4} \equiv 0 \pmod p$.

Examples 6.16, 6.25, 6.27, and 6.29 demonstrate the codes of such almost difference sets.

Remark 6.3. If $n = 4v^2 + 1$ for some odd v, then $D = C_1^{(4,n)} \cup C_2^{(4,n)} \cup C_3^{(4,n)} \cup \{0\}$ is a difference set in $(\mathbb{Z}_n, +)$ with parameters

$$\left(n, \frac{3n+1}{4}, \frac{3(3n+1)}{16} \right).$$

Due to Corollary 6.1, only when p divides the order of this difference set, which is $\frac{3n+1}{16} = \frac{1+3v^2}{4} = v^2 - \frac{v^2-1}{4}$, the code $\mathcal{C}_{\mathrm{GF}(q)}(D)$ of the difference set D is nontrivial and thus interesting. In Theorem 6.18, we have the restriction that p divides $\frac{n-1}{4} = v^2$. It is thus impossible that p divides both $\frac{1+3v^2}{4}$ and v^2. Hence, Theorem 6.18 does not cover the cyclic code $\mathcal{C}_{\mathrm{GF}(q)}(D)$ of the difference set D for the case that $n = 4v^2 + 1$, where v is odd. Hence, the following problem is open and interesting.

Problem 6.7. Let $n = 4v^2 + 1$ for some odd v, and let $D = C_1^{(4,n)} \cup C_2^{(4,n)} \cup C_3^{(4,n)} \cup \{0\}$. Determine the parameters of the code $\mathcal{C}_{\mathrm{GF}(q)}(D)$ of the difference set D for the case that p divides $\frac{3n+1}{16}$ (Note that $\frac{n-1}{4} \not\equiv 0 \pmod{p}$ in this case).

Regarding Problem 6.7, we have the following example.

Example 6.31. Let $n = 37 = 1^2 + 4 \times 3^2$ and $q = p = 7$. Then $D = C_1^{(4,n)} \cup C_2^{(4,n)} \cup C_3^{(4,n)} \cup \{0\}$ is a $(37, 28, 21)$ difference set in $(\mathbb{Z}_{37}, +)$. The order of D is 7 and $(n-1)/4 \equiv 2 \pmod{p}$. The code $\mathcal{C}_{\mathrm{GF}(7)}(D)$ has parameters $[37, 18, 12]$ and generator polynomial

$$x^{19} + 6x^{18} + x^{17} + 4x^{16} + x^{15} + 3x^{14} + 2x^{13} + 5x^{12} + x^{10}$$
$$+ 2x^9 + 2x^8 + 2x^7 + 3x^6 + 6x^5 + 5x^4 + 5x^3 + x^2 + 6.$$

There are six cyclic codes of length 37 and dimension 18 over GF(7). All of them have minimum weight 12.

Remark 6.4. If $n = 4v^2 + 9$ for some odd v, then $D = C_1^{(4,n)} \cup C_2^{(4,n)} \cup C_3^{(4,n)}$ is a difference set in $(\mathbb{Z}_n, +)$ with parameters

$$\left(n, \frac{3(n-1)}{4}, \frac{3(3n-7)}{16} \right).$$

Due to Corollary 6.1, only when p divides the order of this difference set, which is

$$\frac{3(n+3)}{16} = \frac{3}{4} \times \left(\frac{n-1}{4} + 1 \right),$$

the code $\mathcal{C}_{\mathrm{GF}(q)}(D)$ of the difference set D is nontrivial and thus interesting. In Theorem 6.18, we have the restriction that p divides $\frac{n-1}{4}$. These conditions will force that $p = 3$. Example 6.20 is such a case. Hence, Theorem 6.18 covers the cyclic code $\mathcal{C}_{\mathrm{GF}(q)}(D)$ of the difference set D only for the case that $p = 3$ and $n = 4v^2 + 9$, where v is odd. Hence, the following problem is open and interesting.

Problem 6.8. *Let $n = 4v^2 + 9$ for some odd v, and let $D = C_1^{(4,n)} \cup C_2^{(4,n)} \cup C_3^{(4,n)}$. Determine the parameters of the code $\mathcal{C}_{\mathrm{GF}(q)}(D)$ of the difference set D for the case that p divides $\frac{3(3n+1)}{16}$ and $p \neq 3$ (note that $\frac{n-1}{4} \not\equiv 0 \pmod{p}$ in this case).*

Regarding Problem 6.8, we have the following example.

Example 6.32. Let $n = 109 = 3^2 + 4 \times 5^2$ and $q = p = 7$. Then $D = C_1^{(4,n)} \cup C_2^{(4,n)} \cup C_3^{(4,n)}$ is a $(109, 81, 60)$ difference set in $(\mathbb{Z}_{37}, +)$. The order of D is 7 and $(n-1)/4 \equiv 2 \pmod{p}$. The code $\mathcal{C}_{\mathrm{GF}(7)}(D)$ has parameters $[109, 55]$ and generator polynomial

$$x^{54} + 2x^{53} + x^{52} + 2x^{50} + 4x^{49} + 2x^{48} + 4x^{47} + x^{46} + 4x^{45} + 3x^{43}$$

$$+ 4x^{42} + 5x^{41} + 4x^{40} + 5x^{39} + 4x^{38} + 3x^{36} + 2x^{34} + 6x^{33} + 3x^{32}$$

$$+ 3x^{31} + 4x^{30} + 3x^{29} + 4x^{28} + x^{27} + 6x^{26} + 3x^{24} + 4x^{23} + x^{22}$$

$$+ 6x^{21} + 3x^{20} + 4x^{19} + 4x^{18} + 6x^{17} + 3x^{16} + x^{15} + 4x^{14} + x^{13}$$

$$+ 2x^{12} + 2x^{11} + 3x^9 + 3x^7 + 6x^5 + x^4 + 5x^3 + 2x^2 + 6x + 1.$$

This may be a quadratic residue code.

The p-rank of the difference sets $C_0^{(4,n)}$ and $C_0^{(4,n)} \cup \{0\}$ were determined in Pott (1992). For the cyclic codes of the Lehmer difference sets, we also need to find out their generator or parity check polynomial and information on their minimum weight.

6.6 Cyclic Codes of the Two-Prime (Almost) Difference Sets

6.6.1 *The cyclic codes of the two-prime sets*

Throughout this section, let n_1 and n_2 be two distinct odd primes and let $n = n_1 n_2$. Define $N = \gcd(n_1 - 1, n_2 - 1)$ and $e = (n_1 - 1)(n_2 - 1)/N$. It is well-known that any prime n_1 has $\phi(n_1 - 1)$ primitive roots. The Chinese Remainder Theorem guarantees that there are common primitive roots of both n_1 and n_2. Let ς be a fixed common primitive root of both n_1 and n_2, and ϱ be an integer satisfying

$$\varrho \equiv \varsigma \pmod{n_1}, \quad \varrho \equiv 1 \pmod{n_2}.$$

Whiteman proved that [Whiteman (1962)]

$$\mathbb{Z}_n^* = \{\varsigma^s \varrho^i : s = 0, 1, \ldots, e - 1; i = 0, 1, \ldots, N - 1\},$$

where $\mathbb{Z}_n^*$ denotes the set of all invertible elements of the residue class ring $\mathbb{Z}_n$. The generalized cyclotomic classes W_i of order N with respect to n_1 and n_2 are defined by

$$W_i^{(N)} = \{\varsigma^s \varrho^i : s = 0, 1, \ldots, e - 1\}, \quad i = 0, 1, \ldots, N - 1.$$

It was proved in Whiteman (1962) that

$$\mathbb{Z}_n^* = \bigcup_{i=0}^{N-1} W_i^{(N)}, \quad W_i^{(N)} \bigcap W_j^{(N)} = \emptyset \text{ for } i \neq j.$$

Properties of Whiteman's generalized cyclotomy of order N are given in Section 1.5.

Define

$$D_0^{(2)} = \bigcup_{i=0}^{(N-2)/2} W_{2i}^{(N)} \quad \text{and} \quad D_1^{(2)} = \bigcup_{i=0}^{(N-2)/2} W_{2i+1}^{(N)}.$$

Clearly $D_0^{(2)}$ is a subgroup of $\mathbb{Z}_n^*$ and $D_1^{(2)} = \varrho D_0^{(2)}$. The sets $D_0^{(2)}$ and $D_1^{(2)}$ are called the generalized cyclotomic classes of order two and are identical to Whiteman's cyclotomic classes of order N when and only when $N = 2$. In other words, the cyclotomy $\{D_0^{(2)}, D_1^{(2)}\}$ is different from Whiteman's cyclotomy when $N > 2$.

Let

$$C_0 = \{0\} \cup N_2 \cup D_0^{(2)}, \quad C_1 = N_1 \cup D_1^{(2)},$$

where

$$N_1 = \{n_1, 2n_1, 3n_1, \ldots, (n_2 - 1)n_1\}, \quad N_2 = \{n_2, 2n_2, 3n_2, \ldots, (n_1 - 1)n_2\}.$$

Then

$$C_0 \cup C_1 = \mathbb{Z}_n, \quad C_0 \cap C_1 = \emptyset.$$

Throughout this section, let $D = C_1 = N_1 \cup D_1^{(2)}$. When $n_2 = n_1 + 2$, D is the twin-prime difference set. When $n_2 = n_1 + 4$, D is the two-prime almost difference set. In this section, we investigate the complement code $\mathcal{C}_{GF(q)}(D)^c$ of the code $\mathcal{C}_{GF(q)}(D)$ of D and always assume that $\gcd(q, n) = 1$. The dimension and generator polynomial of the code $\mathcal{C}_{GF(q)}(D)$ follow from those of $\mathcal{C}_{GF(q)}(D)^c$ automatically.

Recall that the code $\mathcal{C}_{GF(q)}(D)^c$ of D has generator polynomial

$$g^c(x) = \frac{x^n - 1}{\gcd(x^n - 1, D(x))}, \tag{6.24}$$

where $D(x)$ is the Hall polynomial of D and is defined by

$$D(x) = \sum_{i \in D} x^i = \sum_{i \in N_1 \cup D_1^{(2)}} x^i \in GF(q)[x]. \tag{6.25}$$

6.6.2 *Properties of the generalized cyclotomy of order 2*

In this subsection, we will prove a number of results regarding the generalized cyclotomy of order two, which will be required later when we deal with the cyclic codes $\mathcal{C}_{GF(q)}(D)^c$.

The conclusions in the following lemma are straightforward but useful.

Lemma 6.7. *Let symbols and notations be the same as before. Then*

(1) $\mathrm{ord}_n(\varsigma) = e$, *where* $\mathrm{ord}_n(\varsigma)$ *denotes the order of* ς *modulo* n.
(2) $D_0^{(2)}$ *is a subgroup of* $\mathbb{Z}_n^*$.
(3) *If* $a \in D_j^{(2)}$, $aD_i^{(2)} = D_{(i+j) \bmod 2}^{(2)}$.

The next lemma follows from Lemma 1.13 and the definition of $D_0^{(2)}$ and $D_1^{(2)}$.

Lemma 6.8. *The following four statements are equivalent*:

• $-1 \in D_1^{(2)}$.
• $(n_1 - 1)(n_2 - 1)/N^2$ *is even and* $N/2$ *is odd*.
• *One of the following sets of equations is satisfied*:

$$\begin{cases} n_1 \equiv 1 \pmod 4, \\ n_2 \equiv 3 \pmod 4, \end{cases} \quad \begin{cases} n_1 \equiv 3 \pmod 4, \\ n_2 \equiv 1 \pmod 4. \end{cases}$$

• $n_1 n_2 \equiv 3 \pmod 4$.

The cyclotomic numbers of order two corresponding to the generalized cyclotomic classes of order two are defined by

$$(i, j)_2 = \left| \left(D_i^{(2)} + 1 \right) \cap D_j^{(2)} \right|$$

for any pair of i and j with $0 \le i \le 1$ and $0 \le j \le 1$.

The following lemma summarizes a number of properties of the generalized cyclotomic numbers of order 2 [Ding (2012)].

Lemma 6.9. *Define* $\vartheta = (n_1 - 1)(n_2 - 1)/N^2$. *Then the generalized cyclotomic numbers of order two have the following properties*:

(D0) $(i, j)_2 = (i', j')_2$ *when* $i \equiv i'$ (mod 2) *and* $j \equiv j'$ (mod 2).
(D1) $(i, j)_2 = (2 - i, j - i)_2$.
(D2) $(i, j)_2 = \begin{cases} (j, i)_2 & \textit{if } \vartheta \textit{ odd,} \\ (j + 1, i + 1)_2 & \textit{if } \vartheta \textit{ even and } N/2 \textit{ odd.} \end{cases}$
(D3) $\sum_{j=0}^{1} (i, j)_2 = \frac{(n_1 - 2)(n_2 - 2) - 1}{2} + \bar{\delta}_i$, *where*

$$\bar{\delta}_i = \begin{cases} i & \textit{if } \vartheta \textit{ even and } N/2 \textit{ odd,} \\ 1 - i & \textit{otherwise.} \end{cases}$$

(D4) $\sum_{i=0}^{1} (i, j)_2 = \frac{(n_1 - 2)(n_2 - 2) - 1}{2} + \epsilon_j$, *where*

$$\epsilon_j = \begin{cases} 1 & \textit{if } j = 0, \\ 0 & \textit{if } j = 1. \end{cases}$$

Proof.   Property D0 is straightforward. We first prove Property D1. On one hand, it follows from Property W1 in Lemma 1.14 that

$$(i, j)_2 = \left| \left(D_i^{(2)} + 1 \right) \cap D_j^{(2)} \right|$$

$$= \left| \left(\bigcup_{i_1=0}^{(N-2)/2} W_{2i_1+i}^{(N)} + 1 \right) \cap \left(\bigcup_{j_1=0}^{(N-2)/2} W_{2j_1+j}^{(N)} \right) \right|$$

$$= \sum_{i_1=0}^{(N-2)/2} \sum_{j_1=0}^{(N-2)/2} (2i_1 + i, 2j_1 + j)_N$$

$$= \sum_{i_1=0}^{(N-2)/2} \sum_{j_1=0}^{(N-2)/2} (N - 2i_1 - i, 2(j_1 - i_1) + j - i)_N. \qquad (6.26)$$

On the other hand, since N is even, we have

$$
\begin{aligned}
(2-i, j-i)_2 &= (N-i, j-i)_2 \\
&= \left| \left(D_{N-i}^{(2)} + 1 \right) \bigcap D_{j-i}^{(2)} \right| \\
&= \left| \left(\bigcup_{i_1=0}^{(N-2)/2} W_{N-i-2i_1}^{(N)} + 1 \right) \bigcap \left(\bigcup_{j_1=0}^{(N-2)/2} W_{2j_1+j-i}^{(N)} \right) \right| \\
&= \sum_{i_1=0}^{(N-2)/2} \sum_{j_1=0}^{(N-2)/2} \left| \left(W_{N-i-2i_1}^{(N)} + 1 \right) \bigcap W_{2j_1+j-i}^{(N)} \right| \\
&= \sum_{i_1=0}^{(N-2)/2} \sum_{j_1=0}^{(N-2)/2} \left| \left(D_{N-i-2i_1}^{(N)} + 1 \right) \bigcap D_{2(j_1-i_1)+j-i}^{(N)} \right| \\
&= \sum_{i_1=0}^{(N-2)/2} \sum_{j_1=0}^{(N-2)/2} (N - 2i_1 - i, 2(j_1 - i_1) + j - i)_N. \quad (6.27)
\end{aligned}
$$

Comparing (6.26) and (6.27) proves Property D1.

Property D2 is proved in a similar way. We omit the details here.

We then prove Property D3. It follows from the proof of (6.26) and Property W3 in Lemma 1.14 that

$$
\begin{aligned}
\sum_{j=0}^{1} (i, j)_2 &= \sum_{i_1=0}^{(N-2)/2} \sum_{j=0}^{1} \sum_{j_1=0}^{(N-2)/2} (2i_1 + i, 2j_1 + j)_N \\
&= \sum_{i_1=0}^{(N-2)/2} \sum_{j_1=0}^{N-1} (2i_1 + i, j_1)_N \\
&= \sum_{i_1=0}^{(N-2)/2} \left(\frac{(n_1-1)(n_2-1) - 1}{N} + \delta_{2i_1+i} \right) \\
&= \frac{(n_1-1)(n_2-1) - 1}{2} + \bar{\delta}_i.
\end{aligned}
$$

Property D4 is similarly proved. The detail is left to the reader. $\quad\square$

The generalized cyclotomic numbers of order two are given in the following lemma [Ding (2012)].

Lemma 6.10. *If $n_1 n_2 \equiv 3 \pmod{4}$, we have $(0,0)_2 = (1,0)_2 = (1,1)_2$ and*

$$(0,0)_2 = \frac{(n_1-2)(n_2-2)+1}{4}, \quad (0,1)_2 = \frac{(n_1-2)(n_2-2)-3}{4}.$$

If $n_1 n_2 \equiv 1 \pmod{4}$, we have $(0,1)_2 = (1,0)_2 = (1,1)_2$ and

$$(0,0)_2 = \frac{(n_1-2)(n_2-2)+3}{4}, \quad (0,1)_2 = \frac{(n_1-2)(n_2-2)-1}{4}.$$

Proof. We prove the conclusion of only the first part of this lemma. The conclusion in the second part can be similarly proved.

Assume now that $n_1 n_2 \equiv 3 \pmod{4}$. By Lemma 6.8, $(n_1-1)(n_2-1)/N^2$ is even and $N/2$ is odd. It follows from Property D2 in Lemma 6.9 that $(0,0)_2 = (1,1)_2$. By Property D1 in Lemma 6.9, $(1,0)_2 = (1,1)_2$. Hence

$$(0,0)_2 = (1,0)_2 = (1,1)_2.$$

By Property D4 in Lemma 6.9,

$$(0,0)_2 + (1,0)_2 = \frac{(n_1-2)(n_2-2)+1}{2}.$$

Hence,

$$(0,0)_2 = \frac{(n_1-2)(n_2-2)+1}{4}.$$

By Property D4 in Lemma 6.9 again,

$$(0,1)_2 + (1,1)_2 = \frac{(n_1-2)(n_2-2)-1}{2}.$$

Hence,

$$(0,1)_2 = \frac{(n_1-2)(n_2-2)-3}{4}.$$

This completes the proof of the first part. $\qquad\square$

Remark 6.5. The cyclotomic numbers given in Lemma 6.10 are the same as those given in Whiteman (1962) when $N = 2$, and are new when $N > 2$.

Lemma 6.11. *For any $y \in N_1$, we have*

$$\left|\left(D_1^{(2)} + y\right) \cap D_1^{(2)}\right| = \frac{(n_1 - 1)(n_2 - 3)}{4},$$

$$\left|\left(D_0^{(2)} + y\right) \cap D_1^{(2)}\right| = \frac{(n_1 - 1)(n_2 - 1)}{4}.$$

For any $y \in N_2$, we have

$$\left|\left(D_1^{(2)} + y\right) \cap D_1^{(2)}\right| = \frac{(n_1 - 3)(n_2 - 1)}{4},$$

$$\left|\left(D_0^{(2)} + y\right) \cap D_1^{(2)}\right| = \frac{(n_1 - 1)(n_2 - 1)}{4}.$$

Proof. We prove only two of the four equalities. The other two can be similarly proved.

Let $y \in N_i$ for any $i \in \{0, 1\}$. It follows from Lemma 1.15 that

$$\left|\left(D_0^{(2)} + y\right) \bigcap D_1^{(2)}\right|$$

$$= \left|\left(\bigcup_{i=0}^{(N-2)/2} W_{2i}^{(N)} + y\right) \cap \bigcup_{j=0}^{(N-2)/2} W_{2j+1}^{(N)}\right|$$

$$= \sum_{i=0}^{(N-2)/2} \sum_{j=0}^{(N-2)/2} \left|\left(W_{2i}^{(N)} + y\right) \bigcap W_{2j+1}^{(N)}\right|$$

$$= \left(\frac{N}{2}\right)^2 \frac{(n_1 - 1)(n_2 - 1)}{N^2}$$

$$= \frac{(n_1 - 1)(n_2 - 1)}{4}.$$

This completes the proof of the two equalities. □

6.6.3 *Parameters of the cyclic codes $\mathcal{C}_{\mathrm{GF}(q)}(D)^{\mathrm{c}}$*

Recall that we assume $\gcd(q, n) = 1$. Let ℓ be the order of q modulo n. Then the field $\mathrm{GF}(q^\ell)$ has a primitive nth root of unity θ.

To determine the generator polynomial of the code $\mathcal{C}_{\mathrm{GF}(q)}(D)^{\mathrm{c}}$, we need to compute the polynomial $\gcd(x^n - 1, D(x))$ over $\mathrm{GF}(q)$. To this end, we need a number of auxiliary results.

We have obviously

$$\sum_{i=1}^{n_2-1} \theta^{in_1} = -1 \quad \text{and} \quad \sum_{i=1}^{n_1-1} \theta^{in_2} = -1. \tag{6.28}$$

Lemma 6.12. *Let symbols be the same as before. Then*

$$\sum_{i \in D_1^{(2)}} \theta^{ai} = \begin{cases} -\left(\dfrac{n_1-1}{2} \bmod p\right) & \text{if } a \in N_1, \\[2ex] -\left(\dfrac{n_2-1}{2} \bmod p\right) & \text{if } a \in N_2. \end{cases}$$

Proof. Suppose that $a \in N_2$. Since ς is a common primitive root of both n_1 and n_2 and the order of ς modulo n is e, by the definition of ϱ we have the following multiset equalities:

$$\begin{aligned} D_1^{(2)} \bmod n_1 &= \{\varsigma^s \varrho^i \bmod n_1 : s = 0, 1, \ldots, e-1, i \text{ odd}\} \\ &= \{\varsigma^{s+i} \bmod n_1 : s = 0, 1, \ldots, e-1, i \text{ odd}\} \\ &= \frac{n_2-1}{2} * \{1, 2, \ldots, n_1-1\}, \end{aligned}$$

where $\frac{n_2-1}{2}$ denotes the multiplicity of each element in the set $\{1, 2, \ldots, n_1-1\}$. In other words, when s ranges over $\{0, 1, \ldots, e-1\}$ and i ranges over all even integers between 0 and $N-1$, $\varsigma^s \varrho^i \bmod n_1$ takes on each element of $\{1, 2, \ldots, n_1-1\}$ exactly $(n_2-1)/2$ times. It follows from (6.28) that

$$\sum_{j \in D_1^{(2)}} \theta^{aj} = \left(\frac{n_2-1}{2}\right) \sum_{j \in N_2} \theta^j = -\left(\frac{n_2-1}{2} \bmod p\right).$$

The rest of the conclusions of this lemma can be similarly proved. □

Lemma 6.13. *Let symbols be the same as before. Then*

$$D(\theta^a) = \begin{cases} D(\theta) & \text{if } a \in D_0^{(2)}, \\[1ex] -D(\theta)-1 & \text{if } a \in D_1^{(2)}, \\[1ex] -\left(\dfrac{n_1+1}{2} \bmod p\right) & \text{if } a \in N_1, \\[1ex] \left(\dfrac{n_2-1}{2} \bmod p\right) & \text{if } a \in N_2. \end{cases}$$

Proof. By Lemma 6.7, $aD_0^{(2)} = D_0^{(2)}$ if $a \in D_0^{(2)}$. If $a \in D_0^{(2)}$, $aN_1 = N_1$ since $\gcd(a, n_2) = 1$. Hence

$$D(\theta^a) = \sum_{i \in N_1} \theta^{ai} + \sum_{i \in D_1^{(2)}} \theta^{ai} = \sum_{i \in N_1} \theta^i + \sum_{j \in D_1^{(2)}} \theta^j = D(\theta).$$

If $a \in D_1^{(2)}$, by Lemma 6.7 $aD_1^{(2)} = D_0^{(2)}$. Note that

$$\left(\sum_{i \in D_0^{(2)}} + \sum_{i \in D_1^{(2)}} + \sum_{i \in N_1} + \sum_{i \in N_2} \right) \theta^i + 1 = \sum_{i=0}^{n-1} \theta^i = 0.$$

By (6.28) and Lemma 6.12, we obtain

$$D(\theta^a) = \sum_{i \in N_1} \theta^{ai} + \sum_{i \in D_1^{(2)}} \theta^{ai}$$

$$= \sum_{i \in N_1} \theta^i + \sum_{j \in aD_1^{(2)}} \theta^j$$

$$= \sum_{i \in N_1} \theta^i + \sum_{j \in D_0^{(2)}} \theta^j$$

$$= -D(\theta) - 1.$$

If $a \in N_1$, then $aN_1 = N_1$ since $\gcd(n_1, n_2) = 1$. Then by Lemma 6.12

$$D(\theta^a) = \sum_{i \in N_1} \theta^{ai} + \sum_{i \in D_1^{(2)}} \theta^{ai}$$

$$= \sum_{i \in N_1} \theta^i + \sum_{i \in D_1^{(2)}} \theta^{ai}$$

$$= -\left(\frac{n_1 + 1}{2} \mod p \right).$$

If $a \in N_2$, then $aN_1 = \{0\}$. Then by Lemma 6.12

$$D(\theta^a) = \sum_{i \in N_1} \theta^{ai} + \sum_{i \in D_1^{(2)}} \theta^{ai}$$

$$= [(n_2 - 1) \mod p] + \sum_{i \in D_1^{(2)}} \theta^{ai}$$

$$= \frac{n_2 - 1}{2} \bmod p.$$

This completes the proof of this lemma. □

Note also that

$$D(1) = [n_2 - 1 + (n_1 - 1)(n_2 - 1)/2] \bmod p = \frac{(n_1 + 1)(n_2 - 1)}{2} \bmod p.$$

$$(6.29)$$

Lemma 6.14. *If* $q \in D_0^{(2)}$, *we have* $D(\theta) \in GF(q)$ *and* $D(\theta)^q = (-1)^q + 1 + D(\theta)$.

If $q \in D_1^{(2)}$, *we have* $D(\theta)^q = -1 - D(\theta)$.

Proof. By (6.28) and definition

$$D(\theta) = \sum_{i \in N_1} \theta^i + \sum_{i \in D_1^{(2)}} \theta^i = -1 + \sum_{i \in D_1^{(2)}} \theta^i.$$

If $q \in D_0^{(2)}$ then $qD_i^{(2)} = D_i^{(2)}$ for each i. By Lemma 6.13

$$D(\theta)^q = D(\theta^q) = (-1)^q + \sum_{i \in D_1^{(2)}} \theta^{qi} = (-1)^q + 1 + D(\theta).$$

Hence, $(D(\theta) + 1)^q = D(\theta) + 1$ for both q being odd and even.

If $q \in D_1^{(2)}$, then by Lemma 6.13 we have similarly

$$D(\theta)^q = D(\theta^q) = -1 - D(\theta).$$

In this case, $D(\theta) \notin GF(q)$ if q is even, and $D(\theta)$ could be in $GF(q)$ if q is odd. This completes the proof. □

Lemma 6.15. *If* $n_1 n_2 \equiv 3 \pmod 4$, *we have*

$$D(\theta)(D(\theta) + 1) = -\frac{n+1}{4}.$$

If $n_1 n_2 \equiv 1 \pmod 4$, *we have*

$$D(\theta)(D(\theta) + 1) = \frac{n-1}{4}.$$

Proof. Note that

$$D(\theta) = -1 + \sum_{i \in D_1^{(2)}} \theta^i.$$

We have then

$$D(\theta)(D(\theta) + 1) = D(\theta) + \left(-1 + \sum_{i \in D_1^{(2)}} \theta^i\right)^2$$

$$= -1 - \sum_{i \in D_1^{(2)}} \theta^i + \sum_{i \in D_1^{(2)}} \sum_{j \in D_1^{(2)}} \theta^{i+j}. \qquad (6.30)$$

Suppose that $n_1 n_2 \equiv 3 \pmod 4$. By Lemma 6.8, $-1 \in D_1^{(2)}$. It follows from Lemmas 6.10 and 6.11 that

$$\sum_{i \in D_1^{(2)}} \sum_{j \in D_1^{(2)}} \theta^{i+j}$$

$$= \sum_{i \in D_1^{(2)}} \sum_{j \in D_0^{(2)}} \theta^{i-j}$$

$$= \frac{(n_1 - 1)(n_2 - 1)}{4} \sum_{i \in N_1 \cup N_2} \theta^i + (0,1)_2 \sum_{i \in D_0^{(2)}} \theta^i + (1,0)_2 \sum_{i \in D_1^{(2)}} \theta^i$$

$$= \frac{(n_1 - 1)(n_2 - 1)}{4} \sum_{i \in N_1 \cup N_2} \theta^i$$

$$+ \frac{(n_1 - 2)(n_2 - 2) - 3}{4} \sum_{i \in D_0^{(2)}} \theta^i + \frac{(n_1 - 2)(n_2 - 2) + 1}{4} \sum_{i \in D_1^{(2)}} \theta^i$$

$$= \frac{n_1 + n_2}{4} \sum_{i \in N_1 \cup N_2} \theta^i + \frac{(n_1 - 2)(n_2 - 2) - 3}{4}$$

$$\times \sum_{i \in N_1 \cup N_2 \cup D_0^{(2)} \cup D_1^{(2)}} \theta^i + \sum_{i \in D_1^{(2)}} \theta^i$$

$$= \frac{-n + 3}{4} + D(\theta). \qquad (6.31)$$

Combining (6.30) and (6.31) proves the first conclusion of this lemma.

Suppose now that $n_1 n_2 \equiv 1 \pmod 4$. By Lemma 6.8, $-1 \in D_0^{(2)}$. It follows from Lemmas 6.10 and 6.11 that

$$\sum_{i \in D_1^{(2)}} \sum_{j \in D_1^{(2)}} \theta^{i+j}$$

$$= \sum_{i \in D_1^{(2)}} \sum_{j \in D_1^{(2)}} \theta^{i-j}$$

$$= \frac{(n_1 - 1)(n_2 - 3)}{4} \sum_{i \in N_1} \theta^i + \frac{(n_1 - 3)(n_2 - 1)}{4} \sum_{i N_2} \theta^i$$

$$+ \frac{(n_1 - 1)(n_2 - 1)}{2} + (1, 1)_2 \sum_{i \in D_0^{(2)}} \theta^i + (0, 0)_2 \sum_{i \in D_1^{(2)}} \theta^i$$

$$= \frac{(n_1 - 1)(n_2 - 3)}{4} \sum_{i \in N_1} \theta^i + \frac{(n_1 - 3)(n_2 - 1)}{4} \sum_{i N_2} \frac{(n_1 - 1)(n_2 - 1)}{2}$$

$$+ \frac{(n_1 - 2)(n_2 - 2) - 1}{4} \sum_{i \in D_0^{(2)}} \theta^i + \frac{(n_1 - 2)(n_2 - 2) + 3}{4} \sum_{i \in D_1^{(2)}} \theta^i$$

$$= \frac{n - 1}{4} + \sum_{i \in D_1^{(2)}} \theta^i$$

$$= \frac{n + 3}{4} + D(\theta). \tag{6.32}$$

Combining (6.30) and (6.32) proves the desired second conclusion of this lemma. $\qquad\square$

Lemma 6.16. *If $n \equiv 3 \pmod 4$ and $\frac{n+1}{4} \bmod p = 0$, then $p \in D_0^{(2)}$ and $q \bmod n \in D_0^{(2)}$.*

If $n \equiv 1 \pmod 4$ and $\frac{n-1}{4} \bmod p = 0$, then $p \in D_0^{(2)}$ and $q \bmod n \in D_0^{(2)}$.

Proof. Since $D_0^{(2)}$ is a multiplicative group and q is a power of p, $q \in D_0^{(2)}$ if $p \in D_0^{(2)}$. So we need to prove the conclusions for p only. We prove only the first conclusion of this lemma. The second part can be proved similarly. Hence we assume that $n \equiv 3 \pmod 4$. By Lemma 6.8, $-1 \in D_1^{(2)}$.

We first consider the case that $p = 2$. If $\frac{n+1}{4} \bmod p = 0$, then 8 divides $n + 1$. It is easily checked that in this case, we have only the following four possibilities:

$$\begin{cases} n_1 \equiv 1 & (\bmod 8), \\ n_2 \equiv -1 & (\bmod 8), \end{cases} \quad \begin{cases} n_1 \equiv -1 & (\bmod 8), \\ n_2 \equiv 1 & (\bmod 8), \end{cases}$$
$$\begin{cases} n_1 \equiv 3 & (\bmod 8), \\ n_2 \equiv -3 & (\bmod 8), \end{cases} \quad \begin{cases} n_1 \equiv -3 & (\bmod 8), \\ n_2 \equiv 3 & (\bmod 8). \end{cases} \tag{6.33}$$

Suppose on the contrary that $2 \in D_1^{(2)}$. By definition, 2 can be expressed as $2 = \varsigma^e \varrho^s$ for an odd s. It then follows from the definition of ϱ and ς that

$$2 \equiv \varsigma^{e+s} \pmod{n_1} \quad \text{and} \quad 2 \equiv \varsigma^e \pmod{n_2}.$$

Hence, 2 must be a quadratic residue (nonresidue respectively) modulo n_1 if it is a quadratic nonresidue (residue, respectively) modulo n_2. It follows that none of the four cases in (6.33) could happen. The contradiction proves the conclusion of this lemma for the case $p = 2$.

We now prove the first part of this lemma for the case that p is an odd prime. Note that $-1 \in D_1^{(2)}$. By Lemma 6.8 $(n_1 + n_2)/2$ is even. If $\frac{n+1}{4} \bmod p = 0$, then $n = n_1 n_2 \equiv -1 \pmod{p}$. By the Law of quadratic Reciprocity,

$$\left(\frac{p}{n_i} \right) = (-1)^{\frac{p-1}{2} \frac{n_i-1}{2}} \left(\frac{n_i}{p} \right),$$

where $(\frac{\cdot}{\cdot})$ is the Legendre symbol. Note that

$$\left(\frac{-1}{p} \right) = (-1)^{\frac{p-1}{2}}.$$

It follows then that

$$\left(\frac{p}{n_1} \right) \left(\frac{p}{n_2} \right) = (-1)^{\frac{p-1}{2} \frac{n_1+n_2-2}{2}} \left(\frac{n_1}{p} \right) \left(\frac{n_2}{p} \right)$$

$$= (-1)^{\frac{p-1}{2} \frac{n_1+n_2-2}{2}} \left(\frac{n_1 n_2}{p} \right)$$

$$= (-1)^{\frac{p-1}{2} \frac{n_1+n_2-2}{2}} \left(\frac{-1}{p} \right)$$

$$= (-1)^{\frac{p-1}{2} \frac{n_1+n_2}{2}}$$

$$= (-1)^{p-1}$$

$$= 1. \tag{6.34}$$

Suppose on the contrary that $p \in D_1^{(2)}$. By definition, $p = \varsigma^e \varrho^s$ for some e and odd s. We have then

$$p \equiv \varsigma^{e+s} \pmod{n_1} \quad \text{and} \quad p \equiv \varsigma^e \pmod{n_2}.$$

Since s is odd, we have

$$\left(\frac{p}{n_1}\right)\left(\frac{p}{n_2}\right) = -1. \tag{6.35}$$

This is contrary to (6.34). Hence, the desired conclusion follows. $\qquad\square$

After the preparations above, we are ready to compute the generator polynomial of the code $\mathcal{C}_{\mathrm{GF}(q)}(D)^c$. To this end, we need to discuss the factorization of $x^n - 1$ over $\mathrm{GF}(q)$.

Let θ be the same as before. Among the nth roots of unity θ^i, where $0 \le i \le n - 1$, the n_2 elements θ^i, $i \in N_1 \cup \{0\}$, are n_2th roots of unity, the n_1 elements θ^i, $i \in N_2 \cup \{0\}$, are n_1th roots of unity. Hence,

$$x^{n_1} - 1 = \prod_{i \in N_2 \cup \{0\}} (x - \theta^i), \quad x^{n_2} - 1 = \prod_{i \in N_1 \cup \{0\}} (x - \theta^i).$$

We define

$$d_i(x) = \prod_{i \in D_i^{(2)}} (x - \theta^i)$$

for $i \in \{0, 1\}$. If $q \in D_0^{(2)}$, it is easily proved that $d_i(x) \in \mathrm{GF}(q)[x]$ for all i. Let $d(x) = d_0(x)d_1(x) \in \mathrm{GF}(q)[x]$. We have then

$$x^n - 1 = \prod_{i=0}^{n-1} (x - \theta^i) = \frac{(x^{n_1} - 1)(x^{n_2} - 1)}{x - 1} d(x).$$

Theorem 6.19 ([Ding (2012)]). *Define for $i \in \{1, 2\}$*

$$\Delta_i = \frac{n_i + (-1)^{i-1}}{2} \bmod p, \quad \Delta = \frac{(n_1 + 1)(n_2 - 1)}{2} \bmod p.$$

The cyclic code $\mathcal{C}_{\mathrm{GF}(q)(D)}^c$ has parameters $[n, \kappa, d]$ and generator polynomial $g^c(x)$, where $\kappa = n - \deg(g^c(x))$, and $g^c(x)$ is given below.

(1) *When $n \equiv 3$ (mod 4) and $\frac{n+1}{4}$ mod $p \neq 0$ or $n \equiv 1$ (mod 4) and $\frac{n-1}{4}$ mod $p \neq 0$, we have*

$$
g^c(x) = \begin{cases}
x^n - 1, & \text{if } \Delta_1 \neq 0, \ \Delta_2 \neq 0, \ \Delta \neq 0, \\[2mm]
\dfrac{x^n - 1}{x - 1}, & \text{if } \Delta_1 \neq 0, \ \Delta_2 \neq 0, \ \Delta = 0, \\[2mm]
\dfrac{x^n - 1}{x^{n_2} - 1}, & \text{if } \Delta_1 = 0, \ \Delta_2 \neq 0, \\[2mm]
\dfrac{x^n - 1}{x^{n_1} - 1}, & \text{if } \Delta_1 \neq 0, \ \Delta_2 = 0, \\[2mm]
\dfrac{(x^n - 1)(x - 1)}{(x^{n_1} - 1)(x^{n_2} - 1)}, & \text{if } \Delta_1 = 0, \ \Delta_2 = 0.
\end{cases}
\tag{6.36}
$$

(2) *When $n \equiv 3$ (mod 4) and $\frac{n+1}{4}$ mod $p = 0$ or $n \equiv 1$ (mod 4) and $\frac{n-1}{4}$ mod $p = 0$, by Lemma 6.16, $d_i(x) \in \mathrm{GF}(q)[x]$ for each i. In this case, $D(\theta) \in \{0, 1\}$ and*

$$
g^c(x) = \begin{cases}
\dfrac{x^n - 1}{d_0(x)} & \text{if } \Delta_1 \neq 0, \ \Delta_2 \neq 0, \ \Delta \neq 0, \ D(\theta) = 0, \\[2mm]
\dfrac{x^n - 1}{d_1(x)} & \text{if } \Delta_1 \neq 0, \ \Delta_2 \neq 0, \ \Delta \neq 0, \ D(\theta) = -1, \\[2mm]
\dfrac{x^n - 1}{(x - 1)d_0(x)} & \text{if } \Delta_1 \neq 0, \ \Delta_2 \neq 0, \ \Delta = 0, \ D(\theta) = 0, \\[2mm]
\dfrac{x^n - 1}{(x - 1)d_1(x)} & \text{if } \Delta_1 \neq 0, \Delta_2 \neq 0, \Delta = 0, D(\theta) = -1, \\[2mm]
\dfrac{x^n - 1}{(x^{n_2} - 1)d_0(x)} & \text{if } \Delta_1 = 0, \ \Delta_2 \neq 0, \ D(\theta) = 0, \\[2mm]
\dfrac{x^n - 1}{(x^{n_2} - 1)d_1(x)} & \text{if } \Delta_1 = 0, \ \Delta_2 \neq 0, \ D(\theta) = -1, \\[2mm]
\dfrac{x^n - 1}{(x^{n_1} - 1)d_0(x)} & \text{if } \Delta_1 \neq 0, \ \Delta_2 = 0, \ D(\theta) = 0, \\[2mm]
\dfrac{x^n - 1}{(x^{n_1} - 1)d_1(x)} & \text{if } \Delta_1 \neq 0, \ \Delta_2 = 0, \ D(\theta) = -1, \\[2mm]
\dfrac{(x^n - 1)(x - 1)}{d_0(x)\prod_{i=1}^{2}(x^{n_i} - 1)} & \text{if } \Delta_1 = 0, \ \Delta_2 = 0, \ D(\theta) = 0, \\[2mm]
\dfrac{(x^n - 1)(x - 1)}{d_1(x)\prod_{i=1}^{2}(x^{n_i} - 1)} & \text{if } \Delta_1 = 0, \ \Delta_2 = 0, D(\theta) = -1.
\end{cases}
$$

$$\tag{6.37}$$

Proof. The desired conclusions on the dimension and generator polynomial of the code $\mathcal{C}^c_{\mathrm{GF}(q)(D)}$ then follow from (6.29), Lemmas 6.13–6.16. $\qquad\square$

The following corollary follows directly from Theorem 6.19 and Lemma 6.16.

Corollary 6.3 ([Ding (2012)]). *Let* $q = 2$. *The cyclic code* $\mathcal{C}^c_{\mathrm{GF}(q)(D)}$ *has the following parameters* $[n, \kappa, d]$ *and generator polynomial* $g^c(x)$:

(1) *If* $n_1 \equiv 1 \pmod 8$ *and* $n_2 \equiv 3 \pmod 8$ *or* $n_1 \equiv -3 \pmod 8$ *and* $n_2 \equiv -1 \pmod 8$, *we have*

$$\kappa = 1, \quad g^c(x) = \frac{x^n - 1}{x - 1},$$

and $d = n - 1$.

(2) *If* $n_1 \equiv -1 \pmod 8$ *and* $n_2 \equiv 3 \pmod 8$ *or* $n_1 \equiv 3 \pmod 8$ *and* $n_2 \equiv -1 \pmod 8$, *we have*

$$\kappa = n_2, \quad g^c(x) = \frac{x^n - 1}{x^{n_2} - 1},$$

and $d = n_1$, *which follows from Theorem* 6.20.

(3) *If* $n_1 \equiv -3 \pmod 8$ *and* $n_2 \equiv 1 \pmod 8$ *or* $n_1 \equiv 1 \pmod 8$ *and* $n_2 \equiv -3 \pmod 8$, *we have*

$$\kappa = n_1, \quad g^c(x) = \frac{x^n - 1}{x^{n_1} - 1},$$

and $d = n_2$, *which follows from Theorem* 6.20.

(4) *If* $n_1 \equiv -1 \pmod 8$ *and* $n_2 \equiv -3 \pmod 8$ *or* $n_1 \equiv 3 \pmod 8$ *and* $n_2 \equiv 1 \pmod 8$, *we have*

$$\kappa = n_1 + n_2 - 1, \quad g^c(x) = \frac{(x^n - 1)(x - 1)}{(x^{n_2} - 1)(x^{n_2} - 1)}$$

and d *has the lower bound of Theorem* 6.21.

(5) *If* $n_1 \equiv 1 \pmod 8$ *and* $n_2 \equiv -1 \pmod 8$ *or* $n_1 \equiv -3 \pmod 8$ *and* $n_2 \equiv 3 \pmod 8$, *we have*

$$\kappa = \frac{(n_1 - 1)(n_2 - 1) + 2}{2},$$

$$g^c(x) = \begin{cases} \dfrac{x^n - 1}{(x - 1)d_0(x)} & \text{if } D(\theta) = 0, \\[2mm] \dfrac{x^n - 1}{(x - 1)d_1(x)} & \text{if } D(\theta) = 1. \end{cases}$$

(6) *If* $n_1 \equiv -1 \pmod 8$ *and* $n_2 \equiv 1 \pmod 8$ *or* $n_1 \equiv 3 \pmod 8$ *and* $n_2 \equiv -3$ (mod 8), *we have*

$$\kappa = \frac{(n_1 + 1)(n_2 + 1) - 2}{2}$$

and

$$g^c(x) = \begin{cases} \dfrac{(x^n - 1)(x - 1)}{(x^{n_1} - 1)(x^{n_2} - 1)d_1(x)} & \text{if } D(\theta) = 0, \\[3mm] \dfrac{(x^n - 1)(x - 1)}{(x^{n_1} - 1)(x^{n_2} - 1)d_0(x)} & \text{if } D(\theta) = 1, \end{cases}$$

and d *has the lower bound of Theorem 6.22.*

(7) *If* $n_1 \equiv -1 \pmod 8$ *and* $n_2 \equiv -1 \pmod 8$ *or* $n_1 \equiv 3 \pmod 8$ *and* $n_2 \equiv 3$ (mod 8), *we have*

$$\kappa = \frac{(n_1 + 1)(n_2 - 1) + 2}{2},$$

$$g^c(x) = \begin{cases} \dfrac{x^n - 1}{(x^{n_2} - 1)d_0(x)} & \text{if } D(\theta) = 0, \\[3mm] \dfrac{x^n - 1}{(x^{n_2} - 1)d_1(x)} & \text{if } D(\theta) = 1, \end{cases}$$

and d *has the lower bound of Theorem 6.23.*

(8) *If* $n_1 \equiv 1 \pmod 8$ *and* $n_2 \equiv 1 \pmod 8$ *or* $n_1 \equiv -3 \pmod 8$ *and* $n_2 \equiv -3$ (mod 8), *we have*

$$\kappa = \frac{(n_1 - 1)(n_2 + 1) + 2}{2},$$

$$g^c(x) = \begin{cases} \dfrac{x^n - 1}{(x^{n_1} - 1)d_0(x)} & \text{if } D(\theta) = 0, \\[3mm] \dfrac{x^n - 1}{(x^{n_1} - 1)d_1(x)} & \text{if } D(\theta) = 1, \end{cases}$$

and d *has the lower bound of Theorem 6.23.*

Example 6.33. Let $(q, n_1, n_2) = (2, 3, 5)$. Then $n = 15$, and $\mathcal{C}^c_{\mathrm{GF}(q)(D)}$ is a [15, 11, 3] cyclic code over GF(q) with generator polynomial $x^4 + x^3 + 1$. This code is optimal.

Example 6.34. Let $(q, n_1, n_2) = (2, 5, 3)$. Then $n = 15$, and $\mathcal{C}^{\mathrm{c}}_{\mathrm{GF}(q)(D)}$ is a $[15, 5, 7]$ cyclic code over GF(q) with generator polynomial $x^{10} + x^9 + x^8 + x^6 + x^5 + x^2 + 1$. This code is optimal.

Example 6.35. Let $(q, n_1, n_2) = (2, 3, 7)$. Then $n = 21$, and $\mathcal{C}^{\mathrm{c}}_{\mathrm{GF}(q)(D)}$ is a $[21, 7, 3]$ cyclic code over GF(q) with generator polynomial $x^{14} + x^7 + 1$. This is a bad cyclic code due to its poor minimum distance. The code in this case is bad because $q \notin D_0^{(2)}$.

Example 6.36. Let $(q, n_1, n_2) = (2, 7, 5)$. Then $n = 35$, and $\mathcal{C}^{\mathrm{c}}_{\mathrm{GF}(q)(D)}$ is a $[35, 11, 5]$ cyclic code over GF(q) with generator polynomial

$$x^{24} + x^{23} + x^{19} + x^{18} + x^{17} + x^{16} + x^{14} + x^{13}$$
$$+ x^{12} + x^{11} + x^{10} + x^8 + x^7 + x^6 + x^5 + x + 1.$$

This is the best binary cyclic code of length 35 and dimension 11 according to Table A.15.

Example 6.37. Let $(q, n_1, n_2) = (2, 11, 13)$. Then $q = 2$, $n = 143$, and $\mathcal{C}^{\mathrm{c}}_{\mathrm{GF}(q)(D)}$ is a $[143, 83, 11]$ cyclic code over GF(q) with generator polynomial

$$x^{60} + x^{59} + x^{56} + x^{54} + x^{52} + x^{49} + x^{48} + x^{47} + x^{46} + x^{44}$$
$$+ x^{43} + x^{42} + x^{41} + x^{40} + x^{35} + x^{33} + x^{32} + x^{29} + x^{28} + x^{27}$$
$$+ x^{25} + x^{24} + x^{22} + x^{20} + x^{17} + x^{15} + x^{11} + x^9 + x^5 + x^4 + 1.$$

The best binary linear code known of length 143 and dimension 83 has minimum distance 18.

The following corollary follows directly from Theorem 6.19 and Lemma 6.16.

Corollary 6.4 ([Ding (2012)]). *Let* $q = 3$. *The cyclic code* $\mathcal{C}^{\mathrm{c}}_{\mathrm{GF}(q)(D)}$ *has the following parameters* $[n, \kappa, d]$ *and generator polynomial* $g^{\mathrm{c}}(x)$:

(1) *If* $n_1 \equiv 1 \pmod{12}$ *and* $n_2 \equiv -1 \pmod{12}$ *or* $n_1 \equiv -5 \pmod{12}$ *and* $n_2 \equiv 5 \pmod{12}$, *we have*

$$\kappa = \frac{(n_1 - 1)(n_2 - 1)}{2},$$

$$g^{\mathrm{c}}(x) = \begin{cases} \dfrac{x^n - 1}{d_0(x)} & \text{if } D(\theta) = 0, \\[3mm] \dfrac{x^n - 1}{d_1(x)} & \text{if } D(\theta) = -1. \end{cases}$$

(2) *If $n_1 \equiv -1$ (mod 12) and $n_2 \equiv 5$ (mod 12) or $n_1 \equiv 5$ (mod 12) and $n_2 \equiv -1$ (mod 12), we have*

$$\kappa = n_2, \quad g^c(x) = \frac{x^n - 1}{x^{n_2} - 1},$$

and $d = n_1$, which follows from Theorem 6.20.

(3) *If $n_1 \equiv -5$ (mod 12) and $n_2 \equiv 1$ (mod 12) or $n_1 \equiv 1$ (mod 12) and $n_2 \equiv -5$ (mod 12), we have*

$$\kappa = n_1, \quad g^c(x) = \frac{x^n - 1}{x^{n_1} - 1},$$

and $d = n_2$, which follows from Theorem 6.20.

(4) *If $n_1 \equiv -1$ (mod 12) and $n_2 \equiv 1$ (mod 12) or $n_1 \equiv 5$ (mod 12) and $n_2 \equiv -5$ (mod 12), we have*

$$\kappa = \frac{(n_1 + 1)(n_2 + 1) - 2}{2},$$

$$g^c(x) = \begin{cases} \dfrac{(x^n - 1)(x - 1)}{(x^{n_1} - 1)(x^{n_2} - 1)d_0(x)} & \text{if } D(\theta) = 0, \\[4mm] \dfrac{(x^n - 1)(x - 1)}{(x^{n_1} - 1)(x^{n_2} - 1)d_1(x)} & \text{if } D(\theta) = -1, \end{cases}$$

and d has the lower bound of Theorem 6.22.

(5) *If $n_1 \equiv 1$ (mod 12) and $n_2 \equiv 5$ (mod 12) or $n_1 \equiv -5$ (mod 12) and $n_2 \equiv -1$ (mod 12), we have*

$$\kappa = 0, \quad g^c(x) = x^n - 1.$$

(6) *If $n_1 \equiv -1$ (mod 12) and $n_2 \equiv -5$ (mod 12) or $n_1 \equiv 5$ (mod 12) and $n_2 \equiv 1$ (mod 12), we have*

$$\kappa = n_1 + n_2 - 1, \quad g^c(x) = \frac{(x^n - 1)(x - 1)}{(x^{n_1} - 1)(x^{n_2} - 1)},$$

and d has the lower bound of Theorem 6.21.

(7) *If $n_1 \equiv 5$ (mod 12) and $n_2 \equiv 5$ (mod 12) or $n_1 \equiv -1$ (mod 12) and $n_2 \equiv -1$ (mod 12), we have*

$$\kappa = \frac{(n_1 + 1)(n_2 - 1) + 2}{2},$$

$$g^c(x) = \begin{cases} \dfrac{x^n - 1}{(x^{n_2} - 1)d_0(x)} & \text{if } D(\theta) = 0, \\[3ex] \dfrac{x^n - 1}{(x^{n_2} - 1)d_1(x)} & \text{if } D(\theta) = -1, \end{cases}$$

and d has the lower bound of Theorem 6.23.

(8) *If* $n_1 \equiv 1$ (mod 12) *and* $n_2 \equiv 1$ (mod 12) *or* $n_1 \equiv -5$ (mod 12) *and* $n_2 \equiv -5$ (mod 12), *we have*

$$\kappa = \frac{(n_1 - 1)(n_2 + 1) + 2}{2},$$

$$g^c(x) = \begin{cases} \dfrac{x^n - 1}{(x^{n_1} - 1)d_0(x)} & \text{if } D(\theta) = 0, \\[3ex] \dfrac{x^n - 1}{(x^{n_1} - 1)d_1(x)} & \text{if } D(\theta) = -1, \end{cases}$$

and d has the lower bound of Theorem 6.23.

Example 6.38. Let $(q, n_1, n_2) = (3, 7, 5)$. Then $n = 35$, and $\mathcal{C}^c_{\text{GF}(q)(D)}$ is a [35, 12, 12] cyclic code over GF(q) with generator polynomial

$$x^{23} + 2x^{21} + 2x^{19} + 2x^{18} + x^{16} + 2x^{14} + x^{13} + 2x^{12}$$
$$+ 2x^{11} + 2x^{10} + x^8 + 2x^6 + 2x^5 + 2x^4 + x^3 + x^2 + 2x + 2.$$

This is the best ternary cyclic code of length 35 and dimension 12 according to Table A.98.

Example 6.39. Let $(q, n_1, n_2) = (3, 5, 7)$. Then $n = 35$, and $\mathcal{C}^c_{\text{GF}(q)(D)}$ is a [35, 23, 5] cyclic code over GF(q) with generator polynomial $x^{12} + 2x^{11} + 2x^{10} + x^9 + 2x^8 + x^7 + x^5 + 2x^4 + x^2 + 1$. This is the best ternary cyclic code of length 35 and dimension 23 according to Table A.98. The best ternary linear code known of length 35 and dimension 23 has minimum distance 6.

Remark 6.6. Note that $C_1 = N_1 \cup D_1^{(2)}$ is a difference set when $n_2 = n_1 + 2$ and n_1 are primes.

Examples 6.39 and 6.33 show that a difference set may give an optimal cyclic code.

Remark 6.7. Note that $D = N_1 \cup D_1^{(2)}$ is an almost difference set when $n_2 = n_1 + 4$ and n_1 both are primes.

Example 6.35 shows that this difference set gives a bad cyclic code over GF(2). However, the following example demonstrates that the same almost difference set could give an almost optimal cyclic code over a different field.

Example 6.40. Let $(q, n_1, n_2) = (5, 3, 7)$. Then $n = 21$, and $\mathcal{C}_{\mathrm{GF}(q)}(D)^c$ is a $[21, 6, 12]$ cyclic code over GF(q) with generator polynomial $x^{15} + 3x^{13} + 3x^{12} + 2x^{11} + 3x^{10} + 3x^9 + 4x^8 + x^7 + 2x^6 + 2x^5 + 3x^4 + 2x^3 + 2x^2 + 4$. This code is almost optimal, as the upper bound on the minimum distance of any linear code over GF(q) of length 21 and dimension 6 is 13. The best linear code known of length 21 and dimension 6 over GF(5) has minimum distance 12.

The reader should be informed that the p-rank of the twin-prime difference sets was determined in Pott (1992).

6.6.4 *Minimum weight of the codes* $\mathcal{C}^c_{\mathrm{GF}(q)(D)}$

In this subsection, we determine the minimum distance of some of the cyclic codes and derive lower bounds for other cyclic codes $\mathcal{C}^c_{\mathrm{GF}(q)(D)}$.

Theorem 6.20 ([Ding (2012)]). *The cyclic code over* GF(q) *with the generator polynomial* $g(x) = (x^n - 1)/(x^{n_i} - 1)$ *has parameters* $[n, n_i, d_i]$, *where*

$$d_i = n_{i - (-1)^i}, \tag{6.38}$$

where $i \in \{1, 2\}$.

Proof. Without loss of generality, let $i = 1$. Let η be an nth primitive root of unity over the splitting field of $x^n - 1$. Then $g(\eta^j) = 0$ for all $j \in \{1, 2, \ldots, n - 1\} \backslash \{n_1, 2n_1, \ldots, (n_2 - 1)n_1\}$. By the Pigeonhole Principle, there must exist at least $n_2 - 1$ consecutive j such that $g(\eta^j) = 0$. It then follows from the BCH bound that $d_1 \geq n_2$.

On the other hand,

$$g(x) = x^{n_1(n_2 - 1)} + x^{n_1(n_2 - 2)} + \cdots + x^{n_1} + 1$$

is a codeword of Hamming weight n_2 in the code. It then follows that $d_1 = n_2$. $\qquad\square$

Example 6.41. Let $q = 2$ and $(n_1, n_2) = (7, 5)$. Then the cyclic code over GF(q) with the generator polynomial $g(x) = (x^n - 1)/(x^{n_1} - 1)$ has parameters $[35, 7, 5]$.

Theorem 6.21 ([Ding (2012)]). *Let* $\mathcal{C}_{(n_1, n_2, q)}$ *denote the cyclic code over* GF(q) *with the generator polynomial*

$$g(x) = (x - 1)(x^n - 1)/(x^{n_1} - 1)(x^{n_2} - 1).$$

Then the code $\mathcal{C}_{(n_1,n_2,q)}$ has parameters $[n, n_1 + n_2 - 1, d_{(n_1,n_2)}]$, where

$$d_{(n_1,n_2)} = \min(n_1, n_2). \tag{6.39}$$

Proof. Without loss of generality, we assume that $n_1 < n_2$. Let η be an nth primitive root of unity over the splitting field of $x^n - 1$. Then $g(\eta^i) = 0$ for all i with $1 \le i \le n_1 - 1$. It then follows from the BCH bound that

$$d_{(n_1,n_2)} \ge n_1.$$

On the other hand,

$$g(x)\frac{x^{n_1} - 1}{x - 1} = x^{n_2(n_1-1)} + x^{n_2(n_1-2)} + \cdots + x^{n_2} + 1$$

is a codeword of Hamming weight n_1 in this code. It then follows that $d_{(n_1,n_2)} = n_1$. $\qquad\square$

Example 6.42. Let $q = 2$ and $(n_1, n_2) = (7, 5)$. The cyclic code over GF(q) with the generator polynomial $g(x) = (x^n - 1)/(x^{n_1} - 1)(x^{n_2} - 1)$ has parameters $[35, 11, 5]$.

Theorem 6.22 ([Ding (2012)]). *Assume that $q \in D_0^{(2)}$. Let $\mathcal{C}_{(i,q)}$ denote the cyclic code over* GF(q) *with the generator polynomial $d_i(x)$ for $i = 0$ and $i = 1$. Then the code $\mathcal{C}_{(i,q)}$ has parameters $[n, ((n_1 + 1)(n_2 + 1) - 2)/2, d_i]$, where*

$$d_i \ge \left\lceil \sqrt{\min(n_1, n_2)} \right\rceil. \tag{6.40}$$

If $-1 \in D_1^{(2)}$, we have

$$d_i^2 - d_i + 1 \ge \min(n_1, n_2). \tag{6.41}$$

Proof. Let $\mathcal{C}_{(n_1,n_2,q)}$ denote the cyclic code over GF(q) with the generator polynomial $g(x) = (x - 1)(x^n - 1)/(x^{n_1} - 1)(x^{n_2} - 1)$ and let $d_{(n_1,n_2)}$ denote the minimum distance of this code.

Take any $\ell \in D_1^{(2)}$. Let $c(x) \in$ GF$(q)[x]/(x^n - 1)$ be a codeword of Hamming weight w in $\mathcal{C}_{(i,q)}$. Then $c(x^\ell)$ is a codeword of Hamming weight w in $\mathcal{C}_{((i+1) \bmod 2,q)}$. It then follows that $d_1 = d_2$.

Let $c(x) \in$ GF$(q)[x]/(x^n - 1)$ be a codeword of minimum weight d_i in $\mathcal{C}_{(i,q)}$. Then $c(x^\ell)$ is a codeword of the same weight in $\mathcal{C}_{((i+1) \bmod 2,q)}$. Hence, $c(x)c(x^\ell)$ is a codeword of

$$\mathcal{C}_{(i,q)} \cap \mathcal{C}_{((i+1) \bmod 2,q)} = \mathcal{C}_{(n_1,n_2,q)}.$$

It then follows from Theorem 6.21 that $d_i^2 \ge d_{(n_1,n_2)} = \min(n_1, n_2)$ and $d_i^2 - d_i + 1 \ge d_{(n_1,n_2)} = \min(n_1, n_2)$ if $-1 \in D_1^{(2)}$. $\qquad\square$

Example 6.43. Let $(q, n_1, n_2) = (3, 5, 7)$. Then $n = 35$, and $\mathcal{C}_{(0,q)}$ is a $[35, 23, 5]$ cyclic code over GF(q) with generator polynomial $x^{12} + 2x^{11} + 2x^{10} + x^9 + 2x^8 + x^7 + x^5 + 2x^4 + x^2 + 1$. In this case, $-1 \in D_1^{(2)}$ and $d_{(n_1,n_2)} = 5$. We have

$$d_0^2 - d_0 + 1 \geq 5.$$

Hence, the lower bounds of (6.40) and (6.41) are 3. In this case, the two lower bounds are not met.

Theorem 6.23 ([Ding (2012)]). *Let $q \in D_0^{(2)}$. Let $\mathcal{C}_{(n_1,n_2,q)}^{(i,j)}$ denote the cyclic code over GF(q) with the generator polynomial*

$$g^{(i,j)}(x) = \frac{(x^{n_i} - 1)}{x - 1} d_j(x),$$

and let $d_{(n_1,n_2,q)}^{(i,j)}$ denote the minimum distance of this code, where $i \in \{1, 2\}$ and $j \in \{0, 1\}$. Then the code $\mathcal{C}_{(n_1,n_2,q)}^{(i,j)}$ has parameters

$$\left[n, \frac{(n_i - 1)(n_{i-(-1)^i} + 1) + 2}{2}, d_{(n_1,n_2,q)}^{(i,j)} \right],$$

where

$$d_{(n_1,n_2,q)}^{(i,j)} \geq \lceil \sqrt{n_i} \rceil. \tag{6.42}$$

If $-1 \in D_1^{(2)}$, we have

$$\left(d_{(n_1,n_2,q)}^{(i,j)} \right)^2 - d_{(n_1,n_2,q)}^{(i,j)} + 1 \geq n_i. \tag{6.43}$$

Proof. Take any $\ell \in D_1^{(2)}$. Let $c(x) \in$ GF$(q)[x]/(x^n - 1)$ be a codeword of Hamming weight w in $\mathcal{C}_{(n_1,n_2,q)}^{(i,j)}$. Then $c(x^\ell)$ is a codeword of Hamming weight w in $\mathcal{C}_{(n_1,n_2,q)}^{(i,(j+1) \bmod 2)}$. It then follows that $d_{(n_1,n_2,q)}^{(i,0)} = d_{(n_1,n_2,q)}^{(i,1)}$.

Let $c(x) \in$ GF$(q)[x]/(x^n - 1)$ be a codeword of minimum weight in $\mathcal{C}_{(n_1,n_2,q)}^{(i,j)}$. Then $c(x^\ell)$ is a codeword of the same weight in $\mathcal{C}_{(n_1,n_2,q)}^{(i,(j+1) \bmod 2)}$. Hence, $c(x)c(x^\ell)$ is a codeword of

$$\mathcal{C}_{(n_1,n_2,q)}^{(i,j)} \cap \mathcal{C}_{(n_1,n_2,q)}^{(i,(j+1) \bmod 2)},$$

which is the cyclic code with generator polynomial $(x^n - 1)/(x^{n_{i-(-1)^i}} - 1)$ and minimum weight n_i by Theorem 6.20. It then follows that

$$\left(d_{(n_1,n_2,q)}^{(i,j)} \right)^2 \geq n_i$$

and

$$\left(d^{(i,j)}_{(n_1,n_2,q)}\right)^2 - d^{(i,j)}_{(n_1,n_2,q)} + 1 \geq n_i$$

if $-1 \in D_1^{(2)}$. □

Example 6.44. Let $(q, n_1, n_2) = (2, 3, 11)$. Then $n = 33$, and $\mathcal{C}^c_{\mathrm{GF}(q)(D)}$ is a $[33, 21, 3]$ cyclic code over $\mathrm{GF}(q)$ with generator polynomial

$$x^{12} + x^9 + x^7 + x^6 + x^5 + x^3 + 1 = \frac{x^{n_1} - 1}{x - 1} d_0(x).$$

In this case, the lower bound of (6.42) is 2, while the actual minimum distance is 3.

6.7 Cyclic Codes of Singer Difference Sets

Let q be a prime power and let $m \geq 3$ be a positive integer. Let α be a generator of $\mathrm{GF}(q^m)^*$. Put $n = (q^m - 1)/(q - 1)$. Recall that

$$D = \{0 \leq i < n : \mathrm{Tr}_{q^m/q}(\alpha^i) = 0\} \subset \mathbb{Z}_n$$

is the Singer difference set in $(\mathbb{Z}_n, +)$ with parameters

$$\left(\frac{q^m - 1}{q - 1}, \frac{q^{m-1} - 1}{q - 1}, \frac{q^{m-2} - 1}{q - 1}\right).$$

The cyclic code $\mathcal{C}_{\mathrm{GF}(q)}(D)$ from the Singer difference set D is interesting, as it is decodable with majority logic [Goethals and Delsarte (1968)]. The following theorem gives parameters of the code $\mathcal{C}_{\mathrm{GF}(q)}(D)$.

Theorem 6.24. *Let* $q = p^s$ *for some positive integer* s *and let* $m \geq 2$ *be an integer. Let* α *be a generator of* $\mathrm{GF}(q^m)^*$ *and put* $\beta = \alpha^{q-1}$. *Let* S *be the set of* $b, 0 < b < q^m - 1$, *such that the* p-adic representation of b is

$$b = \sum_{h=0}^{m-1} \sum_{j=0}^{s-1} b_{j,h} p^j q^h \tag{6.44}$$

with

$$0 \leq b_{j,h} \leq p - 1 \quad and \quad \sum_{h=0}^{m-1} b_{j,h} = p - 1 \text{ for all } j \text{ with } 0 \leq j \leq s - 1. \tag{6.45}$$

Then the cyclic code $\mathcal{C}_{\mathrm{GF}(q)}(D)$ of the Singer difference set D has parameters

$$\left[\frac{q^m - 1}{q - 1}, \binom{p + m - 2}{m - 1}^s + 1, \frac{q^{m-1} - 1}{q - 1}\right] \tag{6.46}$$

and parity check polynomial

$$h(x) = (x - 1) \prod_{\substack{1 \le u \le n-1 \\ q^m - 1 - u(q-1) \in S}} (x - \beta^u), \tag{6.47}$$

where $n = (q^m - 1)/(q - 1)$.

Proof. For simplicity, we write $\mathrm{Tr}(a)$ for $\mathrm{Tr}_{q^m/q}(a)$ throughout the proof. Recall that the Hall polynomial of D is defined by

$$D(x) = \sum_{i \in D} x^i \in \mathrm{GF}(q)[x].$$

By Theorem 6.1, the code $\mathcal{C}_{\mathrm{GF}(q)}(D)$ has generator polynomial $g(x) = \gcd(x^n - 1, D(x))$, which is computed over $\mathrm{GF}(q)$. The parity check polynomial $h(x)$ of the code $\mathcal{C}_{\mathrm{GF}(q)}(D)$ is then

$$h(x) = \frac{x^n - 1}{g(x)} = \frac{x^n - 1}{\gcd(x^n - 1, D(x))}. \tag{6.48}$$

Let $\gamma_u = \beta^u$ for all u with $0 \le u \le n - 1$, where β and α are defined in the statement of this theorem. Then these γ_u are the distinct nth roots of unity. Put

$$T_u = \sum_{i \in D} \gamma_u^i = \sum_{i \in D} \alpha^{iu(q-1)}, \quad u = 0, 1, \ldots, n - 1.$$

It then follows from (6.48) that

$$g(x) = \sum_{\substack{0 \le u \le n-1 \\ T_u = 0}} (x - \beta^u) \quad \text{and} \quad h(x) = \sum_{\substack{0 \le u \le n-1 \\ T_u \ne 0}} (x - \beta^u). \tag{6.49}$$

The degree of $h(x)$, i.e., the number of u such that $T_u \ne 0$, is the dimension of the code and also the p-rank of the Singer difference set. We now determine $h(x)$ and its degree.

First, for $u = 0$, we have

$$T_0 = |D| \bmod p = \frac{q^{m-1} - 1}{q - 1} \equiv 1 \pmod{p}.$$

Hence $T_0 \ne 0$.

We now restrict u such that $1 \le u \le n - 1$. By definition,

$$
\begin{aligned}
T_u &= \sum_{i=0}^{n-1} (1 - \text{Tr}(\alpha^i)^{q-1}) \alpha^{iu(q-1)} \\
&= \sum_{i=0}^{n-1} \beta^{iu} - \sum_{i=0}^{n-1} \text{Tr}(\alpha^i)^{q-1} \alpha^{iu(q-1)} \\
&= -\sum_{i=0}^{n-1} \text{Tr}(\alpha^i)^{q-1} \alpha^{iu(q-1)},
\end{aligned} \tag{6.50}
$$

where the last equality follows from the fact that $\sum_{i=0}^{n-1} \beta^{iu} = 0$.

Since $q - 1 = (p - 1) \sum_{i=0}^{s-1} p^i$ and GF(q^m) has characteristic p, we have

$$
[\text{Tr}(\alpha^i)]^{q-1} = \prod_{j=0}^{s-1} \left[\sum_{l=0}^{m-1} \alpha^{ip^j q^l} \right]^{p-1}.
$$

Expanding each term in the product by the multinomial theorem, we obtain for $j \in \{0, 1, \ldots, s - 1\}$,

$$
\left[\sum_{t=0}^{m-1} \alpha^{ip^j q^t} \right]^{p-1} = \sum_{b_j} \frac{(p - 1)!}{b_{j,0}! b_{j,1}! \cdots b_{j,m-1}!} \alpha^{i \left(\sum_{t=0}^{m-1} b_{j,t} p^j q^t \right)}, \tag{6.51}
$$

where the summation is taken over all choices of integers $b_{j,0}, b_{j,1}, \ldots, b_{j,m-1}$ such that

$$
\begin{cases}
0 \le b_{j,t} \le p - 1, \quad t = 0, 1, \ldots, m - 1, \\
\displaystyle\sum_{t=0}^{m-1} b_{j,t} = p - 1.
\end{cases} \tag{6.52}
$$

Note that none of the multinomial coefficients in (6.51) is congruent to zero modulo p. Then

$$
\text{Tr}(\alpha^i)^{q-1} = \sum_{b_0} \cdots \sum_{b_{s-1}} \frac{(p - 1)!}{\prod_{t=0}^{m-1} (b_{0,t}!)} \cdots \frac{(p - 1)!}{\prod_{t=0}^{m-1} (b_{s-1,t}!)} \alpha^{i \left[\sum_{j=0}^{s-1} \sum_{t=0}^{m-1} b_{j,t} p^j q^t \right]}.
$$

Recall the definition of the set S in the statement of this theorem. It follows from (6.44) and (6.45) that $b \equiv 0 \pmod{q-1}$ for $b \in S$. Put

$$m_b = \prod_{j=0}^{s-1} \frac{(p-1)!}{\prod_{t=0}^{m-1} (b_{j,t}!)}.$$

Now m_b is not zero modulo p for $b \in S$ and

$$\mathrm{Tr}(\alpha^i)^{q-1} = \sum_{b \in S} m_b \alpha^{ib}.$$

It then follows from (6.50) that

$$T_u = -\sum_{i=0}^{n-1} \sum_{b \in S} m_b \alpha^{ib+iu(q-1)} = -\sum_{b \in S} m_b \sum_{i=0}^{n-1} \alpha^{i(b+u(q-1))}.$$

As noted earlier, $b \equiv 0 \pmod{q-1}$ for $b \in S$. Suppose that $b = \ell(q-1)$ for some integer ℓ. Then

$$\sum_{i=0}^{n-1} \alpha^{i(b+u(q-1))} = \sum_{i=0}^{n-1} \alpha^{i(q-1)(\ell+u)}$$

$$\equiv \begin{cases} 1 \pmod{p} & \text{if } \ell + u \equiv 0 \pmod{n}, \\ 0 & \text{otherwise.} \end{cases}$$

Since $b = \ell(q-1)$ and $n(q-1) = q^m - 1$, then

$$\sum_{i=0}^{n-1} \alpha^{i(b+u(q-1))} \equiv \begin{cases} 1 \pmod{p} & \text{if } b + u(q-1) \equiv 0 \pmod{q^m - 1}, \\ 0 & \text{otherwise.} \end{cases}$$

For any fixed $b \in S$, there is a unique integer u, $1 \leq u \leq n-1$, such that

$$b + u(q-1) \equiv 0 \pmod{q^m - 1},$$

namely, that integer u such that $u(q-1) = q^m - 1 - b$. Hence,

$$T_u = \begin{cases} -m_b \not\equiv 0 \pmod{p}, & \text{if } q^m - 1 - u(q-1) \in S, \\ 0, & \text{otherwise.} \end{cases}$$

This completes the proof of the conclusion on the parity check polynomial $h(x)$ of the code.

We now determine the degree of $h(x)$, i.e., the dimension of the code and the p-rank of the Singer difference set D. The number of integers u, $1 \leq u \leq n-1$, such that $T_u \neq 0$ is then the number of elements of S. Recall that S is the set of integers b of the form (6.44) satisfying (6.45). If L is the number of ways of choosing m integers in the interval $[0, p-1]$ whose sum is $p-1$, then the number of elements of S is equal to L^s. It is easy to verify that L is the coefficient of $p-1$ in the (real) expansion of

$$(1 + y + \cdots p^{p-1})^m$$

and is equal to $\binom{p+m-2}{m-1}$. Consequently, the number of integers u, $1 \leq u \leq n-1$, such that $T_u \neq 0$ is equal to $\binom{p+m-2}{m-1}^s$. Since $T_0 \neq 0$, the degree of $h(x)$ and the p-rank of the Singer difference set D is equal to

$$\binom{p+m-2}{m-1}^s + 1.$$

This finishes the proof of the conclusion on the dimension of the code.

An elementary proof of the conclusion on the minimum weight of the code is missing. The code $\mathcal{C}_{\mathrm{GF}(q)}(D)$ of the Singer difference set D is actually equivalent to the code $\mathcal{P}(1, m+1)$ of Proposition 5.7.1 in Assmus and Key (1992a) [p. 179], which is a subfield subcode of a nonprimitive generalized Reed–Muller code. The reader is referred to Assmus and Key (1992a) [Section 5.7] for a proof of the minimum weight of the code $\mathcal{C}_{\mathrm{GF}(q)}(D)$. $\qquad\square$

Example 6.45. Let $(q, m) = (3, 3)$ and let α be a generator of $\mathrm{GF}(q^m)^*$ with $\alpha^3 + 2\alpha + 1 = 0$. Then $n = 13$ and the Singer difference set $D = \{0, 1, 3, 9\}$. The ternary cyclic code $\mathcal{C}_{\mathrm{GF}(q)}(D)$ has parameters $[13, 7, 4]$ and generator polynomial $g(x) = x^6 + 2x^5 + 2x^4 + 2x^3 + x^2 + 2x + 1$. The parity check polynomial of this code is $h(x) = x^7 + x^6 + 2x^5 + x^4 + 2x + 2$. The weight enumerator of $\mathcal{C}_{\mathrm{GF}(q)}(D)$ is

$$1 + 26z^4 + 156z^6 + 624z^7 + 494z^9 + 780z^{10} + 78z^{12} + 28z^{13}.$$

The p-rank of the Singer difference set D was determined independently in Goethals and Delsarte (1968), MacWilliams and Mann (1968), and Smith (1969). A proof of the p-rank using Gauss and Jacobi sums, and Stickelberger's theorem on the prime ideal factorization of Gauss sums was developed in Evans, Hollmann, Krattenthaler and Xiang (1999). The parity check polynomial of the code $\mathcal{C}_{\mathrm{GF}(q)}(D)$ and the p-rank of D were also determined in a different way in No, Shin and Helleseth (2004). Our proof of the p-rank result in Theorem 6.24 is a refined version of the proof given in Smith (1969).

Hamada made the following interesting conjecture [Hamada (1973)].

Conjecture 6.1. *Let $\mathbb{D}$ be a symmetric design with parameters*

$$\left(\frac{q^m - 1}{q - 1}, \frac{q^{m-1} - 1}{q - 1}, \frac{q^{m-2} - 1}{q - 1} \right),$$

where $q = p^s$. Then

$$\mathrm{rank}_p(\mathbb{D}) \geq \binom{p + m - 2}{m - 1}^s + 1.$$

This conjecture still remains open. But Hamada and Ohmori made the following progress in this direction.

Theorem 6.25 ([Hamada and Ohmori (1975)]). *Let $\mathbb{D}$ be a symmetric design with parameters $(2^m - 1, 2^{m-1} - 1, 2^{m-2} - 1)$. Then $\mathbb{D}$ is the development of the Singer difference set with these parameters if and only if $\mathrm{rank}_2(\mathbb{D}) = m + 1$.*

6.8 Cyclic Codes of the Hyperoval Difference Sets

Let m be odd, $n = 2^m - 1$ and let α be a generator of $\mathrm{GF}(2^m)$. Maschietti showed that

$$D_k := \log_\alpha[\mathrm{GF}(2^m) \backslash \{x^k + x : x \in \mathrm{GF}(2^m)\}]$$

is a difference set in $(\mathbb{Z}_n, +)$ for the following k:

- $k = 2$ (Singer case),
- $k = 6$ (Segre case),
- $k = 2^\sigma + 2^\pi$ with $\sigma = (m + 1)/2$ and $4\pi \equiv 1 \bmod m$ (Glynn I case), and
- $k = 3 \cdot 2^\sigma + 4$ with $\sigma = (m + 1)/2$ (Glynn II case).

In this section, we deal with the dual $\mathcal{C}_{\mathrm{GF}(2)}(D_k)^\perp$ of the code of the Hyperoval difference set D_k. The dimension and the generator polynomial of the code $\mathcal{C}_{\mathrm{GF}(2)}(D_k)$ follow automatically from those of $\mathcal{C}_{\mathrm{GF}(2)}(D_k)^\perp$. The Singer case was dealt with in Section 6.7. We treat only the remaining three cases in this section.

6.8.1 *The Segre Case*

The linear span and a trace representation of the characteristic sequence of the Segre difference set were given in Chang, Golomb, Gong and Kumar (1998); Evans, Hollmann, Krattenthaler and Xiang (1999); Dobbertin (1999b). Below is a summary of the linear span and trace characterization.

Lemma 6.17. *Let s^∞ denote the characteristic sequence of the difference set D_6. Then the linear span of this sequence equals $m(2F_{(m-1)/2} - 1)$ for all odd $m \geq 1$, where F_n is the Fibonacci number with the initial state $F_0 = F_1 = 1$.*

This sequence has the following trace representation

$$s_t = 1 + \sum_{i \in I} (\alpha^t)^{5i}, \tag{6.53}$$

where I consists of all the following exponents e:

$$e = 1,$$

$$e = 2^{2j} + \sum_{i=1}^{j-1} (-1)^{\epsilon_i} 2^{2i} + 1, \quad 1 \leq j \leq (m-3)/2,$$

$$e = 2^{m-2} + \sum_{i=2}^{(n-5)/2} (-1)^{\epsilon_i} 2^{2i+1} + 2^3 - 1,$$

with the binary sequences (ϵ_i) satisfying the condition $(\epsilon_i, \epsilon_{i-1}) \neq (1, 1)$.

The reciprocal of the minimal polynomial of this sequence is then given by

$$g(x) = (x - 1) \prod_{i \in I} \mathbb{M}_{\alpha^{-i}(x)}, \tag{6.54}$$

where I was defined above and $\mathbb{M}_a(x)$ denotes the minimal polynomial of $a \in \mathrm{GF}(2^m)$ over $\mathrm{GF}(2)$.

The following theorem follows from Lemma 6.17 and the definition of the cyclic code $\mathcal{C}_{\mathrm{GF}(2)}(D_6)^\perp$.

Theorem 6.26. *The code $\mathcal{C}_{\mathrm{GF}(2)}(D_6)^\perp$ has parameters $[2^m - 1, 2^m - 1 - m(2F_{(m-1)/2} - 1), d]$ and generator polynomial $g(x)$, where F_m and $g(x)$ are defined in Lemma 6.17.*

Example 6.46. Let $m = 3$ and α be a generator of $\mathrm{GF}(2^m)^*$ with $\alpha^3 + \alpha + 1 = 0$. Then $\mathcal{C}_{\mathrm{GF}(2)}(D_6)^\perp$ is a $[7, 4, 3]$ binary cyclic code with generator polynomial $x^3 + x + 1$. This code is optimal.

Example 6.47. Let $m = 5$ and α be a generator of $\mathrm{GF}(2^m)^*$ with $\alpha^5 + \alpha^2 + 1 = 0$. Then $\mathcal{C}_{\mathrm{GF}(2)}(D_6)^\perp$ is a $[31, 16, 7]$ binary cyclic code with generator polynomial $x^{15} + x^{12} + x^7 + x^6 + x^2 + x + 1$. This code is the best cyclic code of length 31 and dimension 16 according to Table A.13. The upper bound on the minimum weight of any binary linear code with length 31 and dimension 16 is 8.

Example 6.48. Let $m = 7$ and α be a generator of $\mathrm{GF}(2^m)^*$ with $\alpha^7 + \alpha + 1 = 0$. Then $\mathcal{C}_{\mathrm{GF}(2)}(D_6)^\perp$ is a $[127, 92, 9]$ binary cyclic code with generator polynomial

$$x^{35} + x^{34} + x^{33} + x^{32} + x^{29} + x^{28} + x^{27} + x^{25} + x^{20} + x^{18}$$

$$+x^{14} + x^{11} + x^{10} + x^8 + x^5 + x^4 + x^2 + x + 1.$$

The best binary linear code in the Database with parameters $[127, 92, 11]$ is not known to be cyclic.

These examples indicate that the codes defined by the Segre difference sets are extremely good.

Remark 6.8. Evans, Hollmann, Krattenthaler and Xiang computed the code $\mathcal{C}_{\mathrm{GF}(2)}(D_6)^\perp$ for the case $m = 9$ in Evans, Hollmann, Krattenthaler and Xiang (1999).

Problem 6.9. *Develop a tight lower bound on the minimum weight of the codes $\mathcal{C}_{\mathrm{GF}(2)}(D_6)^\perp$ and the code $\mathcal{C}_{\mathrm{GF}(2)}(D_6)$.*

The generator polynomial of the code $\mathcal{C}_{\mathrm{GF}(2)}(D_6)^\perp$ looks complex. However, its generating idempotent is simple and given in the following theorem whose proof is left to the reader.

Theorem 6.27. *The generating idempotent of the code $\mathcal{C}_{\mathrm{GF}(2)}(D_6)^\perp$ is $1 + D_6(x)$, where $D_6(x)$ is the Hall polynomial of D_6 over $\mathrm{GF}(2)$. Similarly, the generating idempotent of the code $\mathcal{C}_{\mathrm{GF}(2)}(D_6)$ is $D_6(x)$.*

6.8.2 *The Glynn I Case*

The following lemma is proved in Evans, Hollmann, Krattenthaler and Xiang (1999).

Lemma 6.18. *Let s^∞ denote the characteristic sequence of the hyperoval difference set in the Glynn I Case. The linear span of this sequence is equal to mU_m for all odd m, where U_m satisfies the recurrence*

$$U_i = U_{i-2} + U_{i-4} + U_{i-6} + U_{i-8} - 1$$

for all $i \geq 13$ with the initial state $U_5 = 1$, $U_7 = 3$, $U_9 = 7$, and $U_{11} = 13$.

The following theorem follows from Lemma 6.18 and the definition of the cyclic code $\mathcal{C}_{\mathrm{GF}(2)}(D_k)^\perp$.

Theorem 6.28. *The code $\mathcal{C}_{\mathrm{GF}(2)}(D_k)^\perp$ of the hyperoval difference set in Glynn I case has parameters $[2^m - 1, 2^m - 1 - mU_m, d]$, where U_m is defined in Lemma 6.18.*

Example 6.49. Let $m = 3$ and α be a generator of $GF(2^m)^*$ with $\alpha^3 + \alpha + 1 = 0$. Then $\mathcal{C}_{GF(2)}(D_k)^\perp$ is a $[7, 4, 3]$ binary cyclic code with generator polynomial $x^3 + x + 1$. This code is optimal.

Example 6.50. Let $m = 5$ and α be a generator of $GF(2^m)^*$ with $\alpha^5 + \alpha^2 + 1 = 0$. Then $\mathcal{C}_{GF(2)}(D_k)^\perp$ is a $[31, 26, 3]$ binary cyclic code with generator polynomial $x^5 + x^4 + x^2 + x + 1$. This code is optimal.

Example 6.51. Let $m = 7$ and α be a generator of $GF(2^m)^*$ with $\alpha^7 + \alpha + 1 = 0$. Then $\mathcal{C}_{GF(2)}(D_k)^\perp$ is a $[127, 106, 6]$ binary cyclic code with generator polynomial

$$x^{21} + x^{20} + x^{18} + x^{17} + x^{15} + x^{13} + x^8 + x^6 + x^2 + x + 1.$$

The best binary linear code with parameters $[127, 106, 7]$ in the Database is not known to be cyclic.

The examples above indicate that the codes $\mathcal{C}_{GF(2)}(D_k)^\perp$ are extremely good.

Problem 6.10. *Determine the generator polynomial of the code $\mathcal{C}_{GF(2)}(D_k)^\perp$ and develop a tight lower bound on the minimum weight of this code.*

The generator polynomial of the code $\mathcal{C}_{GF(2)}(D_k)^\perp$ is unknown. However, its generating idempotent is simple and given in the following theorem whose proof is left to the reader.

Theorem 6.29. *The generating idempotent of the code $\mathcal{C}_{GF(2)}(D_k)^\perp$ is $1 + D_k(x)$, where $D_k(x)$ is the Hall polynomial of D_k over $GF(2)$. Similarly, the generating idempotent of the code $\mathcal{C}_{GF(2)}(D_k)$ is $D_k(x)$.*

6.8.3 The Glynn II Case

The following lemma is proved in Evans, Hollmann, Krattenthaler and Xiang (1999).

Lemma 6.19. *Let s^∞ denote the characteristic sequence of the hyperoval difference set in the Glynn II case. Then the linear span of this sequence is equal to mV_m for all odd m, where V_m satisfies the recurrence*

$$V_i = V_{i-2} + 3V_{i-4} - V_{i-6} - V_{i-8} + 1$$

for all $i \geq 11$ with the initial state $V_3 = 1$, $V_5 = 1$, $V_7 = 5$, and $V_9 = 7$.

The following theorem follows from Lemma 6.19 and the definition of the cyclic code $\mathcal{C}_{GF(2)}(D_k)^\perp$.

Theorem 6.30. *The code* $\mathcal{C}_{\mathrm{GF}(2)}(D_k)^{\perp}$ *of the hyperoval difference set in the Glynn II case has parameters* $[2^m - 1, 2^m - 1 - mV_m, d]$, *where* V_m *is defined in Lemma 6.19.*

Example 6.52. Let $m = 3$ and α be a generator of $\mathrm{GF}(2^m)^*$ with $\alpha^3 + \alpha + 1 = 0$. Then $\mathcal{C}_{\mathrm{GF}(2)}(D_k)^{\perp}$ is a $[7, 4, 3]$ binary cyclic code with generator polynomial $x^3 + x + 1$. This code is optimal.

Example 6.53. Let $m = 5$ and α be a generator of $\mathrm{GF}(2^m)^*$ with $\alpha^5 + \alpha^2 + 1 = 0$. Then $\mathcal{C}_{\mathrm{GF}(2)}(D_k)^{\perp}$ is a $[31, 26, 3]$ binary cyclic code with generator polynomial $x^5 + x^4 + x^2 + x + 1$. This code is optimal.

Example 6.54. Let $m = 7$ and α be a generator of $\mathrm{GF}(2^m)^*$ with $\alpha^7 + \alpha + 1 = 0$. Then $\mathcal{C}_{\mathrm{GF}(2)}(D_k)^{\perp}$ is a $[127, 92, 9]$ binary cyclic code with generator polynomial

$$x^{35} + x^{34} + x^{33} + x^{29} + x^{27} + x^{26} + x^{22} + x^{17}$$
$$+ x^{14} + x^{12} + x^{11} + x^{10} + x^7 + x^5 + 1.$$

The best binary linear code with parameters $[127, 92, 11]$ in the Database is not cyclic.

The examples above show that the codes $\mathcal{C}_{\mathrm{GF}(2)}(D_k)^{\perp}$ are extremely good.

Problem 6.11. *Determine the generator polynomial of the code* $\mathcal{C}_{\mathrm{GF}(2)}(D_k)^{\perp}$ *and develop a tight lower bound on the minimum weight of this code.*

The generator polynomial of the code $\mathcal{C}_{\mathrm{GF}(2)}(D_k)^{\perp}$ is unknown. However, its generating idempotent is simple and given in the following theorem whose proof is left to the reader.

Theorem 6.31. *The generating idempotent of the code* $\mathcal{C}_{\mathrm{GF}(2)}(D_k)^{\perp}$ *is* $1 + D_k(x)$, *where* $D_k(x)$ *is the Hall polynomial of* D_k *over* $\mathrm{GF}(2)$. *Similarly, the generating idempotent of the code* $\mathcal{C}_{\mathrm{GF}(2)}(D_k)$ *is* $D_k(x)$.

6.9 Cyclic Codes of the N_k Type of Difference Sets

Let $m \not\equiv 0 \pmod 3$ be a positive integer and let $n = 2^m - 1$. Define $\delta_k(x) = x^d + (x + 1)^d \in \mathrm{GF}(2^m)[x]$, where $d = 4^k - 2^k + 1$ and $k = (m \pm 1)/3$. Put

$$N_k = \begin{cases} \log_\alpha[\delta_k(\mathrm{GF}(2^m))], & \text{if } m \text{ is odd,} \\ \log_\alpha[\mathrm{GF}(2^m) \backslash \delta_k(\mathrm{GF}(2^m))], & \text{if } m \text{ is even,} \end{cases}$$

where α is a generator of $\mathrm{GF}(2^m)^*$. Then N_k is a difference set with Singer parameters in $(\mathbb{Z}_n, +)$. This family of cyclic difference sets were conjectured in

No, Chung and Yun (1998); No, Golomb, Gong, Lee and Gaal (1998), and the conjecture was confirmed in Dillon (1999); Dillon and Dobbertin (2004).

In this section, we deal with the dual $\mathcal{C}_{\mathrm{GF}(2)}(N_k)^{\perp}$ of the code $\mathcal{C}_{\mathrm{GF}(2)}(N_k)$ of the difference set N_k and related difference sets. Obviously, the dimension and the generator polynomial of the code $\mathcal{C}_{\mathrm{GF}(2)}(N_k)$ follow automatically from those of $\mathcal{C}_{\mathrm{GF}(2)}(N_k)^{\perp}$.

6.9.1 The first class of cyclic codes

The following lemma gives the trace representation of the characteristic sequences of a class of cyclic difference sets with Singer parameters [No, Golomb, Gong, Lee and Gaal (1998); Dillon and Dobbertin (2004)].

Lemma 6.20. *Let* $m = 2t + 1 > 3$, *and define*

$$s_i = \mathrm{Tr}_{2^m/2}(\alpha^i + (\alpha^i)^{2^t+1} + (\alpha^i)^{2^t+2^{t-1}+1}) \tag{6.55}$$

for all $i \geq 0$. *Then the linear span of the sequence equals* $3m$ *and the reciprocal of the minimal polynomial of the sequence is given by*

$$g(x) = \mathbb{M}_{\alpha^{-1}}(x)\mathbb{M}_{\alpha^{-(2^t+1)}}(x)\mathbb{M}_{\alpha^{-(2^t+2^{t-1}+1)}}(x), \tag{6.56}$$

where $\mathbb{M}_{\alpha^i}(x)$ *denotes the minimal polynomial of* $\alpha^i \in \mathrm{GF}(2^m)$ *over* $\mathrm{GF}(2)$.

Let D denote the support of the binary sequence in Lemma 6.20. It is known that D is a difference set with Singer parameters [Chang, Gaal, Golomb, Gong and Kumar (1998)]. The following theorem was proved in Chang, Gaal, Golomb, Gong and Kumar (1998).

Theorem 6.32. *Let* $m > 3$ *and let* D *denote the support of the sequence in Lemma* 6.20. *Then the code* $\mathcal{C}_{\mathrm{GF}(2)}(D)^{\perp}$ *of the difference set* D *has parameters* $[n, n - 3m, 7]$ *and generator polynomial* $g(x)$ *of* (6.56).

Example 6.55. Let $m = 3$ and α be a generator of $\mathrm{GF}(2^m)^*$ with $\alpha^3 + \alpha + 1 = 0$. Then $\mathcal{C}_{\mathrm{GF}(2)}(D)^{\perp}$ is a $[7, 4, 3]$ binary cyclic code with generator polynomial $x^3 + x + 1$. This code is optimal.

Example 6.56. Let $m = 5$ and α be a generator of $\mathrm{GF}(2^m)^*$ with $\alpha^5 + \alpha^2 + 1 = 0$. Then $\mathcal{C}_{\mathrm{GF}(2)}(D)^{\perp}$ is a $[31, 16, 7]$ binary cyclic code with generator polynomial

$$x^{15} + x^{12} + x^7 + x^6 + x^2 + x + 1.$$

This code is the best binary cyclic code of length 31 and dimension 16 according to Table A.13. The optimal binary linear code of length 31 and dimension 16 has minimum weight 8, and is not cyclic.

Example 6.57. Let $m = 7$ and α be a generator of $\mathrm{GF}(2^m)^*$ with $\alpha^7 + \alpha + 1 = 0$. Then $\mathcal{C}_{\mathrm{GF}(2)}(D)^\perp$ is a $[127, 106, 7]$ binary cyclic code with generator polynomial

$$x^{21} + x^{20} + x^{19} + x^{15} + x^{12} + x^{11} + x^9 + x^8 + x^7 + x^5 + x^4 + x + 1.$$

This cyclic code has the same parameters as the best linear code known in the Database, which is not known to be cyclic. An upper bound on the minimum weight of any binary linear code of length 127 and dimension 106 is 8.

6.9.2 *The second class of cyclic codes*

The following lemma gives the trace representation of the characteristic sequences of a class of cyclic difference sets with Singer parameters [No, Chung and Yun (1998); No, Golomb, Gong, Lee and Gaal (1998); Dillon and Dobbertin (2004)].

Lemma 6.21. *Let $t \geq 3$ be a positive integer and $m = 3t - 1$. Define*

$$s_i = \mathrm{Tr}_{2^m/2}(\alpha^i + (\alpha^i)^{2^t+1} + (\alpha^i)^{2^{2t-1}+2^{t-1}+1})$$

$$+ \mathrm{Tr}_{2^m/2}((\alpha^i)^{2^{2t-1}-2^{t-1}+1} + (\alpha^i)^{2^{2t-1}+2^t-1}) \qquad (6.57)$$

for all $i \geq 0$. Then the linear span of this sequence is equal to $5m$ and the reciprocal of the minimal polynomial of this sequence is given by

$$g(x) = \mathbb{M}_{\alpha^{-1}}(x) \times \mathbb{M}_{\alpha^{-(2^t+1)}}(x) \times \mathbb{M}_{\alpha^{-(2^{2t-1}+2^{t-1}+1)}}(x)$$

$$\times \mathbb{M}_{\alpha^{-(2^{2t-1}-2^{t-1}+1)}}(x) \times \mathbb{M}_{\alpha^{-(2^{2t-1}+2^t-1)}}(x), \qquad (6.58)$$

where $\mathbb{M}_{\alpha^i}(x)$ denotes the minimal polynomial of $\alpha^i \in \mathrm{GF}(2^m)$ over $\mathrm{GF}(2)$.

Proof.   Since $t \geq 3$, it can be easily proved that

$$|C_1| = |C_{2^t+1}| = |C_{2^{2t-1}+2^{t-1}+1}| = |C_{2^{2t-1}-2^{t-1}+1}| = |C_{2^{2t-1}+2^{t-1}-1}| = m,$$

where C_i is the 2-cyclotomic coset modulo n containing i. The desired conclusion on the minimal polynomial of s^∞ then follows from Lemma 1.17. □

Theorem 6.33. *Let $t \geq 3$ and let D denote the support of the sequence of Lemma 6.21. The code $\mathcal{C}_{\mathrm{GF}(2)}(D)^\perp$ of the difference set D has parameters $[n, n - 5m, d]$ and generator polynomial $g(x)$ of (6.58), where $d \geq 5$.*

Proof.   The desired conclusions on the generator polynomial and the dimension of the code follow from Lemma 6.21 and the definition of the code.

Note that

$$2^{2t} + 2^t - 1 = 2^{2t}(2^{2t-1} - 2^{t-1} + 1) \bmod n \in C_{2^{2t-1}-2^{t-1}+1},$$

$$2^{2t} + 2^t = 2^t(2^t + 1) \bmod n \in C_{2^t+1},$$

$$2^{2t} + 2^t + 1 = 2^{2t}(2^{2t-1} + 2^{t-1} + 1) \bmod n \in C_{2^{2t-1}+2^{t-1}+1},$$

$$2^{2t} + 2^t + 2 = 2(2^{2t-1} + 2^{t-1} + 1) \bmod n \in C_{2^{2t-1}+2^{t-1}+1}.$$

By Lemma 6.21, the reciprocal of $g(x)$ has the zeros α^i for all i in the following set

$$\{2^{2t} + 2^t - 1, 2^{2t} + 2^t, 2^{2t} + 2^t + 1, 2^{2t} + 2^t + 2\}.$$

It then follows from the BCH bound that $d \geq 5$. □

Example 6.58. Let $t = 2$ and α be a generator of GF$(2^m)^*$ with $\alpha^5 + \alpha^2 + 1 = 0$. Then $\mathcal{C}_{\mathrm{GF}(2)}(D)^\perp$ is a [31, 16, 5] binary cyclic code with generator polynomial $x^{15} + x^{13} + x^{11} + x^8 + x^7 + x^6 + x^4 + x^3 + 1$. This code is not the best cyclic code of length 31 and dimension 16.

Example 6.59. Let $t = 3$ and α be a generator of GF$(2^m)^*$ with $\alpha^8 + \alpha^4 + \alpha^3 + \alpha^2 + 1 = 0$. Then $\mathcal{C}_{\mathrm{GF}(2)}(D)^\perp$ is a [255, 215, 5] binary cyclic code with generator polynomial

$$x^{40} + x^{39} + x^{38} + x^{35} + x^{33} + x^{32} + x^{28} + x^{27} + x^{26} + x^{24} + x^{21} + x^{19}$$

$$+ x^{18} + x^{17} + x^{16} + x^{13} + x^{12} + x^{11} + x^{10} + x^9 + x^6 + x^5 + x^4 + x + 1.$$

The best linear code known of length 255 and dimension 215 has minimum weight 11.

The two examples above indicate that the lower bound on d given in Theorem 6.33 is tight.

6.9.3 *The third class of cyclic codes*

The following lemma gives the trace representation of the characteristic sequences of a class of cyclic difference sets with Singer parameters [No, Chung and Yun (1998); No, Golomb, Gong, Lee and Gaal (1998); Dillon and Dobbertin (2004)].

Lemma 6.22. *Let $t \geq 3$ be a positive integer and $m = 3t - 2$. Define*

$$s_i = \mathrm{Tr}_{2^m/2}\left(\alpha^i + (\alpha^i)^{2^{t-1}+1} + (\alpha^i)^{2^{2t-2}+2^{t-1}+1}\right)$$

$$+ \mathrm{Tr}_{2^m/2}\left((\alpha^i)^{2^{2t-2}-2^{t-1}+1} + (\alpha^i)^{2^{2t-1}-2^{t-1}+1}\right) \qquad (6.59)$$

for all $i \geq 0$. Then the linear span of this sequence equals $5m$ and the reciprocal of the minimal polynomial of this sequence is given by

$$g(x) = \mathbb{M}_{\alpha^{-1}}(x) \times \mathbb{M}_{\alpha^{-(2^t-1}+1)}(x) \times \mathbb{M}_{\alpha^{-(2^{2t-2}+2^{t-1}+1)}}(x)$$

$$\times \mathbb{M}_{\alpha^{-(2^{2t-2}-2^{t-1}+1)}}(x) \times \mathbb{M}_{\alpha^{-(2^{2t-1}-2^t+1)}}(x), \tag{6.60}$$

where $m_{\alpha^i}(x)$ denotes the minimal polynomial of $\alpha^i \in \mathrm{GF}(2^m)$ over $\mathrm{GF}(2)$.

Proof.   Since $t \geq 3$, it can be easily proved that

$$|C_1| = |C_{2^{t-1}+1}| = |C_{2^{2t-2}+2^{t-1}+1}| = |C_{2^{2t-2}-2^{t-1}+1}| = |C_{2^{2t-1}-2^{t-1}+1}| = m.$$

The desired conclusion on the minimal polynomial of the sequence s^∞ then follows from Lemma 1.17.   □

Theorem 6.34. *Let $t \geq 3$ and let D denote the support of the sequence of Lemma 6.22. Then the code $\mathcal{C}_{\mathrm{GF}(2)}(D)^\perp$ of the difference set D has parameters $[n, n - 5m, d]$ and generator polynomial $g(x)$ of (6.60), where $d \geq 5$.*

Proof.   The desired conclusions on the generator polynomial and the dimension of the code follow from Lemma 6.22 and the definition of the code.

Note that

$$2^{2t-1} + 2^t - 1 = 2^{2t-1}(2^{2t-1} - 2^{t-1} + 1) \bmod n \in C_{2^{2t-1}-2^{t-1}+1},$$

$$2^{2t-1} + 2^t = 2^t(2^{t-1} + 1) \bmod n \in C_{2^{t-1}+1},$$

$$2^{2t-1} + 2^t + 1 = 2^t(2^{2t-2} + 2^{t-1} + 1) \bmod n \in C_{2^{2t-2}+2^{t-1}+1},$$

$$2^{2t-1} + 2^t + 2 = 2(2^{2t-2} + 2^{t-1} + 1) \bmod n \in C_{2^{2t-2}+2^{t-1}+1}.$$

By Lemma 6.22, the reciprocal of $g(x)$ has the zeros α^i for all i in the following set

$$\{2^{2t-1} + 2^t - 1, 2^{2t-1} + 2^t, 2^{2t-1} + 2^t + 1, 2^{2t-1} + 2^t + 2\}.$$

It then follows from the BCH bound that $d \geq 5$.   □

Example 6.60. Let $t = 2$ and α be a generator of $\mathrm{GF}(2^m)^*$ with $\alpha^4 + \alpha + 1 = 0$. Then $\mathcal{C}_{\mathrm{GF}(2)}(D)^\perp$ is a $[15, 11, 3]$ binary cyclic code with generator polynomial $x^4 + x^3 + 1$. This is an optimal linear code.

Example 6.61. Let $t = 3$ and α be a generator of $GF(2^m)^*$ with $\alpha^7 + \alpha + 1 = 0$. Then $\mathcal{C}_{GF(2)}(D)^\perp$ is a $[127, 92, 9]$ binary cyclic code with generator polynomial

$$x^{35} + x^{34} + x^{33} + x^{29} + x^{27} + x^{26} + x^{22} + x^{17} + x^{14} + x^{12}$$

$$+ x^{11} + x^{10} + x^7 + x^5 + 1.$$

The best linear code known has parameters $[127, 92, 11]$.

Example 6.61 indicates that the lower bound on d given in Theorem 6.34 is not tight. Therefore, the following problem is interesting.

Problem 6.12. *Improve the lower bound on the minimum weight of the code $\mathcal{C}_{GF(2)}(D)^\perp$ of Theorem 6.34.*

6.9.4 *The fourth class of cyclic codes*

The following lemma gives the trace representation of the characteristic sequences of a class of cyclic difference sets with Singer parameters [No, Golomb, Gong, Lee and Gaal (1998); Dillon and Dobbertin (2004)].

Lemma 6.23. *Let $t \geq 2$ be a positive integer and $m = 3t - 1$. Define*

$$s_i = \mathrm{Tr}_{2^m/2}(\alpha^i + (\alpha^i + 1)^{2^t+1} + (\alpha^i + 1)^{2^{2t-1}+2^{t-1}+1})$$

$$+ \mathrm{Tr}_{2^m/2}((\alpha^i + 1)^{2^{2t-1}-2^{t-1}+1} + (\alpha^i + 1)^{2^{2t-1}+2^t-1}) \qquad (6.61)$$

for all $i \geq 0$. Then the linear span of the sequence is equal to $m(2^{\lceil m/3 \rceil} - 3)$ and the reciprocal of the minimal polynomial of this sequence is given by

$$g(x) = \prod_{i \in I_3 \cup I_4} \mathbb{M}_{\alpha^{-i}}(x), \qquad (6.62)$$

where $m_{\alpha^i}(x)$ denotes the minimal polynomial of $\alpha^i \in GF(2^m)$ over $GF(2)$ and

$$I_3 = \{2^{2t-1} + 2^{t-1} + 2 + i : 0 \leq i \leq 2^{t-1} - 3\},$$

$$I_4 = \{2^{2t} + 3 + 2i : 0 \leq i \leq 2^{t-1} - 2\}.$$

Proof. It was shown in No, Golomb, Gong, Lee and Gaal (1998) that

$$s_i = \sum_{j \in I_3 \cup I_4} \mathrm{Tr}_{2^m/2}(\alpha^i)^j,$$

and all the elements in $I_3 \cup I_4$ belong to different 2-cyclotomic cosets modulo n. It was proved in Gong and Youssef (2002) that the size $|C_i| = m$ for all $i \in I_3 \cup I_4$

and that the linear span is equal to $m(2^{\lceil m/3 \rceil} - 3)$. The desired conclusion on the minimal polynomial of the sequence s^{∞} then follows from Lemma 1.17. $\qquad\square$

Theorem 6.35. *Let $t \geq 2$ and let D be the support of the sequence of Lemma 6.23. Then the code $\mathcal{C}_{GF(2)}(D)^{\perp}$ of the difference set D has parameters $[n, n-m(2^{\lceil m/3 \rceil} - 3), d]$ and generator polynomial $g(x)$ given in Lemma 6.23, and $d \geq 2^{t-1} + 1$.*

Proof. The desired conclusions on the generator polynomial and the dimension of the code follow from Lemma 6.23 and the definition of the code.

By Lemma 6.23, the reciprocal of $g(x)$ has the zeros α^i for all i in the following set:

$$2^{3t-2}I_4 \bmod n = \{2^{2t+1} + 2^{3t-2} + 1 + i : 0 \leq i \leq 2^{t-1} - 2\}.$$

It then follows from the BCH bound that $d \geq 2^{t-1}$. Since $g(1) = 1$, d must be odd. Hence, $d \geq 2^{t-1}+1$. $\qquad\square$

Example 6.62. Let $t = 2$ and α be a generator of GF$(2^m)^*$ with $\alpha^5 + \alpha^2 + 1 = 0$. Then $\mathcal{C}_{GF(2)}(D)^{\perp}$ is a $[31, 26, 3]$ binary cyclic code with generator polynomial $x^5 + x^4 + x^3 + x^2 + 1$. This code is not optimal, while the optimal linear code with the same parameters in the Database is not cyclic.

Example 6.63. Let $t = 3$ and α be a generator of GF$(2^m)^*$ with $\alpha^8 + \alpha^4 + \alpha^3 + \alpha^2 + 1 = 0$. Then $\mathcal{C}_{GF(2)}(D)^{\perp}$ is a $[255, 215, 5]$ binary cyclic code with generator polynomial

$$x^{40} + x^{38} + x^{34} + x^{33} + x^{28} + x^{24} + x^{22} + x^{21} + x^{15} + x^{14}$$
$$+ x^{13} + x^{12} + x^{11} + x^{10} + x^9 + x^3 + x^2 + x + 1.$$

The best linear code known has parameters $[255, 215, 11]$. The upper bound on d is 14.

The two examples above indicate that the lower bound on d given in Theorem 6.35 is tight.

6.9.5 *The fifth class of cyclic codes*

The following lemma gives the trace representation of the characteristic sequences of a class of cyclic difference sets with Singer parameters [No, Golomb, Gong, Lee and Gaal (1998); Dillon and Dobbertin (2004)].

Lemma 6.24. *Let* $t \geq 3$ *be a positive integer and* $m = 3t - 2$. *Define*

$$s_i = \mathrm{Tr}_{2^m/2}(\alpha^i + (\alpha^i + 1)^{2^{t-1}+1} + (\alpha^i + 1)^{2^{2t-2}+2^{t-1}+1})$$

$$+ \mathrm{Tr}_{2^m/2}((\alpha^i + 1)^{2^{2t-2}-2^{t-1}+1} + (\alpha^i + 1)^{2^{2t-1}-2^{t-1}+1}) \qquad (6.63)$$

for all $i \geq 0$. *Then the linear span of the sequence equals* $m(2^{\lceil m/3 \rceil} - 3)$ *and the reciprocal of the minimal polynomial of this sequence is given by*

$$g(x) = \prod_{i \in \{1\} \cup I_1 \cup I_2} \mathbb{M}_{\alpha^{-i}}(x), \qquad (6.64)$$

where $\mathbb{M}_{\alpha^i}(x)$ *denotes the minimal polynomial of* $\alpha^i \in \mathrm{GF}(2^m)$ *over* $\mathrm{GF}(2)$ *and*

$$I_1 = \{2^{t-1} + 2 + i : 0 \leq i \leq 2^{t-1} - 3\},$$
$$I_2 = \{2^{2t-1} + 2^{t-1} + 2 + i : 0 \leq i \leq 2^{t-1} - 3\}.$$

Proof. It was shown in No, Golomb, Gong, Lee and Gaal (1998) that

$$s_i = \sum_{j \in \{1\} \cup I_1 \cup I_2} \mathrm{Tr}_{2^m/2}(\alpha^i)^j,$$

and all the elements in $\{1\} \cup I_1 \cup I_2$ belong to different 2-cyclotomic cosets modulo n. It was proved in Gong and Youssef (2002) that the size $|C_i| = m$ for all $\{1\} \cup I_1 \cup I_2$ and that the linear span is equal to $m(2^{\lceil m/3 \rceil} - 3)$. The desired conclusion on the minimal polynomial of the sequence s^∞ then follows from Lemma 1.17. □

Theorem 6.36. *Let* $t \geq 2$ *and let* D *denote the support of the sequence of Lemma* 6.24. *Then the code* $\mathcal{C}_{\mathrm{GF}(2)}(D)^\perp$ *of the difference set* D *has parameters* $[n, n - m(2^{\lceil m/3 \rceil} - 3), d]$ *and generator polynomial* $g(x)$ *given in Lemma* 6.24, *and* $d \geq 2^{t-1} + 1$.

Proof. The desired conclusions on the generator polynomial and the dimension of the code follow from Lemma 6.24 and the definition of the code.

Note that the 2-cyclotomic coset C_1 contains 2^{t-1}. By Lemma 6.24, the reciprocal of $g(x)$ has the zeros α^i for all i in the following set:

$$\{2^{t-1}\} \cup I_1 = \{2^{t-1} + j : 2 \leq i \leq 2^{t-1}\}.$$

It then follows from the BCH bound that $d \geq 2^{t-1}$. Since $g(1) = 1, d$ must be odd. Hence, $d \geq 2^{t-1}+1$. □

Example 6.64. Let $t = 2$ and α be a generator of $\mathrm{GF}(2^m)^*$ with $\alpha^4 + \alpha + 1 = 0$. Then $\mathcal{C}_{\mathrm{GF}(2)}(D)^\perp$ is a $[15, 11, 3]$ binary cyclic code with generator polynomial $x^4 + x^3 + 1$. This cyclic code is an optimal linear code.

Example 6.65. Let $t = 3$ and α be a generator of $GF(2^m)^*$ with $\alpha^7 + \alpha + 1 = 0$. Then $\mathcal{C}_{GF(2)}(D)^\perp$ is a $[127, 92, 9]$ binary cyclic code with generator polynomial

$$x^{35} + x^{34} + x^{33} + x^{32} + x^{30} + x^{27} + x^{25} + x^{24} + x^{22} + x^{20}$$

$$+ x^{19} + x^{18} + x^{15} + x^{13} + x^9 + x^8 + x^7 + x^6 + 1.$$

The best linear code known has parameters $[127, 92, 11]$.

The two examples above indicate that the lower bound on d given in Theorem 6.36 is tight.

6.10 Cyclic Codes of Dillon–Dobbertin Difference Sets

Let m be a positive integer and let $n = 2^m - 1$. For each h with $\gcd(h, m) = 1$, define $\Delta_h(x) = (x + 1)^d + x^d + 1$, where $d = 4^h - 2^h + 1$. Then

$$B_k := \log_\alpha[GF(2^m) \backslash \Delta_k(GF(2^m))]$$

is a difference set with Singer parameters in $\mathbb{Z}_{2^m-1}$, where α is a generator of $GF(2^m)^*$. Furthermore, for each fixed m, the $\phi(m)/2$ difference sets B_h are pairwise inequivalent, where ϕ is the Euler function. This family of cyclic difference sets were described in Dillon and Dobbertin (2004). When $h = 1$, the Dillon–Dobbertin difference set coincides with the Singer difference set.

In this section, we deal with the dual $\mathcal{C}_{GF(2)}(B_k)^\perp$ of the code $\mathcal{C}_{GF(2)}(B_k)$ of the difference set B_k. Obviously, the dimension and the generator polynomial of the code $\mathcal{C}_{GF(2)}(B_k)$ follow automatically from those of $\mathcal{C}_{GF(2)}(B_k)^\perp$.

We first define

$$A_1(x) = x,$$

$$A_2(x) = x^{2^h+1},$$

$$A_{i+2}(x) = x^{2^{(i+1)h}} A_{i+1}(x) + x^{2^{(i+1)h}-2^{ih}} A_i(x), \quad i \geq 1$$

and

$$B_1(x) = 0,$$

$$B_2(x) = x^{2^h-1},$$

$$B_{i+2}(x) = x^{2^{(i+1)h}} B_{i+1}(x) + x^{2^{(i+1)h}-2^{ih}} B_i(x), \quad i \geq 1.$$

Let $hh' \equiv 1 \pmod{m}$. We then define

$$R_{(h,h')}(x) = \sum_{i=1}^{h'} A_i(x) + B_{h'}(x). \tag{6.65}$$

The following lemma is proved in Dillon and Dobbertin (2004).

Lemma 6.25. *Let s^∞ denote the characteristic sequence of the Dillon–Dobbertin difference set B_k. The linear span of the sequence is equal to $m(2F_{h_1} - 1)$ for all odd m, where $h_1 = \min\{h', m - h'\}$ with $hh' \equiv 1 \pmod{m}$ and F_i is the ith Fibonacci number with the initial state $F_0 = F_1 = 1$. Furthermore, the trace representation of the sequence is given by*

$$s_t = \mathrm{Tr}_{2^m/2}(R_{(h,h')}(\alpha^t)) \tag{6.66}$$

for all $t \geq 0$.

It was claimed in Dillon and Dobbertin (2004) that the trace representation of (6.66) is *reduced* in the sense that all the exponents of the monomials belong to different 2-cyclotomic cosets modulo $n = 2^m - 1$. However, it will be shown later that this is true only when m is large enough.

Theorem 6.37. *The cyclic code $\mathcal{C}_{\mathrm{GF}(2)}(B_k)^\perp$ has parameters $[2^m - 1, 2^m - 1 - m(2F_{h_1} - 1), d]$, where h_1 and F_{h_1} are defined in Lemma 6.25.*

Proof. The desired conclusion on the dimension of this code follows from the definition of $\mathcal{C}_{\mathrm{GF}(2)}(B_k)^\perp$ and Lemma 6.25. $\qquad\square$

When $h = 1$, the Dillon–Dobbertin difference set coincides with the Singer difference set. In this case the code $\mathcal{C}_{\mathrm{GF}(2)}(B_k)^\perp$ of Theorem 6.37 becomes the Hamming code. When $h = 2$, the Dillon–Dobbertin difference set coincides with the Segre difference set [Dillon and Dobbertin (2004)]. In this case, the code $\mathcal{C}_{\mathrm{GF}(2)}(B_k)^\perp$ of Theorem 6.37 becomes the code of Theorem 6.26.

In order to study the code $\mathcal{C}_{\mathrm{GF}(2)}(B_k)^\perp$, we need information on the minimal polynomial of the characteristic sequence of B_k. Unfortunately, so far we do not have an explicit expression of this minimal polynomial. However, we are able to find out parameters of the code $\mathcal{C}_{\mathrm{GF}(2)}(B_k)^\perp$ of Theorem 6.37 case by case. To this end, we need to compute the polynomial $R_{(h,h')}(x)$ case by case. We have the following special cases:

$$\begin{cases} R_{(h,1)}(x) = x, \\ R_{(h,2)}(x) = x^{2^h+1} + x^{2^h-1} + x, \\ R_{(h,3)}(x) = x^{2^{2h}+2^h+1} + x^{2^{2h}+2^h-1} + x^{2^{2h}-2^h+1} + x^{2^h+1} + x \end{cases} \tag{6.67}$$

and

$$R_{(h,4)}(x) = x^{2^{3h}+2^{2h}+2^h+1} + x^{2^{3h}+2^{2h}-2^h+1} + x^{2^{3h}-2^{2h}+2^h+1}$$

$$+ x^{2^{3h}+2^{2h}+2^h-1} + x^{2^{3h}-2^{2h}+2^h-1} + x^{2^{2h}+2^h+1}$$

$$+ x^{2^{2h}-2^h+1} + x^{2^h+1} + x. \tag{6.68}$$

Corollary 6.5. *Let m be odd and $h = (m+1)/2$. Then the code $\mathcal{C}_{\mathrm{GF}(2)}(B_k)^\perp$ has parameters $[2^m - 1, 2^m - 1 - 3m, 7]$ and generator polynomial*

$$\mathbb{M}_{\alpha^{-(2^{(m+1)/2}+1)}}(x)\mathbb{M}_{\alpha^{-(2^{(m+1)/2}-1)}}(x)\mathbb{M}_{\alpha^{-1}}(x).$$

Proof. Let $m \geq 5$ and $h = (m+1)/2$. Then $h' = 2$. The desired conclusion on the dimension of this code follows from the definition of $\mathcal{C}_{\mathrm{GF}(2)}(B_k)^\perp$ and Lemma 6.25. The conclusion on the generator polynomial follows from the definition of $\mathcal{C}_{\mathrm{GF}(2)}(B_k)^\perp$, Lemma 1.17 and the trace representation

$$s_t = \mathrm{Tr}_{2^m/2}(R_{(h,2)}(\alpha^t)) = \mathrm{Tr}_{2^m/2}((\alpha^t)^{2^{(m+1)/2}+1} + (\alpha^t)^{2^{(m+1)/2}-1} + \alpha^t)$$

for all $t \geq 0$.

It was pointed out in Dobbertin (1999b) that the difference set defined by $h = (m+1)/2$ is equivalent to the difference set defined by the sequence of Lemma 6.20. Hence, the code of this theorem is equivalent to that of Theorem 6.32. Therefore $d = 7$. □

Corollary 6.6. *Let $m \geq 7$ such that $3 \mid (m+1)$ or $3 \mid (2m+1)$. Define $h = (m+1)/3$ or $h = (2m+1)/3$. Then the set $\mathcal{C}_{\mathrm{GF}(2)}(B_k)^\perp$ is a $[2^m - 1, 2^m - 1 - 5m, d]$ binary cyclic code with generator polynomial*

$$\mathbb{M}_{\alpha^{-(2^{2h}+2^h+1)}}(x)\mathbb{M}_{\alpha^{-(2^{2h}+2^h-1)}}(x)\mathbb{M}_{\alpha^{-(2^{2h}-2^h+1)}}(x)\mathbb{M}_{\alpha^{-(2^h+1)}}(x)\mathbb{M}_{\alpha^{-1}}(x),$$

where $d \geq 5$.

Proof. Let $h = (m+1)/3$ or $h = (2m+1)/3$. Then $h' = 3$. The desired conclusion on the dimension of this code follows from the definition of $\mathcal{C}_{\mathrm{GF}(2)}(B_k)^\perp$

and Lemma 6.25. The conclusion on the generator polynomial follows from the definition of $\mathcal{C}_{\mathrm{GF}(2)}(B_k)^{\perp}$, Lemma 1.17 and the trace representation

$$s_t = \mathrm{Tr}_{2^m/2}(R_{(h,3)}(\alpha^t))$$

for all $t \geq 0$, where $R_{(h,3)}(x)$ was defined in (6.67).

We now prove the conclusions about the minimum weight d of this code. When $h = (m+1)/3$ or $h = (2m+1)/3$, we have $2^h(2^{2h} + 2^h + 1) \pmod{n} = 2^{2h} + 2^h + 2$. In addition, $2^h(2^h + 1) = 2^{2h} + 2^h$. Hence,

$$\{2^{2h} + 2^h, 2^{2h} + 2^h + 1, 2^{2h} + 2^h + 2\} \subset C_1 \cup C_{2^{2h}+2^h+1}.$$

It then follows from the BCH bound that $d \geq 4$. However, since d is odd. we have then $d \geq 5$. $\qquad\square$

Example 6.66. Let $(m, h, h') = (5, 3, 3)$ and α be a generator of $\mathrm{GF}(2^m)^*$ with $\alpha^5 + \alpha^2 + 1 = 0$. Then the set $\mathcal{C}_{\mathrm{GF}(2)}(B_k)^{\perp}$ of Corollary 6.6 is a $[31, 16, 7]$ binary cyclic code with generator polynomial $x^{15} + x^{12} + x^7 + x^6 + x^2 + x + 1$. This code is the best binary cyclic code of length 31 and dimension 16 according to Table A.13. An upper bound on the minimum distance of any binary linear code with length 31 and dimension 16 is 8. Note that the parameters of this code do not match with those of the code of Corollary 6.6.

Example 6.67. Let $(m, h, h') = (8, 3, 3)$ and α be a generator of $\mathrm{GF}(2^m)^*$ with $\alpha^8 + \alpha^4 + \alpha^3 + \alpha^2 + 1 = 0$. Then the code $\mathcal{C}_{\mathrm{GF}(2)}(B_k)^{\perp}$ of Corollary 6.6 is a $[255, 215, 5]$ binary cyclic code with generator polynomial

$$x^{40} + x^{39} + x^{33} + x^{32} + x^{28} + x^{22} + x^{20} + x^{19}$$
$$+ x^{14} + x^{12} + x^8 + x^6 + x^5 + x^4 + x^2 + x + 1.$$

The best binary linear code known with parameters $[255, 215, 11]$ in the Database is not known to be cyclic.

Example 6.68. Let $(m, h, h') = (4, 3, 3)$ and α be a generator of $\mathrm{GF}(2^m)^*$ with $\alpha^4 + \alpha + 1 = 0$. Then the code $\mathcal{C}_{\mathrm{GF}(2)}(B_k)^{\perp}$ of Corollary 6.6 is a $[15, 11, 3]$ binary cyclic code with generator polynomial $x^4 + x^3 + 1$. This code is optimal. Note that the parameters of this code do not match with those of the code of Corollary 6.6.

Example 6.69. Let $(m, h, h') = (7, 5, 3)$ and α be a generator of $\mathrm{GF}(2^m)^*$ with $\alpha^7 + \alpha + 1 = 0$. Then the code $\mathcal{C}_{\mathrm{GF}(2)}(B_k)^{\perp}$ of Corollary 6.6 is a $[127, 92, 9]$

binary cyclic code with generator polynomial

$$x^{35} + x^{34} + x^{33} + x^{29} + x^{27} + x^{26} + x^{22}$$
$$+ x^{17} + x^{14} + x^{12} + x^{11} + x^{10} + x^7 + x^5 + 1.$$

The best binary linear code known in the Database with parameters $[127, 92, 11]$ is not known to be cyclic.

Corollary 6.7. *Let $m \geq 9$ such that $4 \mid (m+1)$ or $4 \mid (3m+1)$. Define $h = (m+1)/4$ or $h = (3m+1)/4$. Then the set $\mathcal{C}_{\mathrm{GF}(2)}(B_k)^{\perp}$ is a $[2^m - 1, 2^m - 1 - 9m, d]$ binary cyclic code with generator polynomial*

$$\mathbb{M}_{\alpha^{-(2^{3h}+2^{2h}+2^h+1)}}(x)\mathbb{M}_{\alpha^{-(2^{3h}+2^{2h}-2^h+1)}}(x)\mathbb{M}_{\alpha^{-(2^{3h}-2^{2h}+2^h+1)}}(x)$$

$$\times \mathbb{M}_{\alpha^{-(2^{3h}+2^{2h}+2^h-1)}}(x)\mathbb{M}_{\alpha^{-(2^{3h}-2^{2h}+2^h-1)}}(x)\mathbb{M}_{\alpha^{-(2^{2h}+2^h+1)}}(x)$$

$$\times \mathbb{M}_{\alpha^{-(2^{2h}-2^h+1)}}(x)\mathbb{M}_{\alpha^{-(2^h+1)}}(x)\mathbb{M}_{\alpha^{-1}}(x),$$

where $d \geq 5$.

Proof. Let $h = (m + 1)/4$ or $h = (3m + 1)/4$. Then $h' = 4$. The desired conclusion on the dimension of this code follows from the definition of $\mathcal{C}_{\mathrm{GF}(2)}(B_k)^{\perp}$ and Lemma 6.25. The conclusion on the generator polynomial follows from the definition of $\mathcal{C}_{\mathrm{GF}(2)}(B_k)^{\perp}$, Lemma 1.17 and the trace representation

$$s_t = \mathrm{Tr}_{2^m/2}(R_{(h,4)}(\alpha^t)) \tag{6.69}$$

for all $t \geq 0$, where $R_{(h,4)}(x)$ was defined in (6.68). One can prove that the trace representation of (6.69) is reduced if $m \geq 9$.

We now prove the conclusions about the minimum weight d of this code. When $h = (m + 1)/4$ or $h = (3m + 1)/4$, we have $2^h(2^{3h} + 2^{2h} + 2^h + 1)$ (mod n) $= 2^{3h} + 2^{2h} + 2^h + 2$. In addition, $2^h(2^{2h} + 2^h + 1) = 2^{3h} + 2^{2h} + 2^h$. Hence,

$$\{e, e + 1, e + 2, e + 3\} \subset C_{2^{2h}+2^h+1} \cup C_e \cup C_{e+2},$$

where $e = 2^{3h} + 2^{2h} + 2^h - 1$. It then follows from the BCH bound that the minimum weight $d \geq 5$. $\qquad \square$

Example 6.70. Let $(m, h, h') = (3, 1, 4)$ and α be a generator of $\mathrm{GF}(2^m)^*$ with $\alpha^3 + \alpha + 1 = 0$. Then the code $\mathcal{C}_{\mathrm{GF}(2)}(B_k)^{\perp}$ of Corollary 6.7 is a $[7, 4, 3]$ binary cyclic code with generator polynomial $x^3 + x^2 + 1$. This code is optimal. Note that the parameters of this code do not match with those of the code of Corollary 6.7.

Problem 6.13. *Let $m \geq 9$. Improve the lower bound on the minimum distance d of the code of Corollary* 6.7.

Example 6.71. Let $(m, h, h') = (7, 2, 4)$ and α be a generator of $GF(2^m)^*$ with $\alpha^7 + \alpha + 1 = 0$. Then the code $\mathcal{C}_{GF(2)}(B_k)^{\perp}$ of Corollary 6.7 is a $[127, 92, 9]$ binary cyclic code with generator polynomial

$$x^{35} + x^{34} + x^{33} + x^{29} + x^{27} + x^{26} + x^{22}$$
$$+ x^{17} + x^{14} + x^{12} + x^{11} + x^{10} + x^7 + x^5 + 1.$$

The best binary linear code known in the Database with parameters $[127, 92, 11]$ is not known to be cyclic. Note that the parameters of this code do not match with those of the code of Corollary 6.7.

Example 6.72. Let $(m, h, h') = (5, 4, 4)$ and α be a generator of $GF(2^m)^*$ with $\alpha^5 + \alpha^2 + 1 = 0$. Then the code $\mathcal{C}_{GF(2)}(B_k)^{\perp}$ of Corollary 6.7 is a $[31, 26, 3]$ binary cyclic code with generator polynomial $x^5 + x^3 + 1$. This code is optimal. Note that the parameters of this code do not match with those of the code of Corollary 6.7.

In this section, we considered the code $\mathcal{C}_{GF(2)}(B_k)^{\perp}$ of the Dillon–Dobbertin difference set for the cases when $h \in \{1, 2\}$ and $h' \in \{1, 2, 3, 4\}$. Other cases should also be interesting, but are much more technical. The following is an example for the case $h = 3$.

Example 6.73. Let $(m, h) = (7, 3)$ and α be a generator of $GF(2^m)^*$ with $\alpha^7 + \alpha + 1 = 0$. Then $\mathcal{C}_{GF(2)}(B_k)^{\perp}$ is a $[127, 106, 7]$ binary cyclic code with generator polynomial

$$x^{21} + x^{18} + x^{17} + x^{16} + x^{15} + x^{14} + x^8 + x^6 + x^5 + x^4 + 1.$$

The best binary linear code known with parameters $[127, 106, 7]$ in the Database is not known to be cyclic.

The generator polynomial of the code $\mathcal{C}_{GF(2)}(B_k)^{\perp}$ is unknown in most cases. However, its generating idempotent is simple and given in the following theorem whose proof is left to the reader.

Theorem 6.38. *The generating idempotent of the code $\mathcal{C}_{GF(2)}(B_k)^{\perp}$ is $1 + B_k(x)$, where $B_k(x)$ is the Hall polynomial of B_k over $GF(2)$. Similarly, the generating idempotent of the code $\mathcal{C}_{GF(2)}(B_k)$ is $B_k(x)$.*

6.11   Cyclic Codes of GMW Difference Sets

Consider a proper subfield $\mathrm{GF}(2^{m_0})$ of $\mathrm{GF}(2^m)$, where $m_0 \geq 1$ is a divisor of m. Let

$$R := \{x \in \mathrm{GF}(2^m) : \mathrm{Tr}_{m/m_0}(x) = 1\}.$$

If D is any difference set with Singer parameters $(2^{m_0} - 1, 2^{m_0-1}, 2^{m_0-2})$ in $\mathrm{GF}(2^{m_0})^*$, then $U_D := \log_\alpha[R(D^{(\gamma)})]$ is a difference set with Singer parameters in $(\mathbb{Z}_{2^m-1}, +)$, where γ is any representative of the 2-cyclotomic coset modulo $(2^{m_0} - 1)$ with $\gcd(2^{m_0} - 1, \gamma) = 1$, $D^{(\gamma)} := \{y^\gamma : y \in D\}$, and α is a generator of $\mathrm{GF}(2^m)^*$.

The Gordon–Mills–Welch construction is very powerful and generic. Any difference set with Singer parameters $(2^{m_0} - 1, 2^{m_0-1}, 2^{m_0-2})$ in any subfield $\mathrm{GF}(2^{m_0})^*$ can be plugged into it, and may produce new difference set with Singer parameters.

A special case of the Gorden–Mills–Welch construction is the following:

$$U_{(m_0,\gamma)} = \{0 \leq t < 2^m - 1 : \mathrm{Tr}_{2^{m_0}/2}[(\mathrm{Tr}_{2^m/2^{m_0}}(\alpha^t))^\gamma] = 1\}, \qquad (6.70)$$

where $1 \leq \gamma < 2^{m_0} - 1$ and $\gcd(\gamma, 2^{m_0} - 1) = 1$.

When $\gamma = 1$, $U_{(m_0,\gamma)}$ becomes the Singer difference set and the code $\mathcal{C}_{\mathrm{GF}(2)}(U_{(m_0,\gamma)})^\perp$ is equivalent to the Hamming code with parameters $[2^m, 2^m - 1 - m, 3]$, which is optimal. When $m_0 > 1$ and $\gamma > 1$, the code may be equivalent to the Hamming code, and may be bad if its dimension is different from $2^m - 1 - m$. Hence, the codes from the GMW difference sets might not be interesting in general due to their poor error-correcting capability. The following example may illustrate this.

Example 6.74. Let $(m, m_0, \gamma) = (6, 3, 5)$ and α be a generator of $\mathrm{GF}(2^m)^*$ with $\alpha^6 + \alpha^4 + \alpha^3 + \alpha + 1 = 0$. For the special construction of (6.70), the set $\mathcal{C}_{\mathrm{GF}(2)}(U_{(m_0,\gamma)})^\perp$ is a $[63, 51, 3]$ binary cyclic code with generator polynomial

$$x^{12} + x^{11} + x^{10} + x^9 + x^8 + x^6 + x^5 + x^4 + x^2 + x + 1$$

The best binary cyclic code in Table A.33 has parameters $[63, 51, 5]$.

Example 6.75. Let $(m, m_0, \gamma) = (9, 3, 3)$ and α be a generator of $\mathrm{GF}(2^m)^*$ with $\alpha^9 + \alpha^4 + 1 = 0$. For the special construction of (6.70), the set $\mathcal{C}_{\mathrm{GF}(2)}(U_{(m_0,\gamma)})^\perp$ is a $[511, 484, 3]$ binary cyclic code with generator polynomial $x^{27} + x^{26} + x^{22} + x^{19} + x^{17} + x^{15} + x^{10} + x^8 + x^7 + x^6 + x^4 + x^3 + x^2 + x + 1$.

6.12 Cyclic Codes of the HKM and Lin Difference Sets

The 3-rank of the HDM difference set defined in Theorem 4.33 and the minimal polynomial of its characteristic sequence are presented in the following theorem.

Theorem 6.39 ([No, Shin and Helleseth (2004)]). *Let* $m = 3k > 3$ *for some positive integer* k *and* $d = 3^{2k} - 3^k + 1$. *Define* $n = (3^m - 1)/2$ *and*

$$D = \{t : \mathrm{Tr}_{3^m/3}(\alpha^t + \alpha^{td}) = 0,\ 0 \le t \le n - 1\}, \tag{6.71}$$

where α *is a generator of* $\mathrm{GF}(3^m)^*$. *Then the 3-rank of the difference set* D *is equal to* $2m^2 - 2m + 1$ *and the minimal polynomial of its characteristic sequence is*

$$h(x) := (x - 1)\mathbb{M}_{\alpha^{2d}}(x) \prod_{i=1}^{h} \mathbb{M}_{\alpha^{1+3i}}(x) \prod_{\substack{i=1 \\ i \ne k}}^{h} \mathbb{M}_{\alpha^{(1+3i)d}}(x) \prod_{\substack{i=0 \\ i \ne k}}^{m-1} \mathbb{M}_{\alpha^{(d+3i)d}}(x),$$

$$\tag{6.72}$$

where $h = \lfloor m/2 \rfloor$ *and* $\mathbb{M}_\beta(x)$ *denotes the minimal polynomial of* $\beta \in \mathrm{GF}(3^m)$ *over* $\mathrm{GF}(3)$.

The following theorem then follows from Theorem 6.39.

Theorem 6.40. *The cyclic code* $\mathcal{C}_{\mathrm{GF}(3)}(D)$ *of the HDM cyclic difference set* D *has parameters* $[(3^m - 1)/2, 2m^2 - 2m + 1]$ *and parity check polynomial* $h(x)$ *given in* (6.72).

Problem 6.14. *Determine the minimum weight of the cyclic code* $\mathcal{C}_{\mathrm{GF}(3)}(D)$ *of the HDM difference set* D.

Example 6.76. When $m = 3$, the cyclic code $\mathcal{C}_{\mathrm{GF}(3)}(D)$ of the HDM difference set D has parameters $[13, 7, 4]$ and parity check polynomial $x^7 + 2x^5 + x^3 + 2x^2 + x + 2$.

Example 6.77. When $m = 3$, the cyclic code $\mathcal{C}_{\mathrm{GF}(3)}(D)$ of the HDM difference set D has parameters $[121, 41, 34]$ and parity check polynomial

$$x^{41} + 2x^{40} + 2x^{39} + x^{38} + 2x^{37} + 2x^{36} + x^{34} + 2x^{31} + 2x^{29} + 2x^{27}$$

$$+ 2x^{24} + 2x^{22} + x^{21} + 2x^{19} + 2x^{16} + x^{15} + x^{14} + 2x^{13} + 2x^{12} + x^{11}$$

$$+ x^{10} + x^9 + 2x^8 + x^6 + 2x^5 + x^4 + 2x^3 + x^2 + 2x + 2$$

The 3-rank of the Lin difference set defined in Theorem 4.34 and the minimal polynomial of its characteristic sequence are given in the following theorem.

Theorem 6.41 ([No, Shin and Helleseth (2004)]). *Let* $m = 2h + 1 > 3$ *for some positive integer h and* $d = 2 \times 3^h + 1$. *Define* $n = (3^m - 1)/2$ *and*

$$D = \{t : \text{Tr}_{3^m/3}(\alpha^t + \alpha^{td}) = 0, 0 \le t \le n - 1\}, \tag{6.73}$$

where α *is a generator of* $\text{GF}(3^m)^*$. *Then the 3-rank of the Lin difference set D is equal to* $2m^2 - 2m + 1$ *and the minimal polynomial of its characteristic sequence is*

$$h(x) := (x - 1)\mathbb{M}_{\alpha^2}(x) \prod_{i=1}^{h} \mathbb{M}_{\alpha^{1+3^i}}(x) \prod_{i=1}^{h} \mathbb{M}_{\alpha^{(1+3^i)d}}(x) \prod_{\substack{i=0 \\ i \ne h}}^{m-2} \mathbb{M}_{\alpha^{(d+3^i)d}}(x),$$

$$\tag{6.74}$$

where $\mathbb{M}_\beta(x)$ *denotes the minimal polynomial of* $\beta \in \text{GF}(3^m)$ *over* $\text{GF}(3)$.

The following theorem then follows from Theorem 6.41.

Theorem 6.42. *The cyclic code* $\mathcal{C}_{\text{GF}(3)}(D)$ *of the Lin cyclic difference set D has parameters* $[(3^m - 1)/2, 2m^2 - 2m + 1]$ *and parity check polynomial h(x) given in* (6.74).

Problem 6.15. *Determine the minimum weight of the cyclic code* $\mathcal{C}_{\text{GF}(3)}(D)$ *of the Lin difference set D.*

Example 6.78. When $m = 3$, the cyclic code $\mathcal{C}_{\text{GF}(3)}(D)$ of the Lin difference set D has parameters $[13, 7, 4]$ and parity check polynomial $x^7 + 2x^5 + x^3 + 2x^2 + x + 2$.

6.13   Two More Constructions of Codes with Difference Sets

There are other ways of constructing linear codes from difference sets. In this section, we will briefly introduce two more constructions.

By extending earlier constructions, Lander introduced a method of constructing linear codes from symmetric designs. Such a linear code is self-dual with respect to an inner product which is different from the ordinary one $\mathbf{x} \cdot \mathbf{y} = \sum_{i=1}^{n} x_i y_i$ in the vector space $\text{GF}(q)^n$ [Lander (1981)]. Plugging this construction, difference sets give automatically self-dual codes. However, little research in this direction is done. Hence, the following problem would be interesting.

Problem 6.16. *Study the self-dual codes from difference sets with the Lander construction.*

Another construction of linear codes from difference sets is the following. Let $(A, +)$ be an abelian group of order n, and let D be a subset of A. Recall the characteristic function $\xi_D(x)$ of D, which is a function from A to $\mathrm{GF}(2)$ and can be viewed as a function from A to $\mathrm{GF}(q)$ for any prime power q. In this way, the following matrix

$$
G_D = \begin{bmatrix}
\xi_D(a_1) & \xi_D(a_2) & \cdots & \xi_D(a_n) \\
L_1(a_1) & L_1(a_2) & \cdots & L_1(a_n) \\
L_2(a_1) & L_2(a_2) & \cdots & L_2(a_n) \\
\vdots & \vdots & \vdots & \vdots \\
L_h(a_1) & L_h(a_2) & \cdots & L_h(a_n)
\end{bmatrix}.
\tag{6.75}
$$

generates a linear code of length n over $\mathrm{GF}(q)$, denoted by $\tilde{C}_D$, where $A = \{a_1, a_2, \ldots, a_n\}$ and $\{L_1, L_2, \ldots, L_h\}$ is the set of all linear functions from $(A, +)$ to $(\mathrm{GF}(q), +)$. With this construction, any subset D of A produces a linear code $\tilde{C}_D$.

The following theorem is easily proved.

Theorem 6.43. *Let D be a subset of $(\mathrm{GF}(2)^n, +)$ with $|D| = 2^n \pm 2^{n/2}$. Then D is a Hadamard difference set in $(\mathrm{GF}(2)^n, +)$ if and only if $\tilde{C}_D$ has parameters $[2^n, n+1, 2^n \pm 2^{n/2}]$.*

Regarding the code $\tilde{C}_D$, little work has been done. It would be good if progress can be made in this direction.

Problem 6.17. *Study the codes $\tilde{C}_D$ from other difference sets.*

Chapter 7

Linear Codes of Almost Difference Sets

7.1 Definitions and Fundamentals

Recall that the difference function defined by a subset D of an abelian group $(A, +)$ is given by

$$\text{diff}_D(x) = |D \cap (D + x)|,$$

where $D + x = \{y + x : y \in D\}$.

A k-subset D of an abelian group $(A, +)$ of order n is an (n, k, λ, t) almost difference set of A if the difference function $\text{diff}_D(x)$ takes on λ altogether t times and $\lambda + 1$ altogether $n - 1 - t$ times when x ranges over all the nonzero elements of A. We treated almost difference sets in Chapter 5. In this chapter, we define and study the linear codes of almost difference sets.

Let D be an almost difference set in an abelian group $(A, +)$ of order n. Similar to difference sets, the *development* of D is a triple $(\mathcal{P}, \mathcal{B}, \mathcal{I})$, where $\mathcal{P}$ consists of all the elements of A, $\mathcal{B} = \{B_a : a \in A\}$ with $B_a = \{a + d : d \in D\}$, and the incidence relation $\mathcal{I}$ is the set membership. By definition, the triple $(\mathcal{P}, \mathcal{B}, \mathcal{I})$ is a 2-adesign if D is an almost difference set. In Section 3.3, we defined t-adesigns.

Note that the 2-adesign of Example 3.8 in Section 3.3 is the development of the following almost difference set D in $(\mathbb{Z}_{13}, +)$:

$$D = \{1, 3, 4, 9, 10, 12\}.$$

Recall that the incidence matrix, p-rank, and the code of t-adesigns are defined in the same way as those of t-designs. So the code $\mathcal{C}_{\text{GF}(q)}(D)$ of an almost difference set D is the linear code over $\text{GF}(q)$ spanned by the rows of the incidence matrix of the development of D. The linear code $\mathcal{C}_{\text{GF}(q)}(D)$ may be cyclic or not, depending on whether A is cyclic or not.

When D is an almost difference set in $(\mathbb{Z}_n, +)$, the code $\mathcal{C}_{\text{GF}(q)}(D)$ is cyclic and has generator polynomial $g(x) = \gcd(x^n - 1, D(x))$ and parity check polynomial

$h(x) = (x^n - 1)/g(x)$, where

$$D(x) = \sum_{i \in D} x^i \in \mathrm{GF}(q)[x],$$

which is called the *Hall polynomial* of D, and $\gcd(x^n - 1, D(x))$ is computed over $\mathrm{GF}(q)$. In this chapter, we will study the codes of only cyclic almost difference sets. The codes of some families of almost difference sets were already treated in Chapter 6 when we studied the codes of some families of difference sets.

7.2 Cyclic Codes of DHL Almost Difference Sets

Throughout this section, let n be a prime such that $n \equiv 1 \pmod 4$, and let $D = C_0^{(4,n)} \cup C_1^{(4,n)}$. The Hall polynomial of D is given by

$$D(x) = \sum_{i \in C_0^{(4,n)} \cup C_1^{(4,n)}} x^i \in \mathrm{GF}(q)[x].$$

The cyclic code $\mathcal{C}_{\mathrm{GF}(q)}(D)$ of the set D has generator and parity check polynomials

$$g(x) = \gcd(x^n - 1, D(x)) \quad \text{and} \quad h(x) = \frac{x^n - 1}{g(x)},$$

where $\gcd(x^n - 1, D(x))$ is computed over $\mathrm{GF}(q)$. In this subsection, we treat the cyclic codes $\mathcal{C}_{\mathrm{GF}(q)}(D)$, and always assume that $q \in C_0^{(4,n)}$. This ensures that the polynomials $\Omega_i^{(4,n)}(x)$ defined in Section 6.5.1 are over $\mathrm{GF}(q)$. In this section, we also assume that $\frac{n-1}{4} \bmod p = 0$.

As an auxiliary polynomial, we define

$$\Gamma(x) = \sum_{i \in C_1^{(4,n)} \cup C_2^{(4,n)}} x^i \in \mathrm{GF}(q)[x].$$

Both $D(x)$ and $\Gamma(x)$ depend on the choice of the generator of $\mathrm{GF}(n)^*$ employed to define the cyclotomic classes of order 4.

Notice that

$$\left(\sum_{i \in C_0^{(4,n)}} + \sum_{i \in C_1^{(4,n)}} + \sum_{i \in C_2^{(4,n)}} + \sum_{i \in C_3^{(4,n)}} \right) \eta^i = -1.$$

We have then

$$
D(\eta^i) = \begin{cases} D(\eta) & \text{if } i \in C_0^{(4,n)}, \\ \Gamma(\eta) & \text{if } i \in C_1^{(4,n)}, \\ -(D(\eta)+1) & \text{if } i \in C_2^{(4,n)}, \\ -(\Gamma(\eta)+1) & \text{if } i \in C_3^{(4,n)}. \end{cases} \tag{7.1}
$$

We have also that

$$
D(\eta^0) = D(1) = \frac{n-1}{2} \bmod p. \tag{7.2}
$$

Theorem 7.1 ([Ding (2013)]). *Let $\frac{n-1}{4} \equiv 0 \pmod{p}$, and let $n = u^2 + 4v^2$ with $u \equiv 1 \pmod 4$. The cyclic code $\mathcal{C}_{\mathrm{GF}(q)}(D)$ has parameters $[n, n - \mathrm{rank}_p(D), d]$ and generator polynomial $g(x)$, where $\mathrm{rank}_p(D)$, $g(x)$ and information on d are given below.*

(a) *The case that $n \equiv 1 \pmod 8$:*
When $\frac{v}{2} \not\equiv 0 \pmod p$, we have $\mathrm{rank}_p(D) = n - 1$ and $g(x) = x - 1$.
When $\frac{v}{2} \equiv 0 \pmod p$ and $q \in C_0^{(4,n)}$, we have $\mathrm{rank}_p(D) = (n-1)/2$ and

$$
g(x) = \begin{cases} (x-1)\Omega_0^{(4,n)}(x)\Omega_1^{(4,n)}(x) & \text{if } D(\eta) = 0 \text{ and } \Gamma(\eta) = 0, \\ (x-1)\Omega_0^{(4,n)}(x)\Omega_3^{(4,n)}(x) & \text{if } D(\eta) = 0 \text{ and } \Gamma(\eta) = -1, \\ (x-1)\Omega_1^{(4,n)}(x)\Omega_2^{(4,n)}(x) & \text{if } D(\eta) = -1 \text{ and } \Gamma(\eta) = 0, \\ (x-1)\Omega_2^{(4,n)}(x)\Omega_3^{(4,n)}(x) & \text{if } D(\eta) = -1 \text{ and } \Gamma(\eta) = -1. \end{cases}
$$

In addition, the minimum odd-like weight $d_{\mathrm{odd}} \geq \sqrt{n}$.

(b) *The case that $n \equiv 5 \pmod 8$:*
When $\frac{u^2+3}{4} \not\equiv 0 \pmod p$, we have $\mathrm{rank}_p(D) = n - 1$ and $g(x) = x - 1$. In this subcase, $d = n$.
When $\frac{u^2+3}{4} \equiv 0 \pmod p$ and $q \in C_0^{(4,n)}$, we have $\mathrm{rank}_p(D) = 3(n-1)/4$ and

$$
g(x) = \begin{cases} (x-1)\Omega_0^{(4,n)}(x) & \text{if } D(\eta) = 0, \\ (x-1)\Omega_2^{(4,n)}(x) & \text{if } D(\eta) = -1, \\ (x-1)\Omega_1^{(4,n)}(x) & \text{if } \Gamma(\eta) = 0, \\ (x-1)\Omega_3^{(4,n)}(x) & \text{if } \Gamma(\eta) = -1. \end{cases}
$$

Furthermore, the minimum weight d has the lower bound of Lemma 6.6 if $\mathrm{ord}_n(q) = (n-1)/4$.

Proof.   To prove this theorem, we need information on cyclotomic numbers of order 4. When $n \equiv 5 \pmod 8$ is odd, the relation between the 16 cyclotomic numbers of order 4 is given in Table 1.2. In this case, by Lemma 1.11, there are five possible different cyclotomic numbers in this case; i.e.,

$$A = \frac{n - 7 + 2u}{16},$$

$$B = \frac{n + 1 + 2u - 8v}{16},$$

$$C = \frac{n + 1 - 6u}{16},$$

$$D = \frac{n + 1 + 2u + 8v}{16},$$

$$E = \frac{n - 3 - 2u}{16}.$$

When $n \equiv 1 \pmod 8$, the relation between the 16 cyclotomic numbers is given in Table 1.3. In this case, by Lemma 1.12, there are five possible different cyclotomic numbers in this case; i.e.,

$$A = \frac{n - 11 - 6u}{16},$$

$$B = \frac{n - 3 + 2u + 8v}{16},$$

$$C = \frac{n - 3 + 2u}{16},$$

$$D = \frac{n - 3 + 2u - 8v}{16},$$

$$E = \frac{n + 1 - 2u}{16}.$$

By definition, we have

$$D(\eta)^2 = \left(\sum_{i \in C_0^{(4,n)} \cup C_1^{(4,n)}} \eta^i \right)^2$$

$$= \left(\sum_{\substack{i \in C_0^{(4,n)} \\ j \in C_0^{(4,n)}}} + \sum_{\substack{i \in C_0^{(4,n)} \\ j \in C_1^{(4,n)}}} + \sum_{\substack{i \in C_1^{(4,n)} \\ j \in C_0^{(4,n)}}} + \sum_{\substack{i \in C_1^{(4,n)} \\ j \in C_1^{(4,n)}}} \right) \eta^{i+j}.$$

We now determine the generator polynomial $g(x) = \gcd(D(x), x^n - 1)$ of the code $\mathcal{C}_{\mathrm{GF}(q)}(D)$ and thus the p-rank of D.

We first consider the case that $n \equiv 1 \pmod 8$. In this case $-1 \in C_0^{(4,n)}$ and v must be even. Note that $\frac{n-1}{4} \equiv 0 \pmod p$. It then follows from the relations of the cyclotomic numbers and the cyclotomic numbers above that

$$
D(\eta)^2 = \left(\sum_{\substack{i \in C_0^{(4,n)} \\ j \in C_0^{(4,n)}}} + \sum_{\substack{i \in C_0^{(4,n)} \\ j \in C_1^{(4,n)}}} + \sum_{\substack{i \in C_1^{(4,n)} \\ j \in C_0^{(4,n)}}} + \sum_{\substack{i \in C_1^{(4,n)} \\ j \in C_1^{(4,n)}}} \right) \eta^{i-j}
$$

$$
= ((0,0)_4 + (1,0)_4 + (0,1)_4 + (1,1)_4)
$$

$$
\times \sum_{i \in C_0^{(4,n)}} \eta^i + ((3,3)_4 + (0,3)_4 + (3,0)_4 + (0,0)_4)
$$

$$
\times \sum_{i \in C_1^{(4,n)}} \eta^i + ((2,2)_4 + (3,2)_4 + (2,3)_4 + (3,3)_4)
$$

$$
\times \sum_{i \in C_2^{(4,n)}} \eta^i + ((1,1)_4 + (2,1)_4 + (1,2)_4 + (2,2)_4)
$$

$$
\times \sum_{i \in C_3^{(4,n)}} \eta^i + \frac{n-1}{2}
$$

$$
= (A + 2B + D) \sum_{i \in C_0^{(4,n)}} \eta^i + (A + B + 2D)
$$

$$
\times \sum_{i \in C_1^{(4,n)}} \eta^i + (B + C + 2E) \sum_{i \in C_2^{(4,n)}} \eta^i + (C + D + 2E)
$$

$$
\times \sum_{i \in C_3^{(4,n)}} \eta^i + \frac{n-1}{2}
$$

$$
= -D(\eta) + \frac{n-1}{4} + \frac{v}{2} \left(2 \sum_{i \in C_0^{(2,n)}} \eta^i + 1 \right)
$$

$$
= -D(\eta) + \frac{v}{2} \left(2 \sum_{i \in C_0^{(2,n)}} \eta^i + 1 \right).
$$

Whence,

$$D(\eta)(D(\eta) + 1) = \frac{\upsilon}{2}\left(2\sum_{i \in C_0^{(2,n)}} \eta^i + 1\right). \qquad (7.3)$$

Note that $\frac{n-1}{4} \equiv 0 \pmod{p}$. By (6.14) we have $\sum_{i \in C_0^{(2,n)}} \eta^i \in \{0, -1\}$. It then follows that

$$2\sum_{i \in C_0^{(2,n)}} \eta^i + 1 \in \{1, -1\}. \qquad (7.4)$$

Similarly, one can show that

$$\Gamma(\eta)(\Gamma(\eta) + 1) = -\frac{\upsilon}{2}\left(2\sum_{i \in C_0^{(2,n)}} \eta^i + 1\right). \qquad (7.5)$$

The desired conclusions on the p-rank of D and the generator polynomial $g(x)$ of the code $\mathcal{C}_{\mathrm{GF}(q)}(D)$ for Case 1 then follow from (7.1)–(7.5).

In the first subcase, it is obvious that the minimum weight $d = n$. In the second subcase, the generator polynomial of the code shows that $\mathcal{C}_{\mathrm{GF}(q)}(D)$ is a duadic code. So, we have the square-root bound on the minimum odd-like weight.

We now prove the conclusions for Case 2. Since $n \equiv 5 \pmod{8}$, $-1 \in C_2^{(4,n)}$. In this case υ must be odd. Note that $\frac{n-1}{4} \equiv 0 \pmod{p}$. It then follows from the relations of the cyclotomic numbers and the cyclotomic numbers above that

$$D(\eta)^2 = \left(\sum_{\substack{i \in C_0^{(4,n)} \\ j \in C_2^{(4,n)}}} + \sum_{\substack{i \in C_0^{(4,n)} \\ j \in C_3^{(4,n)}}} + \sum_{\substack{i \in C_1^{(4,n)} \\ j \in C_2^{(4,n)}}} + \sum_{\substack{i \in C_1^{(4,n)} \\ j \in C_3^{(4,n)}}}\right)\eta^{i-j}$$

$$= ((2, 0)_4 + (3, 0)_4 + (2, 1)_4 + (3, 1)_4)$$

$$\times \sum_{i \in C_0^{(4,n)}} \eta^i + (1, 3)_4 + (2, 3)_4 + (1, 0)_4 + (2, 0)_4)$$

$$\times \sum_{i \in C_1^{(4,n)}} \eta^i + ((0, 2)_4 + (1, 2)_4 + (0, 3)_4 + (1, 3)_4)$$

$$\times \sum_{i \in C_2^{(4,n)}} \eta^i + ((3,1)_4 + (0,1)_4 + (3,2)_4 + (0,2)_4) \sum_{i \in C_3^{(4,n)}} \eta^i$$

$$= -D(\eta) + \frac{v\left(2\sum_{i \in C_0^{(2,n)}} \eta^i + 1\right) - 1}{2}.$$

Whence,

$$D(\eta)(D(\eta) + 1) = \frac{v\left(2\sum_{i \in C_0^{(2,n)}} \eta^i + 1\right) - 1}{2}. \tag{7.6}$$

Similarly, one can show that

$$\Gamma(\eta)(\Gamma(\eta) + 1) = -\frac{v\left(2\sum_{i \in C_0^{(2,n)}} \eta^i + 1\right) + 1}{2}. \tag{7.7}$$

Since $n \equiv 5 \pmod 8$ and p divides $(n-1)/4$, p must be odd. Note that

$$\frac{n-1}{4} = \frac{u^2+3}{4} + (|v| - 1)(|v| + 1).$$

Hence, $\frac{u^2+3}{4} \equiv 0 \pmod p$ if and only if $(|v|-1)(|v|+1) \equiv 0 \pmod p$. However, $(|v|-1)(|v|+1) \equiv 0 \pmod p$ if and only if p divides one and only one of $|v| - 1$ and $|v| + 1$.

The desired conclusions on the p-rank of D and the generator polynomial $g(x)$ of the code $\mathcal{C}_{\mathrm{GF}(q)}(D)$ for Case 2 then follow from (7.1), (7.2), (7.6), (7.4), and (7.7).

In the first subcase, it is obvious that the minimum weight $d = n$. In the second subcase, the generator polynomial of the code shows that the minimum weight d has the lower bound of Lemma 6.6 if $\mathrm{ord}_n(q) = (n-1)/4$. $\qquad \square$

Example 7.1. Let $(q, n) = (2, 73)$. Then $q \in C_0^{(4,n)}$ and $n = u^2 + 4v^2 = (-3)^2 + 4 \times 4^2$. Hence $v/2 \bmod p = 0$. Let γ be a generator of $\mathrm{GF}(q^n)^*$ with $\gamma^{73} + \gamma^4 + \gamma^3 + \gamma^2 + 1 = 0$. Then $\mathcal{C}_{\mathrm{GF}(2)}(D)$ is a $[73, 36, 12]$ cyclic code with generator polynomial

$$x^{37} + x^{36} + x^{34} + x^{32} + x^{30} + x^{27} + x^{25} + x^{24} + x^{23} + x^{21} + x^{19}$$

$$+ x^{18} + x^{16} + x^{14} + x^{13} + x^{12} + x^{10} + x^7 + x^5 + x^3 + x + 1.$$

The best binary cyclic code of length 73 and dimension 36 has minimum weight 16 by Table A.38.

Example 7.2. Let $(q, n) = (2, 89)$. Then $q \in C_0^{(4,n)}$ and $n = u^2 + 4v^2 = 5^2 + 4 \times 4^2$. Hence $v/2 \bmod p = 0$. Then $\mathcal{C}_{\mathrm{GF}(2)}(D)$ is a $[89, 44, 16]$ cyclic code with generator polynomial

$$x^{45} + x^{44} + x^{40} + x^{39} + x^{36} + x^{34} + x^{33} + x^{29} + x^{23}$$
$$+ x^{22} + x^{16} + x^{12} + x^{11} + x^9 + x^6 + x^5 + x + 1.$$

The best binary cyclic code of length 89 and dimension 44 is 18 by Table A.48.

Example 7.3. Let $(q, n) = (4, 17)$. Then $q \in C_0^{(4,n)}$ and $n = u^2 + 4v^2 = 1^2 + 4 \times 2^2$. Hence $v/2 \bmod p = 1$. Then $\mathcal{C}_{\mathrm{GF}(4)}(D)$ is a $[17, 16, 2]$ cyclic code with generator polynomial $x - 1$.

Example 7.4. Let $(q, n) = (3, 13)$. Then $q \in C_0^{(4,n)}$ and $n = u^2 + 4v^2 = (-3)^2 + 4 \times 1^2$. Hence $(u^2 + 3)/4 \bmod p = 0$. Let γ be a generator of $\mathrm{GF}(q^n)^*$ with $\gamma^{13} + \gamma^4 + 2\gamma + 1 = 0$. Then $\mathcal{C}_{\mathrm{GF}(3)}(D)$ is an optimal $[13, 9, 3]$ cyclic code with generator polynomial $x^4 + 2x^2 + 2x + 1$. In this example, D is an almost difference set.

Example 7.5. Let $(q, n) = (7, 29)$. Then $q \in C_0^{(4,n)}$ and $n = u^2 + 4v^2 = 5^2 + 4 \times 1^2$. Hence $(u^2 + 3)/4 \bmod p = 0$. Let γ be a generator of $\mathrm{GF}(q^n)^*$ with $\gamma^{29} + 6\gamma + 4 = 0$. Then $\mathcal{C}_{\mathrm{GF}(7)}(D)$ is a $[29, 21, 6]$ cyclic code with generator polynomial

$$x^8 + 5x^7 + 2x^6 + x^5 + 5x^4 + 2x^3 + 3x^2 + x + 1.$$

The best linear code known over $\mathrm{GF}(7)$ with length 29 and dimension 21 has minimum distance 6.

Remark 7.1. Note that the set $C_0^{(4,n)} \cup C_1^{(4,n)}$ is an $(n, (n-1)/2, (n-5)/4, (n-1)/2)$ almost difference set in $(\mathrm{GF}(n), +)$ when $v = \pm 1$. Examples 7.4 and 7.5 demonstrate that the cyclic code $\mathcal{C}_{\mathrm{GF}(q)}(D)$ of such almost difference sets could have good parameters.

Problem 7.1. *Determine the parameters of the code $\mathcal{C}_{\mathrm{GF}(q)}(D)$ for the case that $\frac{n-1}{4} \not\equiv 0 \pmod{p}$.*

7.3 Linear Codes of Planar Almost Difference Sets

We first investigate the cyclic codes of the planar almost difference sets described in Theorem 5.20. To this end, we prove the following more general result.

Theorem 7.2. *Let $q = p^s$ for some positive integer s and let $m \geq 2$ be an integer. We define $\ddot{n} = q^m - 1$ and*

$$\ddot{D} = \{0 \leq i < \ddot{n} : \mathrm{Tr}_{q^m/q}(\alpha^i) = 0\} \subset \mathbb{Z}_{\ddot{n}}, \tag{7.8}$$

where α is a generator of $\mathrm{GF}(q^m)^*$. Let $\mathcal{C}_{\mathrm{GF}(q)}(\ddot{D})$ be the cyclic code over $\mathrm{GF}(q)$ with length $\ddot{n}$ and generator polynomial $\ddot{g}(x) = \gcd(x^{\ddot{n}} - 1, S_{\ddot{D}}(x))$. Then $\mathcal{C}_{\mathrm{GF}(q)}(\ddot{D})$ has parameters

$$\left[q^m - 1, \binom{p+m-2}{m-1}^s + 1, q^{m-1} - 1 \right] \tag{7.9}$$

and the parity check polynomial of (6.47).

Proof. We first establish some connections between the cyclic code $\mathcal{C}_{\mathrm{GF}(q)}(D)$ of Theorem 6.24 and the code $\mathcal{C}_{\mathrm{GF}(q)}(\ddot{D})$.

Note that any i with $0 \leq i < \ddot{n}$ can be uniquely written as $i = i_1 n + i_2$, where $i_2 \in D$ and $0 \leq i_1 \leq q - 2$ and $n = (q^m - 1)/(q - 1)$. Since $\alpha^n \in \mathrm{GF}(q)$, we obtain $\mathrm{Tr}_{q^m/q}(\alpha^i) = \alpha^{n i_1} \mathrm{Tr}_{q^m/q}(\alpha^{i_2})$. It then follows that $i \in \ddot{D}$ if and only if $i_2 \in D$. We then deduce that

$$\ddot{D}(x) = (1 + x^n + x^{2n} + \cdots + x^{(q-2)n}) D(x) = \frac{x^{(q-1)n} - 1}{x^n - 1} D(x), \tag{7.10}$$

where D and $D(x)$ are defined in the proof of Theorem 6.24. We have then

$$\begin{aligned}
\ddot{g}(x) &= \gcd(x^{\ddot{n}} - 1, \ddot{D}(x)) \\
&= \gcd\left(x^{n(q-1)} - 1, \frac{x^{(q-1)n} - 1}{x^n - 1} D(x) \right) \\
&= \frac{x^{(q-1)n} - 1}{x^n - 1} \gcd\left(x^n - 1, D(x) \right) \\
&= \frac{x^{\ddot{n}} - 1}{x^n - 1} g(x).
\end{aligned} \tag{7.11}$$

This means that the code $\mathcal{C}_{\mathrm{GF}(q)}(\ddot{D})$ is a repetition of $q - 1$ copies of the code $\mathcal{C}_{\mathrm{GF}(q)}(D)$. We have then

$$d(\mathcal{C}_{\mathrm{GF}(q)}(\ddot{D})) = (q - 1) d(\mathcal{C}_{\mathrm{GF}(q)}(D)),$$

where $d(\mathcal{C}_{\mathrm{GF}(q)}(D))$ denotes the minimum weight of the code $\mathcal{C}_{\mathrm{GF}(q)}(D)$.

From (7.11), we have

$$\frac{x^{\ddot{n}} - 1}{\ddot{g}(x)} = \frac{x^n - 1}{g(x)},$$

which means that the two codes have the same parity check polynomial. Hence the two codes have the same dimension.

The conclusions of this theorem then follows from those of Theorem 6.24. The proof is now completed. $\qquad\qquad\square$

Example 7.6. Let $(q, m) = (3, 3)$ and let α be a generator of GF$(q^m)^*$ with $\alpha^3 + 2\alpha + 1 = 0$. By definition, $\ddot{n} = 26$ and $\ddot{D} = \{0, 1, 3, 9, 13, 14, 16, 22\}$. The ternary cyclic code $\mathcal{C}_{\mathrm{GF}(q)}(\ddot{D})$ has parameters $[26, 7, 8]$ and generator polynomial $\ddot{g}(x) = x^{19} + 2x^{18} + 2x^{17} + 2x^{16} + x^{15} + 2x^{14} + x^{13} + x^6 + 2x^5 + 2x^4 + 2x^3 + x^2 + 2x + 1$. The parity check polynomial of this code is $\ddot{h}(x) = x^7 + x^6 + 2x^5 + x^4 + 2x + 2$. The weight enumerator of $\mathcal{C}_{\mathrm{GF}(q)}(\ddot{D})$ is

$$1 + 26z^8 + 156z^{12} + 624z^{14} + 494z^{18} + 780z^{20} + 78z^{24} + 28z^{26}.$$

When $m = 2$, $\mathcal{C}_{\mathrm{GF}(q)}(\ddot{D})$ has parameters $[q^2 - 1, q + 1, q - 1]$ and is the code of the planar almost difference set of Theorem 5.20.

Chapter 8

Codebooks from (Almost) Difference Sets

8.1 A Generic Construction of Complex Codebooks

An (n, K) *codebook* $\mathcal{C}$ is a set $\{\mathbf{c}_0, \ldots, \mathbf{c}_{n-1}\}$ of n unit norm $1 \times K$ complex vectors $\mathbf{c}_i$, which are called *codewords* of the codebook. The *alphabet* of the codebook is the set of all different complex values that the coordinates of all the codewords take. The *alphabet size* is the number of elements in the alphabet.

As basic measures of performance of a codebook $\mathcal{C}$ in CDMA communication systems, the *root-mean-square (RMS) correlation* and the *maximum correlation amplitudes* of $\mathcal{C}$ are introduced and defined as

$$I_{\text{rms}}(\mathcal{C}) := \sqrt{\frac{1}{n(n-1)} \sum_{\substack{0 \le i,j \le n-1 \\ i \ne j}} |\mathbf{c}_i \mathbf{c}_j^{\mathbf{H}}|^2},$$

$$I_{\text{max}}(\mathcal{C}) := \max_{0 \le i < j \le n-1} |\mathbf{c}_i \mathbf{c}_j^{\mathbf{H}}|,$$

here and hereafter $\mathbf{c}^H$ stands for the conjugate transpose of the complex vector $\mathbf{c}$. The following Welch bounds are well known [Welch (1974)].

Lemma 8.1. *For any codebook $\mathcal{C}$ with $n \ge K$,*

$$I_{\text{rms}}(\mathcal{C}) \ge \sqrt{\frac{n-K}{(n-1)K}}, \tag{8.1}$$

with equality if and only if $\sum_{i=0}^{n} \mathbf{c}_i^{\mathbf{H}} \mathbf{c}_i = (n/K)\mathbf{I}_K$, where $\mathbf{I}_K$ denotes the $K \times K$ identity matrix. We have also

$$I_{\text{max}}(\mathcal{C}) \ge \sqrt{\frac{n-K}{(n-1)K}}, \tag{8.2}$$

with equality if and only if for all pairs (i, j) with $i \ne j$

$$|\mathbf{c}_i \mathbf{c}_j^{\mathbf{H}}| = \sqrt{\frac{n-K}{(n-1)K}}. \tag{8.3}$$

If the equality holds in (8.1), the codebook $\mathcal{C}_D$ is referred to as a *Welch-bound-equality* (*WBE*) *codebook*. A codebook meeting the bound of (8.2) is called a *maximum-Welch-bound-equality* (*MWBE*) *codebook*. An MWBE codebook must be a WBE codebook, but a WBE codebook may not be an MWBE codebook. MWBE codebooks form a proper subset of WBE codebooks.

A well rounded treatment of MWBE and WBE codebooks is given in Sarwate (1999). They have applications also in quantum information processing [Wootters and Fields (2005)], packing [Conway, Harding and Sloane (1996)], and coding theory [Calderbank, Cameron, Kantor and Seidel (1997); Delsarte, Goethals and Seidel (1977)].

It is trivially easy to construct WBE codebooks. Every linear error correcting code whose dual code with Hamming distance at least 3 yields a WBE codebook [Massey and Mittelholzer (1993); Sarwate (1999)]. There are other constructions of WBE codebooks. The reader is referred to Sarwate (1999) for a survey of constructions of WBE codebooks.

However, it is very hard to construct MWBE codebooks, as pointed out in Sarwate (1999)[p. 100]. The following are some known classes of MWBE codebooks.

(1) (n, n) orthogonal MWBE codebooks for any $n > 1$ [Sarwate (1999)].

(2) $(n, n - 1)$ MWBE codebooks for any $n > 1$ obtained from the FFT matrix, and some $(n, n - 1)$ MWBE codebooks from the m-sequence codes [Sarwate (1999)].

(3) (n, K) MWBE codebooks based on conference matrices [Conway, Harding and Sloane (1996); Strohmer and Heath (2003)], when $n = 2K = 2^{d+1}$ and d is a positive integer, and $n = 2K = p^d + 1$ with p a prime number and d a positive integer.

In this chapter, we will consider the following general construction of complex codebooks.

Let $(A, +)$ be an abelian group of order n and exponent e_A, and let $\{\chi_0, \chi_1, \ldots, \chi_{n-1}\}$ be the set of all characters of A. Given any K-subset $D := \{d_1, d_2, \ldots, d_K\}$ of A, we define a codebook

$$\mathcal{C}_D := \{\mathbf{c}_i : i = 0, \ldots, n - 1\}, \tag{8.4}$$

where for each i

$$\mathbf{c}_i := \frac{1}{\sqrt{K}}(\chi_i(d_1), \chi_i(d_2), \ldots, \chi_i(d_K)). \tag{8.5}$$

Whether this construction yields good codebooks or not depends on the choice of the abelian group and the subset D of A. In the subsequent sections, we will study the codebooks C_D when D is a difference set or almost difference set.

8.2 Optimal Codebooks from Difference Sets

By definition, MWBE codebooks are optimal in the sense that they meet the Welch bound. The following theorem shows that every difference set gives an optimal codebook.

Theorem 8.1. *The set C_D of (8.4) is an (n, K) MWBE codebook with alphabet size e_A if and only if the set D is an (n, K, λ) difference set in $(A, +)$, where $K > 1$.*

Proof. For any pair (i, j) with $i \neq j$ we have $\chi_i \overline{\chi_j} \neq \chi_0$ because χ_i and χ_j are distinct characters. By definition, we have

$$\left| c_i c_j^H \right| = \frac{1}{K} \left| \sum_{k=1}^{K} \chi_i(d_k) \overline{\chi_j}(d_k) \right| = \frac{1}{K} \left| \sum_{k=1}^{K} (\chi_i \overline{\chi_j})(d_k) \right| = \frac{1}{K} |(\chi_i \overline{\chi_j})(D)|.$$

By Lemma 8.1, C_D is an (n, K) MWBE codebook if and only if for each pair (i, j) with $i \neq j$, $\left| c_i c_j^H \right| = \sqrt{\frac{n-K}{(n-1)K}}$. It then follows from Theorem 4.6 that C_D is an (n, K) MWBE codebook if and only if D is an (n, K, λ) difference set in $(A, +)$. $\square$

Theorem 8.1 comes from Ding and Feng (2007), and is an generalization of the constructions in Ding (2006) and Xia, Zhou and Giannakis (2005). The idea of constructing codebooks with difference sets dates back at least to König (1979).

There is a large number of difference sets. All the difference sets covered in Chapter 4 and Section 6.2 can be plugged into Theorem 8.1 to obtain MWBE complex codebooks.

It is known that Welch's bound on the root-mean-square correlation is not tight for binary codebooks in several cases. Better bounds on binary codebooks are developed in Karystinos and Pados (2003). Binary codebooks meeting the Karystinos–Pados bounds are treated in Karystinos and Pados (2003); Ding, Golin and Kløve (2003); Ipatov (2004).

As pointed out in Love, Heath and Strohmer (2003), constructing optimal codebooks with minimal I_{max} is very difficult in general. This problem is equivalent to line packing in Grassmannian space [Conway, Harding and Sloane (1996)]. In frame theory, such a codebook with I_{max} minimized is referred to as a Grassmannian frame [Strohmer and Heath (2003)]. The optimal codebooks presented in this section and the constructions should have applications in these areas.

8.3 Codebooks from Some Almost Difference Sets

In this section, we consider the codebook $\mathcal{C}_D$ from some almost difference sets D. Since the difference function of almost difference sets takes three values rather than two values, we may have to deal with the codebook of almost difference sets case by case. The idea of using almost difference sets dates back to Ding (2006).

8.3.1 *Codebooks from the two-prime almost difference sets*

In this subsection, we need group characters of $(\mathbb{Z}_n, +)$. Define

$$\psi_n(j) := e^{2j\pi\sqrt{-1}/n}. \tag{8.6}$$

Then all group chracters of $(\mathbb{Z}_n, +)$ are given by

$$\psi_n^{(h)}(j) := \psi_n(hj) = e^{2hj\pi\sqrt{-1}/n}, \quad 0 \le h \le n-1.$$

We first recall the two-prime almost difference sets introduced in Section 8.3.1. Let n_1 and n_2 be two distinct primes, and let $n = n_1 n_2$. Put $N = \gcd(n_1, n_2)$. Recall the generalized cyclotomic classes $W_{2i}^{(N)}$ of order N introduced in Section 1.5. Define

$$D_0^{(2)} = \bigcup_{i=0}^{(N-2)/2} W_{2i}^{(N)} \quad \text{and} \quad D_1^{(2)} = \bigcup_{i=0}^{(N-2)/2} W_{2i+1}^{(N)}.$$

Clearly $D_0^{(2)}$ is a subgroup of $\mathbb{Z}_n^*$ and $D_1^{(2)} = \varrho D_0^{(2)}$, where ϱ was defined in Section 1.5.

Let

$$D = D_1^{(2)} \cup \{n_1, 2n_1, \ldots, (n_2-1)n_1\}.$$

Theorem 5.13 says that D is an almost difference set in $(\mathbb{Z}_{n_1(n_1+4)}, +)$ with parameters

$$\left(n_1(n_1+4), \frac{(n_1+3)(n_1+1)}{2}, \frac{(n_1+3)(n_1+1)}{4}, \frac{(n_1-1)(n_1+5)}{2} \right)$$

when $n_2 - n_1 = 4$ and both n_1 and n_2 are primes.

In this subsection, we consider the codebook $\mathcal{C}_D$ from this almost difference set.

We define

$$N_1 = \{n_1, 2n_1, \ldots, (n_2-1)n_1\}, \quad N_2 = \{n_2, 2n_2, \ldots, (n_1-1)n_2\}.$$

It is straightforward to prove the following lemma.

Lemma 8.2. *Let symbols and notation be as before. We have then*

$$\psi_n^{(m)}(N_1) = \begin{cases} n_2 - 1 & \text{if } \gcd(m, n_2) = n_2, \\ -1 & \text{otherwise,} \end{cases}$$

$$\psi_n^{(m)}(N_2) = \begin{cases} n_1 - 1 & \text{if } \gcd(m, n_1) = n_1, \\ -1 & \text{otherwise.} \end{cases}$$

Lemma 8.3. *Let symbols and notation be as before. We have then*

$$\psi_n^{(m)}(D_1^{(2)}) = \begin{cases} -(n_2 - 1)/2 & \text{if } \gcd(m, n) = n_2, \\ -(n_1 - 1)/2 & \text{if } \gcd(m, n) = n_1. \end{cases}$$

Proof. We need to recall the definition of the generalized cyclotomic classes in Section 1.5. Since ς is a common primitive root of both n_1 and n_2 and the order of ς modulo n is $f = (n_1 - 1)(n_2 - 1)/2$, by the definition of ϱ we have

$$\begin{aligned} D_1^{(2)} \bmod n_1 &= \{\varsigma^s \varrho \bmod n_1 : s = 0, 1, \ldots, f - 1\} \\ &= \{\varsigma^{s+1} \bmod n_1 : s = 0, 1, \ldots, f - 1\} \\ &= \{1, 2, \ldots, n_1 - 1\}. \end{aligned}$$

When s ranges over $\{0, 1, \ldots, f - 1\}$, $\varsigma^s x \bmod n_1$ takes on each element of $\{1, 2, \ldots, n_1 - 1\}$ exactly $(n_2 - 1)/2$ times.

If $\gcd(m, n) = n_2$, then n_2 divides m and $m \neq 0$, but $\gcd(m, n_1) = 1$. Hence $\psi_n^{(m)}$ is a group character of $\mathbb{Z}_n$ with order n_1. It then follows that

$$\psi_n^{(m)}(D_1^{(2)}) = \frac{n_2 - 1}{2} \sum_{j \in N_2} \psi_{n_1}^{(m/n_2)}(j) = -\frac{n_2 - 1}{2}.$$

The second part follows by symmetry. $\qquad\square$

We shall need the following lemma in the sequel.

Lemma 8.4. *Recall the difference function* $\text{diff}_S(x)$ *of a subset S of $\mathbb{Z}_n$ defined before. We have then*

$$\text{diff}_{D_1^{(2)}}(w) = \begin{cases} \dfrac{n_1^2 - 1}{4} & \text{if } w \in N_1, \\[2mm] \dfrac{n_1^2 - 9}{4} & \text{if } w \in N_2. \end{cases}$$

Proof. It follows from Lemma 5.32 in Baumert (1971)[p. 138]. $\qquad\square$

Lemma 8.5. *If* $\gcd(h, n) = 1$*, then*

$$\psi_n^{(h)}(D_1^{(2)}) = \frac{1 \pm \sqrt{n_1(n_1 + 4)}}{2}.$$

Proof. Since n_1 and $n_2 = n_1 + 4$ both are primes, by Lemma 1.13, $-1 \in D_0^{(2)}$. It follows from $\gcd(h, n) = 1$ that $\psi_n^{(h)}(D_1^{(2)}) = 1 - \psi_n^{(h)}(D_0^{(2)})$. Define $\delta := \psi_n^{(h)}(D_1^{(2)})$. By the definition of generalized cyclotomic numbers, Lemmas 8.2, 8.4 and 6.10, we have

$$
\begin{aligned}
\delta^2 &= \left(\sum_{\ell \in D_1^{(2)}} \psi_n^{(h)}(\ell) \right) \left(\sum_{m \in D_1^{(2)}} \psi_n^{(h)}(-m) \right) \\
&= \sum_{\ell \in D_1^{(2)}} \sum_{m \in D_1^{(2)}} \psi_n^{(h)}(\ell - m) \\
&= \left| D_1^{(2)} \right| + (1, 1)_2 \psi_n^{(h)}(D_0^{(2)}) + (0, 0)_2 \psi_n^{(h)}(D_1^{(2)}) \\
&\quad + \frac{n_1^2 - 1}{4} \psi_n^{(h)}(N_1) + \frac{n_1^2 - 9}{4} \psi_n^{(h)}(N_2) \\
&= n_1 + 1 + \frac{n_1^2 - 5}{4} \psi_n^{(h)}(D_0^{(2)}) + \frac{n_1^2 - 1}{4} \psi_n^{(h)}(D_1^{(2)}) \\
&= \frac{n_1^2 + 4n_1 - 1}{4} + \psi_n^{(h)}(D_1^{(2)}) \\
&= \frac{n_1^2 + 4n_1 - 1}{4} + \delta.
\end{aligned}
$$

Solving the quadratic equation yields the two possible values of δ. □

Combining Lemmas 8.2, 8.3 and 8.5, we obtain the following.

Lemma 8.6. *Let symbols and notation be the same as before. Then*

$$
\psi_n^{(h)}(N_2 \cup D_1^{(2)}) = \begin{cases} -\dfrac{n_1 + 1}{2} & \text{if } \gcd(h, n) = n_1, \\[2mm] \dfrac{n_1 + 3}{2} & \text{if } \gcd(h, n) = n_2, \\[2mm] \dfrac{-1 \pm \sqrt{n_1(n_1 + 4)}}{2} & \text{if } \gcd(h, n) = 1. \end{cases}
$$

The following lemma follows from the autocorrelation values of the generalized cyclotomic sequence defined in Ding (1998)[Theorem 2].

Lemma 8.7. *Let symbols and notation be the same as before. Then*

$$\text{diff}_D(w) = \begin{cases} \dfrac{n_1{}^2 + 4n_1 + 7}{4} & \text{if } w \in N_1 \cup D_0^{(2)}, \\[2mm] \dfrac{n_1{}^2 + 4n_1 + 3}{4} & \text{if } w \in N_2 \cup D_1^{(2)}. \end{cases}$$

With the preparations above, we are now ready to state and prove the main result of this subsection below.

Theorem 8.2 ([Ding and Feng (2008)]). *Let n_1 and $n_2 = n_1 + 4$ both be prime. For the $(n_1(n_1 + 4), (n_1 + 3)(n_1 + 1)/2)$ codebook $\mathcal{C}_D$ of the two-prime almost difference set D, we have*

$$I_{\max}(\mathcal{C}_D) = \frac{1}{n_1 + 1}.$$

Proof. Let $0 \le j < i \le n - 1$. Note that $K = (n_1 + 1)(n_1 + 3)/2$ and

$$\mathbf{w}_i \mathbf{w}_j^{\mathbf{H}} = \frac{1}{K} \sum_{\ell=1}^{K} \psi_n((i - j)d_\ell),$$

where the d_ℓ's are all the elements of D. It follows from Lemmas 8.6 and 8.7 that

$$\begin{aligned} \left(\mathbf{w}_i \mathbf{w}_j^{\mathbf{H}}\right)^2 &= \frac{1}{K^2} \sum_{\ell=1}^{K} \sum_{m=1}^{K} \psi_n[(i - j)(d_\ell - d_m)] \\ &= \frac{1}{K^2} \left(K + \sum_{\ell=1}^{K} \sum_{\substack{m=1 \\ m \ne \ell}}^{K} \psi_n[(i - j)(d_\ell - d_m)] \right) \\ &= \frac{1}{K^2} \left(K + \sum_{\ell=1}^{K} \sum_{\substack{m=1 \\ m \ne \ell}}^{K} \psi_n^{(i-j)}[d_\ell - d_m] \right) \\ &= \frac{1}{K^2} \left(K - \frac{n_1{}^2 + 4n_1 + 3}{4} + \psi_n^{(i-j)}\left(N_2 \cup D_1^{(2)}\right) \right) \\ &= \begin{cases} \dfrac{1}{(n_1 + 1)^2} & \text{if } \gcd(i - j, n) = n_2, \\[2mm] \dfrac{1}{(n_1 + 3)^2} & \text{if } \gcd(i - j, n) = n_1, \\[2mm] \dfrac{\left(\sqrt{n_1(n_1 + 4)} \pm 1\right)^2}{(n_1 + 1)^2(n_1 + 3)^2} & \text{if } \gcd(i - j, n) = 1. \end{cases} \end{aligned}$$

This completes the proof of this theorem. $\qquad\square$

Note that the Welch bound for $(n_1(n_1 + 4), (n_1 + 3)(n_1 + 1)/2)$ codebook $\mathcal{C}_D$ is

$$\sqrt{\frac{n - K}{(n - 1)K}} = \sqrt{\frac{n_1{}^2 + 4n_1 - 3}{(n_1{}^2 + 4n_1 - 1)(n_1 + 3)(n_1 + 1)}}.$$

Hence $I_{\max}(\mathcal{C}_D)$ nearly meets the Welch bound.

Examples of $(n_1, n_1 + 4)$ that yield such codebooks are

$$(3, 7), (7, 11), (19, 23), (43, 47), (67, 71), (97, 101).$$

8.3.2 *Codebooks from other almost difference sets*

Let $n_1 > 1$ and $n_2 > 1$ be two relatively prime integers. Let ψ_{n_i} be the group character of the abelian group $\mathbb{Z}_{n_i}$ defined in (8.6). For any $(j_1, j_2) \in \mathbb{Z}_{n_1} \times \mathbb{Z}_{n_2}$, we define

$$\Psi((j_1, j_2)) = \psi_{n_1}(j_1)\psi_{n_2}(j_2). \tag{8.7}$$

Then Ψ is a group character of the abelian group $(\mathbb{Z}_{n_1} \times \mathbb{Z}_{n_2}, +)$.

For any subset $D := \{d_1, d_2, \ldots, d_K\}$ of $\mathbb{Z}_{n_1} \times \mathbb{Z}_{n_2}$, where $d_i = (d_{i,1}, d_{i,2})$, recall that the code from D is defined by

$$\mathcal{C}_D := \{\mathbf{u}_i : i = 0, \ldots, n_1 n_2 - 1\}, \tag{8.8}$$

where

$$\mathbf{u}_{(i_1,i_2)} = \frac{1}{\sqrt{K}}(\Psi((i_1 d_{1,1}, i_2 d_{1,2})), \Psi((i_1 d_{2,1}, i_2 d_{2,2})), \ldots, \Psi((i_1 d_{K,1}, i_2 d_{K,2})))$$

for each $(i_1, i_2) \in \mathbb{Z}_{n_1} \times \mathbb{Z}_{n_2}$.

Note that

$$\mathbf{u}_{(i_1,i_2)}\mathbf{u}_{(j_1,j_2)}^{\mathbf{H}} = \sum_{\ell=1}^{K} \Psi((i_1 - j_1)d_{\ell,1}, (i_2 - j_2)d_{\ell,2}).$$

We obtain

$$K^2\left|\mathbf{u}_{(i_1,i_2)}\mathbf{u}_{(j_1,j_2)}^{\mathbf{H}}\right|^2$$

$$= \sum_{m=1}^{K}\sum_{\ell=1}^{K} \Psi((i_1 - j_1)(d_{m,1} - d_{\ell,1}), (i_2 - j_2)(d_{m,2} - d_{\ell,2}))$$

$$= \sum_{m=1}^{K} \sum_{\ell=1}^{K} \Psi^{(i_1-j_1,i_2-j_2)}((d_{m,1}-d_{\ell,1}),(d_{m,2}-d_{\ell,2}))$$

$$= K + \sum_{m=1}^{K} \sum_{\substack{\ell=1 \\ \ell \neq m}}^{K} \Psi^{(i_1-j_1,i_2-j_2)}((d_{m,1}-d_{\ell,1}),(d_{m,2}-d_{\ell,2})), \qquad (8.9)$$

where $\Psi^{(i,j)}((n_1,n_2)) := \Psi((in_1,jn_2))$ is a group character of $\mathbb{Z}_{n_1} \times \mathbb{Z}_{n_2}$.

Below we study the complex codebook $\mathcal{C}_D$ from some almost difference sets D in $\mathbb{Z}_{n_1} \times \mathbb{Z}_{n_2}$, where n_1 and n_2 are positive integers.

8.3.2.1 *The* $n \equiv 2$ (mod 4) *case*

Let $q \equiv 5$ (mod 8) be a prime. It is known that $q = s^2 + 4t^2$ for some s and t with $s \equiv \pm 1$ (mod 4). Set $n = 2q$. Let $i, j, \ell \in \{0, 1, 2, 3\}$ be three pairwise distinct integers, and define

$$D_{(i,j,\ell)} = \left[\{0\} \times \left(D_i^{(4,q)} \cup D_j^{(4,q)}\right)\right] \cup \left[\{1\} \times \left(D_\ell^{(4,q)} \cup D_j^{(4,q)}\right)\right]. \qquad (8.10)$$

Then $D_{(i,j,\ell)}$ is an $(n, \frac{n-2}{2}, \frac{n-6}{4}, \frac{3n-6}{4})$ almost difference set in $\mathbb{Z}_2 \times \mathbb{Z}_q$ if

(1) $t = 1$ and $(i, j, \ell) = (0, 1, 3)$ or $(0, 2, 1)$; or
(2) $s = 1$ and $(i, j, \ell) = (1, 0, 3)$ or $(0, 1, 2)$

Theorem 8.3 ([Ding and Feng (2008)]). *Let* $t = 1$. *For the almost difference set* $D_{(0,1,3)}$ *of* (8.10), $\mathcal{C}_{D_{(0,1,3)}}$ *is a* $(2q, q-1)$ *codebook with*

$$I_{\max}(\mathcal{C}_{D_{(0,1,3)}}) = \frac{\sqrt{2q+2\sqrt{q}}}{2(q-1)}.$$

Proof. It was proved in Ding (2006) that

$$\psi_q\left(D_0^{(2,q)} \cup \{0\}\right) = \psi_q\left(D_0^{(4,q)} \cup D_2^{(4,q)} \cup \{0\}\right) = (1 \pm \sqrt{q})/2.$$

Recall the definition of the difference function $\text{diff}_D(x)$ of D. It was proved in Ding, Helleseth and Martinsen (2001) that

$$\text{diff}_{D_{(0,1,3)}}(w) = \begin{cases} \dfrac{q-3}{2}, & w \in \left[\{0\} \times \mathbb{Z}_q^*\right] \cup \left[\{1\} \times \left(D_1^{(4,q)} \cup D_3^{(4,q)}\right)\right], \\[2mm] \dfrac{q-1}{2}, & w \in \{1\} \times \left(D_0^{(4,q)} \cup D_2^{(4,q)} \cup \{0\}\right). \end{cases}$$

For any distinct pairs (i_1, i_2) and (j_1, j_2), it follows from (8.9) that

$$
K^2 \left| \mathbf{u}_{(i_1,i_2)} \mathbf{u}_{(j_1,j_2)}^{\mathbf{H}} \right|^2
$$

$$
= K - \frac{q-3}{2} + \Psi^{(i_1-j_1, i_2-j_2)} \left(\{1\} \times \left(D_0^{(4,q)} \cup D_2^{(4,q)} \cup \{0\} \right) \right)
$$

$$
= \begin{cases} K - \dfrac{q-3}{2} - \dfrac{q+1}{2}, & i_2 = j_2, \\[2mm] \psi_q \left(D_0^{(4,q)} \cup D_2^{(4,q)} \cup \{0\} \right), & i_2 \neq j_2, \end{cases}
$$

$$
= \begin{cases} 0, & i_2 = j_2, \\[2mm] \dfrac{q \pm \sqrt{q}}{2}, & i_2 \neq j_2. \end{cases}
$$

This completes the proof. $\qquad\square$

Remark: Note that the Welch bound for $(2p, p-1)$ codebooks is

$$
\sqrt{\frac{n-K}{(n-1)K}} = \frac{\sqrt{2q + 2\frac{2q-1}{q-1}}}{2q-1}.
$$

Hence $I_{\max}(\mathcal{C}_{D_{0,1,3}})$ nearly meets the Welch bound.

Theorem 8.4 ([Ding and Feng (2008)]). *Let $s = 1$. For the almost difference set $D_{(0,1,2)}$ of (8.10), $\mathcal{C}_{D_{(0,1,2)}}$ is a $(2q, q-1)$ codebook with*

$$
I_{\max}(\mathcal{C}_{D_{(0,1,2)}}) = \frac{\sqrt{2q + 2\sqrt{q}}}{2(q-1)}.
$$

Proof. It was proved in Ding (2006) that

$$
\psi_q \left(D_1^{(2,q)} \cup \{0\} \right) = \psi_q \left(D_1^{(4,q)} \cup D_3^{(4,q)} \cup \{0\} \right) = (1 \pm \sqrt{q})/2.
$$

Recall the difference function diff_D of D. It was proved in Ding, Helleseth and Martinsen (2001) that

$$
\mathrm{diff}_{D_{(0,1,2)}}(w) = \begin{cases} \dfrac{q-3}{2}, & w \in \left[\{0\} \times \mathbb{Z}_q^* \right] \cup \left[\{1\} \times \left(D_0^{(4,q)} \cup D_2^{(4,q)} \right) \right], \\[3mm] \dfrac{q-1}{2}, & w \in \{1\} \times \left(D_1^{(4,q)} \cup D_3^{(4,q)} \cup \{0\} \right). \end{cases}
$$

For any distinct pair (i_1, i_2) and (j_1, j_2), it follows from (8.9) that

$$
K^2 \left| \mathbf{u}_{(i_1, i_2)} \mathbf{u}^{\mathbf{H}}_{(j_1, j_2)} \right|^2
$$

$$
= K - \frac{q-3}{2} + \Psi^{(i_1 - j_1, i_2 - j_2)} \left(\{1\} \times \left(D_1^{(4,q)} \cup D_3^{(4,q)} \cup \{0\} \right) \right)
$$

$$
= \begin{cases}
K - \dfrac{q-3}{2} - \dfrac{q+1}{2}, & i_2 = j_2, \\[2ex]
\psi_q \left(D_1^{(4,q)} \cup D_3^{(4,q)} \cup \{0\} \right), & i_2 \neq j_2,
\end{cases}
$$

$$
= \begin{cases}
0, & i_2 = j_2, \\[2ex]
\dfrac{q \pm \sqrt{q}}{2}, & i_2 \neq j_2.
\end{cases}
$$

This completes the proof. $\qquad\qquad\qquad\qquad\qquad\qquad\qquad\qquad\qquad$ □

Let $q \equiv 5 \pmod 8$ be a prime. It is known that $q = s^2 + 4t^2$ for some s and t with $s \equiv \pm 1 \pmod 4$. Set $n = 2q$. Let $i, j, \ell \in \{0, 1, 2, 3\}$ be three pairwise distinct integers, and define

$$
D_{(i,j,\ell)} = \left[\{0\} \times \left(D_i^{(4,q)} \cup D_j^{(4,q)} \right) \right] \cup \left[\{1\} \times \left(D_\ell^{(4,q)} \cup D_j^{(4,q)} \right) \right] \cup \{0, 0\}.
$$

Then $D_{(i,j,\ell)}$ is an $(n, \frac{n}{2}, \frac{n-2}{4}, \frac{3n-2}{4})$ almost difference set of $\mathbb{Z}_2 \times \mathbb{Z}_q$ if

(1) $t = 1$ and $(i, j, \ell) \in \{(0, 1, 3), (0, 2, 3), (1, 2, 0), (1, 3, 0)\}$; or
(2) $s = 1$ and $(i, j, \ell) \in \{(0, 1, 2), (0, 3, 2), (1, 0, 3), (1, 2, 3)\}$.

The almost difference sets $D_{(i,j,\ell)}$ above yield $(2q, q)$ codebooks $\mathcal{C}_{D_{(i,j,\ell)}}$ that nearly meet the Welch bound. The parameter $I_{\max}(\mathcal{C}_{D_{(i,j,\ell)}})$ can be computed similarly.

8.3.2.2 *The $n \equiv 0 \pmod 4$ case*

Recall that difference sets with parameters $(\ell, (\ell - 1)/2, (\ell - 3)/4)$ are called *Paley–Hadamard difference sets*. Their complements are also difference sets with parameters $(\ell, (\ell + 1)/2, (\ell + 1)/4)$.

Known cyclic Paley–Hadamard difference sets are of the following types:

(a) The Paley difference sets with parameters $(p, (p - 1)/2, (p - 3)/4)$, where $p \equiv 3 \pmod 4$ is prime.
(b) Difference sets with Singer parameters $(2^t - 1, 2^{t-1} - 1, 2^{t-2} - 1)$.
(c) Twin-prime difference sets with parameters $(\ell, (\ell - 1)/2, (\ell - 3)/4)$, where $\ell = p(p + 2)$ and both p and $p + 2$ are primes.

(d) Hall difference sets with parameters $(p, (p-1)/2, (p-3)/4)$, where p is a prime of the form $p = 4s^2 + 27$.

Let $n = 4\ell$, where ℓ is an odd positive integer. Then the ring $\mathbb{Z}_n$ is isomorphic to $\mathbb{Z}_4 \times \mathbb{Z}_\ell$ under the morphism $\phi(x) = (x \bmod 4, x \bmod \ell)$. Cyclic $(\ell, (\ell-1)/2, (\ell-3)/4)$ difference sets in $\mathbb{Z}_\ell$ were used to construct almost difference sets in $\mathbb{Z}_4 \times \mathbb{Z}_\ell$, which are stated as follows.

Let E be an $(\ell, (\ell-1)/2, (\ell-3)/4)$ difference set in $\mathbb{Z}_\ell$. Then

$$D = (\{0\} \times E) \cup (\{1, 2, 3\} \times E^c) \tag{8.11}$$

is a $(4\ell, 2\ell+1, \ell, \ell-1)$ almost difference set in $\mathbb{Z}_4 \times \mathbb{Z}_\ell$, where E^c is the complement of E.

We now consider the codebook of (8.8) defined by the almost difference sets of (8.11).

Theorem 8.5 ([Ding and Feng (2008)]). *For the* $(4\ell, 2\ell+1)$ *codebooks* $\mathcal{C}_D$ *of* (8.8) *defined by the almost difference sets D of* (8.11), *we have*

$$I_{\max}(\mathcal{C}_D) = \frac{\sqrt{\ell+1}}{2\ell+1}.$$

Proof. To prove this theorem, we need the notion of multiset. If T is a set, $2 \cdot T$ denotes the multiset which contains two copies of each element of T. If T_1 and T_2 are subsets of an abelian group $(A, +)$, we define $T_1 \ominus T_2$ to be the multiset $\{t_1 - t_2 : t_1 \in T_1, t_2 \in T_2\}$.

Since the set E employed to define the almost difference set is an $(\ell, (\ell-1)/2, (\ell-3)/4$ difference set, and its complement D^c is an $(\ell, (\ell+1)/2, (\ell+1)/4)$ difference set, we have

$$\begin{cases} E \ominus E = \dfrac{\ell+1}{4} \cdot \{0\} \cup \dfrac{\ell-3}{4} \cdot \mathbb{Z}_\ell, \\[2mm] E^c \ominus E^c = \dfrac{\ell+1}{4} \cdot \{0\} \cup \dfrac{\ell+1}{4} \cdot \mathbb{Z}_\ell, \\[2mm] E \ominus E^c = \dfrac{\ell+1}{4} \cdot \mathbb{Z}_\ell^*, \\[2mm] E^c \ominus E = \dfrac{\ell+1}{4} \cdot \mathbb{Z}_\ell^*. \end{cases} \tag{8.12}$$

It then follows from (8.12) that

$$D \ominus D = [(\ell+1) \cdot (\mathbb{Z}_4 \times \mathbb{Z}_\ell) \cup (\ell+1) \cdot \{(0,0)\}] \setminus \{0\} \times \mathbb{Z}_\ell. \tag{8.13}$$

Let (i_1, i_2) and (j_1, j_2) be two distinct elements of $\mathbb{Z}_4 \times \mathbb{Z}_\ell$. Then the group character $\Psi^{((i_1-j_1, i_2-j_2))}$ on $\mathbb{Z}_4 \times \mathbb{Z}_\ell$ is nontrivial. Then, we have

$$\Psi^{((i_1-j_1, i_2-j_2))}(\mathbb{Z}_4 \times \mathbb{Z}_\ell) = 0, \tag{8.14}$$

and

$$\Psi^{((i_1-j_1, i_2-j_2))}(\{0\} \times \mathbb{Z}_\ell) = \begin{cases} \ell & \text{if } i_2 = j_2, \\ 0 & \text{if } i_2 \neq j_2. \end{cases} \tag{8.15}$$

The conclusion on $I_{max}(\mathcal{C}_D)$ then follows from (8.9), (8.13), (8.14) and (8.15).
$\square$

Remark: Note that the Welch bound for $(4\ell, 2\ell + 1)$ codebooks is

$$\sqrt{\frac{n - K}{(n - 1)K}} = \sqrt{\frac{2\ell - 1}{(4\ell - 1)(2\ell + 1)}}.$$

Hence $I_{max}(\mathcal{C}_D)$ nearly meets the Welch bound. With the different types of cyclic difference sets in (a)–(d) above, Theorem (8.5) yields four classes of codebooks nearly meeting the Welch bound.

Other almost difference sets presented in Chapter 5 also give complex codebooks within the framework of the construction of this chapter. It remains to compute their maximum correlation amplitude $I_{max}(\mathcal{C}_D)$.

Codebooks from almost difference sets are also considered in Yu, Feng and Zhang (2014) and Zhang and Feng (2012).

Appendix A

Tables of Best Binary and Ternary Cyclic Codes

In this appendix, we present tables of best binary cyclic codes and best BCH binary cyclic codes of odd length up to 125, and tables of best ternary cyclic codes and best ternary BCH codes of length n up to 79, where $\gcd(3, n) = 1$. The objective is to provide a repository of best binary and ternary cyclic codes, which can be used to benchmark cyclic codes constructed with any tools and methods. All the computations are done with the software package Magma.

A.1. Basic Notation and Symbols in the Tables

Let q be any prime. We identify the generator polynomial $g(x) = \sum_{i=0}^{t} a_i x^i \in \mathrm{GF}(q)[x]$ of a cyclic code over $\mathrm{GF}(q)$ with the integer $\sum_{i=0}^{t} a_i q^i$, where these $a_i \in \mathrm{GF}(q)$ are viewed as integers. In all the tables of this appendix, we express the generator polynomial of a code over $\mathrm{GF}(q)$ with an positive integer, in order to save space.

Let δ be an integer with $2 \leq \delta \leq n$. Recall that a BCH code $\mathcal{C}$ of length n over $\mathrm{GF}(q)$ and design distance δ is a cyclic code with the defining set

$$T = C_b \cup C_{b+1} \cup \cdots \cup C_{b+\delta-2}, \qquad (\mathrm{A.1})$$

where C_i is the q-cyclotomic coset modulo n containing i. By the BCH bound, this code has minimum distance at least δ.

Note that two different pairs (b, δ) and (b', δ') may define the same cyclic code. Hence, a BCH code may have different design distances. The largest design distance of a BCH code is called the *Bose distance*.

In all the tables of this appendix, the generator polynomial of a cyclic code is given as an integer, δ is the Bose distance of a BCH code, the symbol "BKLC" indicates if the cyclic code has the same parameters as the best linear code known in the tables of best linear codes known maintained by Markus Grassl, and the symbol "Opt" means that the cyclic code is optimal according to the available information in the tables maintained by Markus Grassl. In all the tables, we ignore the trivial cyclic code with parameters $[n, n, 1]$. The total number of cyclic codes of length n over $GF(q)$ is equal to $2^t - 1$, where t is the number of q-cyclotomic cosets modulo n. The BCH ratio is defined to be the total number of BCH codes over the total number of cyclic codes of the same length.

The total number of cyclic codes of length n over $GF(q)$ is sometimes too large. Instead of documenting all of them, we compile a table of best cyclic codes of length n over $GF(q)$, where the code of each possible dimension with the largest minimum distance is recorded. To give the reader more information on cyclic codes of length n over $GF(q)$, we actually record the best BCH and the best non-BCH code of each dimension in the table if both exist.

We do not know any formula of the total number of BCH codes over $GF(q)$ of length n. So the BCH ratio is computed case by case with Magma. In general, the BCH ratio is small. However, in some cases, it could be large.

A.2. Tables of Best Binary Cyclic Codes of Length up to 125

Table A.1 Best cyclic codes of length 7 over GF(2), where $\text{ord}_7(2) = 3$. BCH ratio $= 6/7$.

k	d	δ	BCH	Generator polynomial	BKLC	Opt.
1	7	7	Y	127	Y	Y
3	4		N	23	Y	Y
3	4	4	Y	29	Y	Y
4	3	3	Y	11	Y	Y
6	2	2	Y	3	Y	Y

Table A.2 Best cyclic codes of length 9 over GF(2), where $\text{ord}_9(2) = 6$. BCH ratio $= 6/7$.

k	d	δ	BCH	Generator polynomial	BKLC	Opt.
1	9	9	Y	511	Y	Y
2	6	4	Y	219	Y	Y
3	3	3	Y	73	N	N
6	2		N	9	Y	Y
7	2	2	Y	7	Y	Y
8	2	2	Y	3	Y	Y

Table A.3 Best cyclic codes of length 11 over GF(2), where $\text{ord}_{11}(2) = 10$. BCH ratio = 3/3.

k	d	δ	BCH	Generator polynomial	BKLC	Opt.
1	11	11	Y	2047	Y	Y
10	2	2	Y	3	Y	Y

Table A.4 Best cyclic codes of length 13 over GF(2), where $\text{ord}_{13}(2) = 12$. BCH ratio = 3/3.

k	d	δ	BCH	Generator polynomial	BKLC	Opt.
1	13	13	Y	8191	Y	Y
12	2	2	Y	3	Y	Y

Table A.5 Best cyclic codes of length 15 over GF(2), where $\text{ord}_{15}(2) = 4$. BCH ratio = 19/31.

k	d	δ	BCH	Generator polynomial	BKLC	Opt.
1	15	15	Y	32767	Y	Y
2	10		N	14043	Y	Y
3	5	5	Y	4681	N	N
4	8		N	2479	Y	Y
4	8	8	Y	3929	Y	Y
5	3		N	1057	N	N
5	7	7	Y	1335	Y	Y
6	6		N	825	Y	Y
6	6	6	Y	627	Y	Y
7	5	5	Y	465	Y	Y
8	4		N	231	Y	Y
9	4	3	Y	93	Y	Y
10	4		N	43	Y	Y
10	4	4	Y	53	Y	Y
11	3	3	Y	25	Y	Y
12	2		N	9	Y	Y
13	2	2	Y	7	Y	Y
14	2	2	Y	3	Y	Y

Table A.6 Best cyclic codes of length 17 over GF(2), where $\text{ord}_{17}(2) = 8$. BCH ratio = 6/7.

k	d	δ	BCH	Generator polynomial	BKLC	Opt.
1	17	17	Y	131071	Y	Y
8	6		N	843	Y	Y
8	6	4	Y	633	Y	Y
9	5	3	Y	471	Y	Y
16	2	2	Y	3	Y	Y

Table A.7 Best cyclic codes of length 19 over GF(2), where ord$_{19}$(2) = 18. BCH ratio = 2/3.

k	d	δ	BCH	Generator polynomial	BKLC	Opt.
1	19	19	Y	524287	Y	Y
18	2	2	Y	3	Y	Y

Table A.8 All cyclic codes of length 21 over GF(2), where ord$_{21}$(2) = 6. BCH ratio = 32/63.

k	d	δ	BCH	Generator polynomial	BKLC	Opt.
1	21	21	Y	2097151	Y	Y
2	14		N	898779	Y	Y
3	12		N	379799	Y	Y
3	12	10	Y	478877	Y	Y
4	9	9	Y	181643	N	N
5	10		N	128305	Y	Y
5	10	8	Y	72031	Y	Y
6	7	7	Y	63285	N	N
6	8		N	42187	Y	Y
7	3		N	16513	N	N
7	8	5	Y	20195	Y	Y
8	6		N	11223	N	N
9	3	3	Y	7003	N	N
9	8		N	7723	Y	Y
10	5	4	Y	2585	N	N
11	6		N	1237	Y	Y
11	6	6	Y	1369	Y	Y
12	4		N	989	N	N
12	5	5	Y	823	Y	Y
13	3	3	Y	421	N	N
13	4		N	381	Y	Y
14	4		N	249	Y	Y
14	4	4	Y	159	Y	Y
15	3	3	Y	87	N	N
15	4		N	101	Y	Y
16	3	3	Y	35	Y	Y
17	2		N	29	Y	Y
18	2		N	9	Y	Y
18	2	2	Y	11	Y	Y
19	2	2	Y	7	Y	Y
20	2	2	Y	3	Y	Y

Table A.9 Best cyclic codes of length 23 over GF(2), where ord$_{23}$(2) = 11. BCH ratio = 6/7.

k	d	δ	BCH	Generator polynomial	BKLC	Opt.
1	23	23	Y	8388607	Y	Y
11	8		N	5279	Y	Y
11	8	6	Y	7973	Y	Y
12	7	5	Y	2787	Y	Y
22	2	2	Y	3	Y	Y

Table A.10 Best cyclic codes of length 25 over GF(2), where ord$_{25}$(2) = 20. BCH ratio = 6/7.

k	d	δ	BCH	Generator polynomial	BKLC	Opt.
1	25	25	Y	33554431	Y	Y
4	10	6	Y	3247203	N	N
5	5	5	Y	1082401	N	N
20	2		N	33	N	N
21	2	2	Y	31	Y	Y
24	2	2	Y	3	Y	Y

Table A.11 All cyclic codes of length 27 over GF(2), where ord$_{27}$(2) = 18. BCH ratio = 10/15.

k	d	δ	BCH	Generator polynomial	BKLC	Opt.
1	27	27	Y	134217727	Y	Y
2	18	10	Y	57521883	Y	Y
3	9	9	Y	19173961	N	N
6	6		N	2363913	N	N
7	6	6	Y	1838599	N	N
8	6	4	Y	787971	N	N
9	3	3	Y	262657	N	N
18	2		N	513	N	N
19	2		N	511	N	N
20	2		N	219	N	N
21	2	2	Y	73	N	N
24	2		N	9	Y	Y
25	2	2	Y	7	Y	Y
26	2	2	Y	3	Y	Y

Table A.12 All cyclic codes of length 29 over GF(2), where ord$_{29}$(2) = 28. BCH ratio = 3/3.

k	d	δ	BCH	Generator polynomial	BKLC	Opt.
1	29	29	Y	536870911	Y	Y
28	2	2	Y	3	Y	Y

Table A.13 Best cyclic codes of length 31 over GF(2), where $\mathrm{ord}_{31}(2) = 5$. BCH ratio $= 55/127$.

k	d	δ	BCH	Generator polynomial	BKLC	Opt.
1	31	31	Y	2147483647	Y	Y
5	16		N	94957459	Y	Y
5	16	16	Y	91635305	Y	Y
6	15		N	57124209	Y	Y
6	15	15	Y	53340711	Y	Y
10	12		N	4175255	Y	Y
10	12	12	Y	3828607	Y	Y
11	11		N	1808993	Y	Y
11	11	11	Y	1404045	Y	Y
15	8		N	98717	Y	Y
15	8	8	Y	102641	Y	Y
16	7		N	56669	N	N
16	7	5	Y	37063	N	N
20	6		N	3989	Y	Y
20	6	6	Y	2491	Y	Y
21	5		N	1975	Y	Y
21	5	3	Y	1219	Y	Y
25	4		N	89	Y	Y
25	4	4	Y	111	Y	Y
26	3	3	Y	59	Y	Y
30	2	2	Y	3	Y	Y

Table A.14 Best cyclic codes of length 33 over GF(2), where $\mathrm{ord}_{33}(2) = 10$. BCH ratio $= 17/31$.

k	d	δ	BCH	Generator polynomial	BKLC	Opt.
1	33	33	Y	8589934591	Y	Y
2	22	12	Y	3681400539	Y	Y
3	11	11	Y	1227133513	N	N
10	12		N	16443231	Y	Y
11	11		N	5659957	N	N
11	11	8	Y	7522727	N	N
12	10		N	3218979	N	N
12	10	6	Y	2444457	N	N
13	10	5	Y	1111521	Y	Y
20	6		N	8673	Y	Y
21	3		N	8031	N	N
21	4	3	Y	6141	N	N
22	6		N	2661	Y	Y
22	6	4	Y	3579	Y	Y
23	3	3	Y	1571	N	N
30	2		N	9	Y	Y
31	2	2	Y	7	Y	Y
32	2	2	Y	3	Y	Y

Table A.15 Best cyclic codes of length 35 over GF(2), where $\mathrm{ord}_{35}(2) = 12$, Part I. BCH ratio $= 31/63$.

k	d	δ	BCH	Generator polynomial	BKLC	Opt.
1	35	35	Y	34359738367	Y	Y
3	20		N	6222629783	Y	Y
3	20	16	Y	7845924509	Y	Y
4	14		N	3325135971	N	N
4	15	15	Y	2976040331	N	N
5	7	7	Y	1108378657	N	N
6	10		N	811647363	N	N
7	14		N	475770777	N	N
7	14	8	Y	322878279	N	N
8	7	7	Y	197748087	N	N
10	10	6	Y	43091493	N	N
11	5	5	Y	26181091	N	N
12	8		N	11906059	N	N
13	8		N	5237915	N	N
15	8		N	1989395	N	N
16	6		N	951229	N	N
16	7	6	Y	587461	N	N
17	6		N	360811	N	N
18	4		N	189815	N	N
19	4	4	Y	92555	N	N
19	6		N	73621	N	N
20	6	5	Y	62835	N	N
22	4		N	14521	N	N
22	4	4	Y	10055	N	N
23	3	3	Y	6039	N	N
24	4		N	3999	N	N
25	4		N	1397	Y	Y
27	4		N	363	Y	Y
28	2		N	129	N	N
28	4	3	Y	155	Y	Y
29	2		N	127	N	N
30	2		N	33	Y	Y
31	2		N	29	Y	Y
31	2	2	Y	31	Y	Y
32	2	2	Y	11	Y	Y
34	2	2	Y	3	Y	Y

Table A.16 All cyclic codes of length 37 over GF(2), where $\mathrm{ord}_{37}(2) = 36$. BCH ratio $= 3/3 = 1$.

k	d	δ	BCH	Generator polynomial	BKLC	Opt.
1	37	37	Y	137438953471	Y	Y
36	2	2	Y	3	Y	Y

Codes from Difference Sets

Table A.17 Best cyclic codes of length 39 over GF(2), where ord$_{39}$(2) = 12. BCH ratio = 19/31.

k	d	δ	BCH	Generator polynomial	BKLC	Opt.
1	39	39	Y	549755813887	Y	Y
2	26	14	Y	235609634523	Y	Y
3	13	13	Y	78536544841	N	N
12	12		N	190200499	N	N
13	12	7	Y	72138139	Y	?
13	3		N	67117057	N	N
14	10		N	49411983	N	N
14	10	8	Y	63422013	N	N
15	10	7	Y	21140971	N	N
24	6		N	61397	Y	?
25	4	3	Y	24573	N	N
26	6		N	14481	Y	Y
26	6	4	Y	8775	Y	Y
27	2	2	Y	8191	N	N
36	2		N	9	Y	Y
37	2	2	Y	7	Y	Y
38	2	2	Y	3	Y	Y

Table A.18 All cyclic codes of length 41 over GF(2), where ord$_{41}$(2) = 20. BCH ratio = 6/7.

k	d	δ	BCH	Generator polynomial	BKLC	Opt.
1	41	41	Y	2199023255551	Y	Y
20	10		N	3003117	Y	Y
20	10	4	Y	3691143	Y	Y
21	9	4	Y	1560189	Y	?
21	9	6	Y	1789531	Y	?
40	2	2	Y	3	Y	Y

Table A.19 Best cyclic codes of length 43 over GF(2), where ord$_{43}$(2) = 14. BCH ratio = 11/15.

k	d	δ	BCH	Generator polynomial	BKLC	Opt.
1	43	43	Y	8796093022207	Y	Y
14	14		N	975489303	Y	Y
14	14	8	Y	659418297	Y	Y
15	13	5	Y	370951949	Y	?
28	6		N	64959	Y	?
28	6	4	Y	47517	Y	?
29	6	3	Y	21653	Y	Y
42	2	2	Y	3	Y	Y

Table A.20 Best cyclic codes of length 45 over GF(2), where $\mathrm{ord}_{45}(2) = 12$, Part I. BCH ratio $= 72/255$.

k	d	δ	BCH	Generator polynomial	BKLC	Opt.
1	45	45	Y	35184372088831	Y	Y
2	30		N	15079016609499	Y	Y
3	15	15	Y	5026338869833	N	N
4	24		N	2661887216047	Y	Y
4	24	22	Y	4218860375897	Y	Y
5	21	21	Y	2032655304549	N	N
5	9		N	1134979744801	N	N
6	18		N	885864039225	N	N
6	18	16	Y	673256669811	N	N
7	10		N	481977699847	N	N
7	15	15	Y	299583111447	N	N
8	12		N	248041930983	N	N
9	12	9	Y	99861037149	N	N
9	5		N	68853957121	N	N
10	12		N	46172307499	N	N
10	12	10	Y	56910053429	N	N
11	10	10	Y	23943361167	N	N
11	9		N	19989722921	N	N
12	10		N	11105459877	N	N
13	5	5	Y	8251653465	N	N
13	8		N	8575222215	N	N
14	10		N	2586583763	N	N
15	10	8	Y	1452009653	N	N
15	7		N	1225031745	N	N
16	10		N	779577203	N	N
17	8		N	496709389	N	N
17	8	8	Y	349638479	N	N
18	8		N	223174969	N	N
19	7		N	73294447	N	N
19	7	7	Y	122149145	N	N
20	8		N	65027815	N	N
21	8		N	31449721	N	N
21	8	7	Y	18869839	N	N
22	8		N	13919965	N	N
22	8	8	Y	13482549	N	N
23	6		N	7904613	N	N
23	7	7	Y	6567057	N	N
24	6		N	3531065	N	N
25	5		N	2012113	N	N

(*Continued*)

Table A.20 (*Continued*)

k	d	δ	BCH	Generator polynomial	BKLC	Opt.
26	6		N	948191	N	N
27	6		N	496561	N	N
27	6	6	Y	324151	N	N
28	6		N	211509	N	N
28	6	6	Y	217501	N	N
29	4		N	90447	N	N
29	5	5	Y	102609	N	N
30	4		N	54449	N	N
31	4		N	31513	N	N
31	4	4	Y	28735	N	N

Table A.21 All cyclic codes of length 45 over GF(2), where $\text{ord}_{45}(2) = 12$, Part II. BCH ratio = 72/255.

k	d	δ	BCH	Generator polynomial	BKLC	Opt.
32	4		N	9747	N	N
32	4	4	Y	12315	N	N
33	3	3	Y	4609	N	N
33	4		N	4599	N	N
34	4		N	3293	N	N
35	2		N	1893	N	N
35	4	3	Y	1099	Y	Y
36	2		N	513	N	N
37	2		N	279	N	N
38	2		N	139	N	N
39	2		N	93	N	N
39	2	2	Y	73	N	N
40	2		N	33	Y	Y
41	2	2	Y	19	Y	Y
42	2		N	9	Y	Y
43	2	2	Y	7	Y	Y
44	2	2	Y	3	Y	Y

Table A.22 All cyclic codes of length 47 over GF(2), where $\text{ord}_{47}(2) = 23$. BCH ratio = 6/7.

k	d	δ	BCH	Generator polynomial	BKLC	Opt.
1	47	47	Y	140737488355327	Y	Y
23	12		N	18461267	Y	Y
23	12	6	Y	26516273	Y	Y
24	11	5	Y	16215601	Y	Y
24	11	5	Y	9205487	Y	Y
46	2	2	Y	3	Y	Y

Table A.23 Best cyclic codes of length 49 over GF(2), where $\text{ord}_{49}(2) = 21$. BCH ratio = 17/31.

k	d	δ	BCH	Generator polynomial	BKLC	Opt.
1	49	49	Y	562949953421311	Y	Y
3	28		N	101951566367639	Y	Y
3	28	22	Y	128547627159197	Y	Y
4	21	21	Y	48759444784523	N	N
6	14	8	Y	13298030395779	N	N
7	7	7	Y	4432676798593	N	N
21	4		N	270548993	N	N
22	4		N	268419199	N	N
24	4		N	48237463	N	N
25	4	4	Y	23070091	N	N
27	4		N	6340611	N	N
27	4	4	Y	6291843	N	N
28	3	3	Y	2097281	N	N
42	2		N	129	N	N
43	2		N	127	N	N
45	2		N	23	Y	Y
46	2	2	Y	11	Y	Y
48	2	2	Y	3	Y	Y

Table A.24 Best cyclic codes of length 51 over GF(2), where $\text{ord}_{51}(2) = 8$, Part I. BCH ratio = 79/255.

k	d	δ	BCH	Generator polynomial	BKLC	Opt.
1	51	51	Y	2251799813685247	Y	Y
2	34		N	965057063007963	Y	Y
3	17	17	Y	321685687669321	N	N
8	24		N	9701490648263	Y	Y
8	24	20	Y	16744204422115	Y	Y
9	15		N	8091780121047	N	N
9	19	19	Y	5593930057377	N	N
10	18		N	3599973291841	N	N
10	18	18	Y	2634386288205	N	N
11	15 .		N	1749888626131	N	N
11	17	8	Y	2170601688953	N	N
16	16		N	67340981993	Y	?

Table A.25 All cyclic codes of length 51 over GF(2), where ord$_{51}$(2) = 8, Part II. BCH ratio = 79/255.

k	d	δ	BCH	Generator polynomial	BKLC	Opt.
17	16		N	30829603125	Y	?
17	16	8	Y	18283749147	Y	?
18	14		N	17122158115	N	N
18	14	12	Y	16164577725	N	N
19	14		N	7721523681	Y	?
19	14	11	Y	5388197739	Y	?
24	10		N	145086071	N	N
25	10		N	94529999	N	N
25	10	6	Y	124271377	N	N
26	10		N	64718761	Y	?
26	10	10	Y	60282213	Y	?
27	8		N	22937425	N	N
27	9	4	Y	20112911	N	N
32	6		N	998883	N	N
33	6		N	516217	N	N
33	6	5	Y	272709	N	N
34	6		N	232721	N	N
34	6	6	Y	243195	N	N
35	2		N	131071	N	N
35	5	5	Y	93353	N	N
40	4		N	2289	Y	?
41	4		N	1317	Y	?
41	4	3	Y	1967	Y	?
42	4		N	813	Y	Y
42	4	4	Y	533	Y	Y
43	3	3	Y	499	N	N
48	2		N	9	Y	Y
49	2	2	Y	7	Y	Y
50	2	2	Y	3	Y	Y

Table A.26 All cyclic codes of length 53 over GF(2), where ord$_{53}$(2) = 52. BCH ratio = 3/3.

k	d	δ	BCH	Generator polynomial	BKLC	Opt.
1	53	53	Y	9007199254740991	Y	Y
52	2	2	Y	3	Y	Y

Table A.27 Best cyclic codes of length 55 over GF(2), where $\text{ord}_{55}(2) = 20$. BCH ratio $= 20/31$.

k	d	δ	BCH	Generator polynomial	BKLC	Opt.
1	55	55	Y	36028797018963967	Y	Y
4	22	12	Y	3486657776028771	N	N
5	11	11	Y	1162219258676257	N	N
10	10		N	52802340526083	N	N
11	5	5	Y	17600780175361	N	N
14	10	6	Y	2838834429093	N	N
15	5	5	Y	1702192387171	N	N
20	16		N	51462183867	Y	?
21	15	8	Y	20208952491	Y	?
24	12		N	3061967591	Y	?
25	11	7	Y	1563590875	N	N
30	10		N	46960135	Y	?
31	5	5	Y	25117627	N	N
34	8		N	2440119	Y	?
34	8	4	Y	3897129	Y	?
35	5	4	Y	1495783	N	N
40	4		N	63519	N	N
41	4	3	Y	22517	N	N
44	2		N	2049	N	N
45	2	2	Y	2047	N	N
50	2		N	33	Y	Y
51	2	2	Y	31	Y	Y
54	2	2	Y	3	Y	Y

Table A.28 All cyclic codes of length 57 over GF(2), where $\text{ord}_{57}(2) = 18$. BCH ratio $= 17/31$.

k	d	δ	BCH	Generator polynomial	BKLC	Opt.
1	57	57	Y	144115188075855871	Y	Y
2	38	20	Y	61763652032509659	Y	Y
3	19	19	Y	20587884010836553	N	N
18	16		N	760821458061	Y	?
19	16		N	477213891707	Y	?
19	16	9	Y	301427800625	Y	?
20	14		N	197852512797	N	N
20	14	6	Y	265608477423	N	N
21	14	5	Y	88899676581	N	N
36	6		N	3829143	N	N
37	3		N	1451149	N	N
37	4	3	Y	1572861	N	N
38	6		N	935463	N	N
38	6	4	Y	577329	N	N
39	3	3	Y	506607	N	N
54	2		N	9	Y	Y
55	2	2	Y	7	Y	Y
56	2	2	Y	3	Y	Y

Table A.29 All cyclic codes of length 59 over GF(2), where $\mathrm{ord}_{59}(2) = 58$. BCH ratio = 3/3.

k	d	δ	BCH	Generator polynomial	BKLC	Opt.
1	59	59	Y	576460752303423487	Y	Y
58	2	2	Y	3	Y	Y

Table A.30 All cyclic codes of length 61 over GF(2), where $\mathrm{ord}_{61}(2) = 60$. BCH ratio = 3/3.

k	d	δ	BCH	Generator polynomial	BKLC	Opt.
1	61	61	Y	2305843009213693951	Y	Y
60	2	2	Y	3	Y	Y

Table A.31 Best cyclic codes of length 63 over GF(2), where $\mathrm{ord}_{63}(2) = 6$, Part I. BCH ratio = 306/8191.

k	d	δ	BCH	Generator polynomial	BKLC	Opt.
1	63	63	Y	9223372036854775807	Y	Y
2	42		N	3952873730080618203	Y	Y
3	21	21	Y	1317624576693539401	N	N
3	36		N	1670374463367400343	Y	Y
4	27	18	Y	798874743349626251	N	N
5	30		N	316795839301359967	N	N
6	32		N	285247320157033569	Y	Y
6	32	32	Y	253483157574931709	Y	Y
7	31		N	95457791829701599	Y	Y
7	31	31	Y	107096222472614827	Y	Y
8	26		N	71870234979549569	N	N
9	28		N	35515675285765533	Y	Y
9	28	28	Y	23453379044915121	Y	Y
10	27		N	16997755446114345	Y	?
10	27	27	Y	13846799311760751	Y	?
11	26		N	8957504890631599	Y	Y
12	21	13	Y	3010168110699365	N	N
12	24		N	4393038335675501	Y	?
13	24		N	2157631890490727	Y	?
13	24	15	Y	1229441488298961	Y	?
14	22		N	892011697113939	N	N
15	24		N	539084853913561	Y	Y
15	24	24	Y	437229697661393	Y	Y
16	23		N	274994884082331	Y	?
16	23	23	Y	145789046216527	Y	?
17	22		N	139737557747015	Y	?
17	22	22	Y	98833622111019	Y	?
18	21		N	51886244352981	Y	?
18	21	21	Y	41936017350331	Y	?

Table A.32 Best cyclic codes of length 63 over GF(2), where $\text{ord}_{63}(2) = 6$, Part II. BCH ratio = 306/8191.

k	d	δ	BCH	Generator polynomial	BKLC	Opt.
19	19		N	32911747805697	N	N
19	19	14	Y	23871235424827	N	N
20	18		N	16051192599235	Y	?
21	18		N	8500268819081	Y	?
21	9	9	Y	8715302627679	N	N
22	16		N	3452578301145	N	N
22	16	11	Y	3143913836125	N	N
23	16		N	2181028058243	Y	?
23	16	16	Y	1325391789925	Y	?
24	15	8	Y	904272203549	N	N
24	16		N	1003053506425	Y	?
25	14	12	Y	428788834845	N	N
25	15		N	280029485683	N	N
26	14		N	270752879679	N	N
27	16		N	92988893669	Y	?
27	9	7	Y	99156315883	N	N
28	13	11	Y	60943896067	N	N
28	15		N	64036488631	Y	?
29	14		N	34116813743	Y	?
29	14	14	Y	27045420115	Y	?
30	13		N	9274952465	Y	?
30	13	13	Y	9406807921	Y	?
31	12		N	8552129335	Y	?
31	9	9	Y	7706782885	N	N
32	12		N	4257853143	Y	?
33	12		N	2136639529	Y	?
33	9	6	Y	1405957447	N	N
34	10	8	Y	541932693	N	N
34	11		N	983426477	N	N
35	12		N	512274445	Y	?
35	12	12	Y	279551037	Y	?
36	11		N	253098353	Y	?
36	11	11	Y	226300319	Y	?
37	10		N	111472133	Y	?
37	8	7	Y	68241683	N	N

Table A.33 Best cyclic codes of length 63 over GF(2), where ord$_{63}(2) = 6$, Part III. BCH ratio = 306/8191.

k	d	δ	BCH	Generator polynomial	BKLC	Opt.
38	10		N	64585799	Y	?
38	10	10	Y	49702627	Y	?
39	9		N	31312311	Y	?
39	9	5	Y	29039303	Y	?
40	8		N	9960299	Y	?
40	8	7	Y	10861271	Y	?
41	8		N	8128011	Y	?
42	7	5	Y	2588209	N	N
42	8		N	2127061	Y	?
43	6	6	Y	1153057	N	N
43	8		N	2071093	Y	?
44	8		N	970305	Y	Y
44	8	8	Y	698363	Y	Y
45	7	7	Y	418473	N	N
45	8		N	310997	Y	Y
46	6	6	Y	169769	N	N
46	7		N	191031	Y	?
47	6		N	94715	Y	?
48	6		N	63499	Y	Y
48	6	4	Y	43971	Y	Y
49	5		N	32081	N	N
49	5	4	Y	19999	N	N
50	6		N	9653	Y	Y
50	6	6	Y	11113	Y	Y
51	4		N	8001	N	N
51	5	3	Y	4779	Y	Y
52	4		N	3829	Y	Y
53	4		N	2003	Y	Y
54	4		N	991	Y	Y
54	4	3	Y	707	Y	Y
55	3	3	Y	295	N	N
55	3	3	Y	309	N	N
56	4		N	197	Y	Y
56	4	4	Y	237	Y	Y
57	2		N	101	N	N
57	3	3	Y	91	Y	Y
58	2		N	35	Y	Y
59	2		N	23	Y	Y
60	2		N	9	Y	Y
60	2	2	Y	11	Y	Y
61	2	2	Y	7	Y	Y
62	2	2	Y	3	Y	Y

Table A.34 Best cyclic codes of length 65 over GF(2), where $\text{ord}_{65}(2) = 12$. BCH ratio = 52/127.

k	d	δ	BCH	Generator polynomial	BKLC	Opt.
1	65	65	Y	36893488147419103231	Y	Y
4	26	14	Y	35703375626534461603	N	N
5	13	13	Y	1190112520884487201	N	N
12	26		N	16824492343224567	Y	?
13	25		N	7085852433458835	Y	?
13	25	16	Y	7783524653931707	Y	?
16	22		N	688282278231609	N	N
16	22	12	Y	909563130934131	N	N
17	13		N	370697749088533	N	N
17	13	10	Y	400398056564845	N	N
24	16		N	3930817671975	N	N
25	15		N	1659643931779	N	N
25	15	5	Y	1829242261291	N	N
28	14		N	248002977639	N	N
28	14	8	Y	265004544879	N	N
29	13		N	117479197531	N	N
29	13	5	Y	100564667101	N	N
36	10		N	1012924047	N	N
37	8		N	321668761	N	N
37	8	4	Y	274160961	N	N
40	10		N	66734463	Y	?
40	10	6	Y	37139889	Y	?
41	5		N	24083949	N	N
41	8	6	Y	32116911	Y	?
48	6		N	214923	Y	?
49	5		N	109483	N	N
49	5	3	Y	100291	N	N
52	6		N	15663	Y	Y
52	6	4	Y	14823	Y	Y
53	5	4	Y	4593	Y	Y
60	2		N	33	Y	Y
61	2	2	Y	31	Y	Y
64	2	2	Y	3	Y	Y

Table A.35 All binary cyclic codes of length 67, where $\text{ord}_{67}(2) = 66$. BCH ratio = 3/3.

k	d	δ	BCH	Generator polynomial	BKLC	Opt.
1	67	67	Y	147573952589676412927	Y	Y
66	2	2	Y	3	Y	Y

Table A.36 Best cyclic codes of length 69 over GF(2), where $\text{ord}_{69}(2) = 22$. BCH ratio $= 34/63$.

k	d	δ	BCH	Generator polynomial	BKLC	Opt.
1	69	69	Y	590295810358705651711	Y	Y
2	46	24	Y	252983918725159565019	Y	Y
3	23	23	Y	84327972908386521673	N	N
11	24		N	371476644797355167	N	N
12	21	9	Y	196117713402202851	N	N
13	24		N	132846124864858857	N	N
13	24	16	Y	85196433824079831	N	N
14	21	15	Y	50293550536382887	N	N
22	16		N	189707041968955	N	N
23	16	9	Y	83255641018131	N	N
23	3		N	70368752566273	N	N
24	14		N	45733681246311	N	N
25	14	7	Y	25833931604515	N	N
33	12		N	111565140407	N	N
34	12	4	Y	37190610797	N	N
35	12		N	18527079421	N	N
35	8	6	Y	26179633179	N	N
36	11	6	Y	14319914959	N	N
44	6		N	65746253	N	N
45	3	3	Y	28938549	N	N
45	4		N	25165821	N	N
46	6		N	9767599	N	N
46	6	4	Y	16076969	N	N
47	2		N	8388607	N	N
47	3	3	Y	5451879	N	N
55	4		N	24059	N	N
56	4	3	Y	13481	N	N
57	2		N	5279	N	N
58	2	2	Y	2787	N	N
66	2		N	9	Y	Y
67	2	2	Y	7	Y	Y
68	2	2	Y	3	Y	Y

Table A.37 All cyclic codes of length 71 over GF(2), where $\text{ord}_{71}(2) = 35$. BCH ratio $= 6/7$.

k	d	δ	BCH	Generator polynomial	BKLC	Opt.
1	71	71	Y	2361183241434822606847	Y	Y
35	12		N	91952517263	N	N
35	12	8	Y	129403085525	N	N
36	11	7	Y	43469906355	N	N
36	11	7	Y	55167946629	N	N
70	2	2	Y	3	Y	Y

Table A.38 Best cyclic codes of length 73 over GF(2), where ord$_{73}$(2) = 9. BCH ratio = 153/511.

k	d	δ	BCH	Generator polynomial	BKLC	Opt.
1	73	73	Y	9444732965739290427391	Y	Y
9	28		N	36845412074761028353	N	N
9	28	26	Y	30256586336110888865	N	N
10	28	13	Y	12839832200019398407	N	N
18	24		N	718224649358103033	Y	?
18	24	18	Y	58911352575600799	Y	?
19	21		N	35585148978166119	N	N
19	21	10	Y	24272516658514059	N	N
27	20		N	95965327783857	Y	?
27	20	14	Y	112005389694573	Y	?
28	17		N	68732646225139	N	N
28	17	6	Y	68849930092049	N	N
36	14	12	Y	157512300549	N	N
36	16		N	151548932819	Y	?
37	13		N	93742684533	N	N
37	13	5	Y	107142351331	N	N
45	10		N	522878837	Y	?
45	10	10	Y	289413601	Y	?
46	9		N	265062027	N	N
46	9	4	Y	136834107	N	N
54	6		N	952677	N	N
54	6	6	Y	998113	N	N
55	6		N	395011	Y	?
55	6	3	Y	267753	Y	?
63	4		N	1757	Y	Y
63	4	4	Y	1963	Y	Y
64	3	3	Y	929	N	N
72	2	2	Y	3	Y	Y

Table A.39 Best cyclic codes of length 75 over GF(2), where ord$_{75}$(2) = 20, Part I. BCH ratio = 59/255.

k	d	δ	BCH	Generator polynomial	BKLC	Opt.
1	75	75	Y	37778931862957161709567	Y	Y
2	50		N	16190970798410212161243	Y	Y
3	25	25	Y	5396990266136737387081	N	N
4	40		N	2858179634640669084079	Y	Y
4	40	36	Y	4529966835217099165529	Y	Y
5	15		N	12186752213857148933857	N	N
5	35	35	Y	1539197181220368385335	N	N
6	30		N	951189269293486080825	N	N
6	30	26	Y	722903844663049421427	N	N

(Continued)

Table A.39   (*Continued*)

k	d	δ	BCH	Generator polynomial	BKLC	Opt.
7	25	25	Y	536124860874510336465	N	N
8	20		N	266332995402176102631	N	N
9	20	15	Y	107224972174902067293	N	N
10	20		N	49577137672266547243	N	N
10	20	16	Y	61106704572793651253	N	N
11	15	15	Y	28823917251317760025	N	N
12	10		N	10376610210474393609	N	N
13	10	10	Y	8070696830368972807	N	N
14	10		N	3458870070158131203	N	N
15	5	5	Y	1152956690052710401	N	N
20	8		N	37190981884346369	N	N
21	8		N	36027731833551903	N	N
22	8		N	15587147900794587	N	N
23	8		N	5312682958787145	N	N
24	8		N	4319985307254905	N	N
25	8		N	2081377496170437	N	N
26	8		N	907097958809625	N	N
27	7		N	487084116073179	N	N
27	8	8	Y	306764036997111	N	N
28	8		N	253987428466695	N	N
29	7		N	86861501511087	N	N
29	8	8	Y	102254678999037	N	N
30	8		N	58274171900565	N	N
30	8	8	Y	57884403564597	N	N
31	3		N	28347650048825	N	N
31	7	7	Y	28373653913625	N	N
32	8		N	9895614096681	N	N
33	3		N	7937342013671	N	N
33	7	7	Y	7696588741863	N	N
34	6		N	3401718005859	N	N
34	6	6	Y	3404838469635	N	N
35	5	5	Y	1134946156545	N	N
40	4		N	35467035681	N	N
41	4		N	34359705631	N	N
42	4		N	14879143643	N	N
43	4		N	5066719081	N	N
44	4		N	4119979129	N	N
45	4		N	1985014725	N	N

Table A.40 Best cyclic codes of length 75 over GF(2), where ord$_{75}$(2) = 20, Part II. BCH ratio = 59/255.

k	d	δ	BCH	Generator polynomial	BKLC	Opt.
46	4		N	865100825	N	N
47	4		N	503316465	N	N
48	4		N	242228231	N	N
49	4		N	97520637	N	N
50	4		N	55576213	N	N
51	2		N	33554431	N	N
51	4	4	Y	33521695	N	N
52	4		N	9741609	N	N
53	4	3	Y	7576807	N	N
54	4		N	3244035	N	N
54	4	4	Y	3145827	N	N
55	3	3	Y	1081345	N	N
60	2		N	32769	N	N
61	2		N	32767	N	N
62	2		N	14043	N	N
63	2		N	4681	N	N
64	2		N	3929	N	N
65	2		N	1893	N	N
66	2		N	825	N	N
67	2		N	465	N	N
68	2		N	231	N	N
69	2		N	93	Y	Y
70	2		N	53	Y	Y
71	2	2	Y	31	Y	Y
72	2		N	9	Y	Y
73	2	2	Y	7	Y	Y
74	2	2	Y	3	Y	Y

Table A.41 Best cyclic codes of length 77 over GF(2), where ord$_{77}$(2) = 30. BCH ratio = 30/63.

k	d	δ	BCH	Generator polynomial	BKLC	Opt.
1	77	77	Y	15111572745182864683827 1	Y	Y
3	44		N	27367415207811487222679	Y	Y
3	44	34	Y	34506740914197092585117	Y	Y
4	33	33	Y	15468539030502144951949	N	N
6	22	12	Y	35696628531928026812 19	N	N
7	11	11	Y	11898876177309342270 73	N	N
10	14		N	221469068077912037379	N	N
11	7		N	73823022692637345793	N	N
13	14		N	33123513016869688505	N	N

(Continued)

Table A.41   (*Continued*)

k	d	δ	BCH	Generator polynomial	BKLC	Opt.
14	7	7	Y	16847577654504957597	N	N
16	14	8	Y	2906418216109959813	N	N
17	7	7	Y	1743850375392057731	N	N
30	8		N	228810142253069	N	N
31	8		N	87918001520635	N	N
33	8		N	33395293600115	N	N
34	8	6	Y	9557961558725	N	N
36	4		N	4017112721053	N	N
37	4	4	Y	1523732880779	N	N
40	6		N	197568542743	N	N
41	6		N	111669176333	N	N
43	6		N	22786927057	N	N
44	6	5	Y	16300581235	N	N
46	4		N	2636764071	N	N
46	4	4	Y	3855324345	N	N
47	3	3	Y	1555657623	N	N
60	4		N	260223	N	N
61	4		N	88021	N	N
63	4		N	26637	N	N
64	4	3	Y	14329	N	N
66	2		N	2049	N	N
67	2	2	Y	2047	N	N
70	2		N	129	N	N
71	2		N	127	N	N
73	2		N	29	Y	Y
74	2	2	Y	13	Y	Y
76	2	2	Y	3	Y	Y

Table A.42   All cyclic codes of length 79 over GF(2), where $\mathrm{ord}_{79}(2) = 39$. BCH ratio = 6/7.

k	d	δ	BCH	Generator polynomial	BKLC	Opt.
1	79	79	Y	604462909807314587353087	Y	Y
39	16		N	1826187884633	Y	?
39	16	4	Y	1323852699947	Y	?
40	15	7	Y	1013993764633	Y	?
40	15	7	Y	656848807991	Y	?
78	2	2	Y	3	Y	Y

Table A.43 All cyclic codes of length 81 over GF(2), where $\text{ord}_{81}(2) = 54$. BCH ratio $= 15/31$.

k	d	δ	BCH	Generator polynomial	BKLC	Opt.
1	81	81	Y	241785163922925834941235	Y	Y
2	54	28	Y	103622213109825357831957	Y	Y
3	27	27	Y	345407377032751192773193	N	N
6	18		N	425844711410241196569	N	N
7	18	18	Y	331212553319076486220	N	N
8	18	10	Y	141948237136747065523	N	N
9	9	9	Y	47316079045582355174	N	N
18	6		N	924138650421795276	N	N
19	6		N	920535770693055334	N	N
20	6		N	394515330297023714	N	N
21	6	6	Y	131505110099007904	N	N
24	6		N	16212958779329741	N	N
25	6	6	Y	12610079050589799	N	N
26	6	4	Y	5404319593109913	N	N
27	3	3	Y	1801439864369971	N	N
54	2		N	13421772	N	N
55	2		N	13421772	N	N
56	2		N	5752188	N	N
57	2		N	1917396	N	N
60	2		N	236391	N	N
61	2		N	183859	N	N
62	2		N	78797	N	N
63	2	2	Y	26265	N	N
72	2		N	513	N	N
73	2		N	511	N	N
74	2		N	219	N	N
75	2	2	Y	73	Y	Y
78	2		N	9	Y	Y
79	2	2	Y	7	Y	Y
80	2	2	Y	3	Y	Y

Table A.44 All cyclic codes of length 83 over GF(2), where $\text{ord}_{83}(2) = 82$. BCH ratio $= 3/3$.

k	d	δ	BCH	Generator polynomial	BKLC	Opt.
1	83	83	Y	967140655691703339764940	Y	Y
82	2	2	Y	3	Y	Y

Table A.45 Best cyclic codes of length 85 over GF(2), where $\mathrm{ord}_{85}(2) = 8$, Part I. BCH ratio = 330/4095.

k	d	δ	BCH	Generator polynomial	BKLC	Opt.
1	85	85	Y	38685626227668133590597631	Y	Y
4	34		N	37437702800969161539288003	N	N
5	17	17	Y	12479234266989720513096011	N	N
8	40		N	29679404357684875608578301	Y	Y
8	40	38	Y	21310538958370065370295101	Y	Y
9	37		N	772009114945150839607271	N	N
9	37	18	Y	102546909997218437041165	N	N
12	34		N	96478081300885212608251	N	N
13	17		N	7835751611502334756395	N	N
13	17	12	Y	8829517263369214558527	N	N
16	32		N	941061156695903158969	Y	?
16	32	30	Y	924911253159598333693	Y	?
17	29		N	491769230776162394035	N	N
17	29	29	Y	329836964637060724139	N	N
20	28		N	73420783910230902769	Y	?
21	17		N	36573828658968662975	N	N
21	17	10	Y	26902586857633390451	N	N
24	24		N	4562818473854209393	Y	?
24	24	22	Y	3175892628701229729	Y	?
25	23	16	Y	1329986958359413001	N	N
25	24		N	1265721781574261127	Y	?
28	22		N	249566652200487573	N	N
28	22	18	Y	269171993141127043	N	N
29	17		N	99440098942171313	N	N
29	17	10	Y	142870685016291937	N	N
32	20		N	9508086021133571	Y	?
33	20		N	8863157433366061	Y	?
33	20	13	Y	5672483182469407	Y	?
36	16	16	Y	1093789964203275	N	N
36	18		N	1045210201813183	Y	?
37	16	8	Y	473734505848881	N	N
37	17		N	320317207372677	N	N
40	16		N	69949617421035	N	N
41	15	8	Y	29474345065933	N	N
41	16		N	19578997812203	N	N

Table A.46 Best cyclic codes of length 85 over GF(2), where $\text{ord}_{85}(2) = 8$, Part II. BCH ratio $= 330/4095$.

k	d	δ	BCH	Generator polynomial	BKLC	Opt.
44	14		N	4355580751141	Y	?
44	14	14	Y	2305758453507	Y	?
45	13	7	Y	2188273971187	N	N
45	14		N	1133051516975	Y	?
48	12		N	138256045779	Y	?
49	12		N	98894976591	Y	?
49	12	7	Y	121396569913	Y	?
52	10		N	9843141907	N	N
52	10	10	Y	10462986053	N	N
53	10		N	8556021291	Y	?
53	10	5	Y	5380190611	Y	?
56	8		N	996810165	N	N
57	8		N	534741237	N	N
57	8	5	Y	336315543	N	N
60	8		N	66834303	Y	?
60	8	8	Y	61099147	Y	?
61	7		N	33228445	N	N
61	7	7	Y	23315577	N	N
64	6		N	4186603	N	N
65	6		N	2063639	N	N
65	6	4	Y	1387427	N	N
68	6		N	260905	Y	?
68	6	6	Y	134661	Y	?
69	5		N	87783	N	N
69	5	3	Y	100643	N	N
72	4		N	11803	Y	?
73	4		N	6665	Y	?
73	4	3	Y	5559	Y	?
76	4		N	909	Y	Y
76	4	4	Y	921	Y	Y
77	3	3	Y	477	N	N
80	2		N	33	Y	Y
81	2	2	Y	31	Y	Y
84	2	2	Y	3	Y	Y

Table A.47   All cyclic codes of length 87 over GF(2), where ord$_{87}$(2) = 28. BCH ratio = 19/31.

k	d	δ	BCH	Generator polynomial	BKLC	Opt.
1	87	87	Y	154742504910672534362390527	Y	Y
2	58	30	Y	66318216390288229012453083	Y	Y
3	29	29	Y	22106072130096076337484361	N	N
28	24		N	840304425435849571	Y	?
28	24		N	895219256762018141	Y	?
28	6		N	864691130065747971	N	N
29	24	8	Y	298412844869654731	Y	?
29	24	8	Y	475353906092286241	Y	?
29	3		N	288230376688582657	N	N
30	22		N	280102209584692173	Y	?
30	22	6	Y	202517277033882399	Y	?
30	6		N	205878839878277997	N	N
31	22	9	Y	124645179283562229	Y	?
31	22	9	Y	98785796496578235	Y	?
31	3	3	Y	123527304141715163	N	N
56	4		N	3758096391	N	N
56	6		N	2779053031	N	N
56	6		N	3890161061	N	N
57	3	3	Y	1565188963	N	N
57	3	3	Y	1667475805	N	N
57	4	3	Y	1610612733	N	N
58	2		N	536870913	N	N
58	6		N	556338263	N	N
58	6	4	Y	982263073	N	N
59	2	2	Y	536870911	N	N
59	3	3	Y	377217823	N	N
59	3	3	Y	521731021	N	N
84	2		N	9	Y	Y
85	2	2	Y	7	Y	Y
86	2	2	Y	3	Y	Y

Table A.48 Best cyclic codes of length 89 over GF(2), where ord$_{89}$(2) = 11. BCH ratio = 159/511.

k	d	δ	BCH	Generator polynomial	BKLC	Opt.
1	89	89	Y	618970019642690137449562111	Y	Y
11	40		N	332038606377631336716351	Y	Y
12	33	15	Y	179642197798545564674951	N	N
22	28		N	286618348002952157821	Y	?
22	28	20	Y	150448858606915674679	Y	?
23	28		N	86433013351972032061	Y	?
23	28	11	Y	116930169944023911429	Y	?
33	20		N	99596613231852981	Y	?
33	20	14	Y	80413784251617965	Y	?
34	20		N	71091185315397437	Y	?
34	20	7	Y	50367161824013479	Y	?
44	12	12	Y	53850825598133	N	N
44	18		N	65880906413559	Y	?
45	17		N	29660881482731	Y	?
45	17	5	Y	24906380995757	Y	?
55	12		N	33820253097	Y	?
55	12	10	Y	29285831383	Y	?
56	11		N	16951063365	Y	?
56	11	4	Y	10530121679	Y	?
66	8		N	9934509	Y	?
66	8	6	Y	11181039	Y	?
67	7		N	4889769	N	N
67	7	3	Y	4655517	N	N
77	4		N	7001	Y	Y
77	4	4	Y	6705	Y	Y
78	4	2	Y	2359	Y	Y
88	2	2	Y	3	Y	Y

Codes from Difference Sets

Table A.49 Best cyclic codes of length 91 over GF(2), where $\text{ord}_{91}(2) = 12$, Part I. BCH ratio $= 200/2013$.

k	d	δ	BCH	Generator polynomial	BKLC	Opt.
1	91	91	Y	2475880078570760549798248447	Y	Y
3	52		N	448387730764783406656375703	Y	Y
3	52	40	Y	565358443138205164914560669	Y	Y
4	39	39	Y	214446306017939890140005771	N	N
6	26		N	58485356186710879129092483	N	N
7	13	13	Y	19495118728903626376364161	N	N
12	36		N	738075935227986378134225	N	N
13	36		N	572454875101920263687703	N	N
13	36	21	Y	423749253578165232068637	N	N
15	28		N	94642391789071680766631	N	N
15	28	20	Y	135328915254189789620741	N	N
16	28		N	65588097652677115232661	N	N
16	28	11	Y	38109274831420197552505	N	N
18	26		N	9540792271373785196583	N	N
19	13	9	Y	9388019958627836418077	N	N
19	7		N	7139616241758021656963	N	N
24	24		N	278433091628284356199	N	N
25	24		N	92812156523268840925	N	N
25	24	16	Y	112741657464786307523	N	N
27	20	18	Y	26778942036031897321	N	N
27	24		N	18629723142490282901	Y	?
28	24		N	18336953258368518515	Y	?
28	24	15	Y	11509058829997084519	Y	?
30	16	14	Y	2487673629188936079	N	N
30	20		N	4240500739958420501	N	N
31	13		N	2305027393173640599	N	N
31	13	13	Y	1450620065385457615	N	N
36	20		N	62313892191115229	Y	?
37	19		N	29652355370959465	N	N
39	20		N	8927450154810179	Y	?
40	16		N	4113305040022787	N	N
40	16	7	Y	2625983827873999	N	N
42	16		N	980225765835461	N	N
42	16	12	Y	1019567306075469	N	N
43	13		N	451409534662661	N	N
43	13	11	Y	410364815203131	N	N
48	14		N	17481986597381	Y	?
49	12		N	4788728330961	N	N
51	14		N	1819670244875	Y	?
52	12		N	989593298111	Y	?
52	12	5	Y	801101945459	Y	?

Table A.50 Best cyclic codes of length 91 over GF(2), where ord$_{91}$(2) = 12, Part II. BCH ratio = 200/2013.

k	d	δ	BCH	Generator polynomial	BKLC	Opt.
54	10	10	Y	259932522077	N	N
54	12		N	137533886569	Y	?
55	10		N	94641197581	N	N
55	10	4	Y	111366043379	N	N
60	8		N	4093053149	N	N
61	8		N	2134451753	N	N
63	8		N	532218921	N	N
64	8		N	255780363	N	N
64	8	4	Y	156882227	N	N
66	8		N	60643537	Y	?
66	8	8	Y	36465383	Y	?
67	7	7	Y	31729245	N	N
72	6		N	917067	Y	?
73	6		N	498519	Y	?
75	4		N	106023	N	N
76	4		N	47185	N	N
76	4	3	Y	40955	N	N
78	4		N	13427	Y	?
78	4	4	Y	14853	Y	?
79	3	3	Y	7115	N	N
84	2		N	129	N	N
85	2		N	127	Y	Y
87	2		N	23	Y	Y
88	2	2	Y	13	Y	Y
90	2	2	Y	3	Y	Y

Table A.51 Best cyclic codes of length 93 over GF(2), where ord$_{93}$(2) = 10, Part I. BCH ratio = 419/16383.

k	d	δ	BCH	Generator polynomial	BKLC	Opt.
1	93	93	Y	99035203142830421991929937 91	Y	Y
2	62		N	42443658489784466567969973 39	Y	Y
3	31	31	Y	14147886163261488855989991 13	N	N
5	48		N	55846843941834088518096106 7	Y	?
5	48	46	Y	42259325505961463009497866 5	Y	?
6	45		N	26343891608169726274172657 7	N	N
6	45	24	Y	18490291772223172666587823 7	N	N
7	46		N	10380054110371505120442896 1	Y	Y

(*Continued*)

Codes from Difference Sets

Table A.51 *(Continued)*

k	d	δ	BCH	Generator polynomial	BKLC	Opt.
8	31		N	6393129007001251550846085	N	N
8	31	23	Y	6201194849351392489205033	N	N
10	36		N	19254965115835335326021015	N	N
10	36	34	Y	17656333380175097251916671	N	N
11	33		N	894898133461155882346805	N	N
11	33	18	Y	617317984871142018648803	N	N
12	36		N	46336778299004256304208	N	N
12	36	32	Y	26616431511119676574033	N	N
13	31		N	2259165499863523763343181	N	N
13	31	15	Y	1321574155827176604337159	N	N
15	32		N	6004543128927489580388	N	N
16	32		N	29554961463403971816108	N	N
16	32	17	Y	15298817842076042336599	N	N
17	30		N	99126122284858632669421	N	N
18	21	12	Y	6205418826291634315909	N	N
18	30		N	3933082074165788383554	N	N
20	32		N	99114528284064412753	Y	?
21	28	16	Y	89824992780326356649	N	N
21	29		N	48411496709865735233	N	N
22	28	24	Y	39282812084370309244	N	N
22	30		N	25051005111985873264	Y	?
23	27	12	Y	21057873225632080092	N	N
23	29		N	13016321769479692196	Y	?
25	24		N	588594196996218378639	N	N
26	21	15	Y	266837805144576759629	N	N
26	24		N	155169190996213670563	N	N
27	24		N	94160379337888606223	Y	?
27	24	22	Y	145446879570185457441	Y	?
28	21	21	Y	71663365287571205653	N	N
28	22		N	37077583153611184163	N	N
30	24		N	9793072940410256617	Y	?
31	22	15	Y	8441279276743620785	N	N
31	24		N	4815377180922330531	Y	?
32	22		N	4592118461954920139	Y	?
33	15	9	Y	2228056523428068859	N	N
33	22		N	2290860637568330769	Y	?
35	20		N	559854585098668391	Y	?
36	16	14	Y	274507608997827753	N	N
36	20		N	277653133640140679	Y	?
37	18	18	Y	139994628024122565	N	N
37	20		N	111430241490796121	Y	?

Table A.52 Best cyclic codes of length 93 over GF(2), where
ord$_{93}$(2) = 10, Part II. BCH ratio = 419/16383.

k	d	δ	BCH	Generator polynomial	BKLC	Opt.
38	18		N	71774657626631417	N	N
38	18	10	Y	57281939847816123	N	N
40	16		N	9109906018329977	N	N
41	15	12	Y	7514186581067353	N	N
41	16		N	4512469712916063	N	N
42	16		N	4487527136084885	N	N
42	16	16	Y	3348901196164807	N	N
43	14		N	1133697809588451	N	N
43	15	7	Y	1131095242524059	N	N
45	16		N	510041342822917	N	N
46	12	12	Y	233383480951939	N	N
46	16		N	176095926353305	Y	?
47	14		N	93734485873369	N	N
48	12	8	Y	60617264374221	N	N
48	14		N	69673941995675	Y	?
50	14		N	9988344649713	Y	?
51	12	11	Y	8684555416885	N	N
51	13		N	4450779054675	N	N
52	12		N	4395460627197	N	N
52	12	12	Y	2633927799205	N	N
53	12		N	2149609796895	N	N
53	12	7	Y	1584275435611	N	N
55	12		N	492941939509	Y	?
56	12		N	274781202891	Y	?
56	12	9	Y	199290053999	Y	?
57	10		N	99941347195	N	N
57	10	10	Y	119035348025	N	N
58	10		N	68543632227	N	N
58	9	9	Y	62340598633	N	N
60	10		N	9724930871	Y	?
61	10		N	4345178895	Y	?
61	8	8	Y	8291289389	N	N
62	10		N	4152093285	Y	?
63	8	6	Y	1589998457	N	N
63	9		N	1139172009	N	N

Table A.53 Best cyclic codes of length 93 over GF(2), where
$\text{ord}_{93}(2) = 10$, Part III. BCH ratio $= 419/16383$.

k	d	δ	BCH	Generator polynomial	BKLC	Opt.
65	8		N	269771103	N	N
66	6	6	Y	239255115	N	N
66	8		N	268237597	N	N
67	8		N	97690963	Y	?
67	8	8	Y	125822041	Y	?
68	7		N	65841333	N	N
68	7	5	Y	34036123	N	N
70	8		N	9876939	Y	?
71	6	5	Y	5358241	N	N
71	7		N	4453183	N	N
72	6		N	3228413	N	N
73	5	5	Y	1565789	N	N
73	6		N	1063871	N	N
75	6		N	520805	Y	?
76	5		N	255737	N	N
76	5	4	Y	217019	N	N
77	6		N	97215	Y	?
77	6	6	Y	88429	Y	?
78	5		N	61613	N	N
78	5	3	Y	35891	N	N
80	4		N	10233	Y	?
81	3	3	Y	5585	N	N
81	4		N	4321	Y	?
82	4		N	4003	Y	Y
82	4	4	Y	3049	Y	Y
83	2		N	2013	N	N
83	3	2	Y	1077	N	N
85	4		N	483	Y	Y
86	3		N	179	Y	Y
86	3	3	Y	161	Y	Y
87	2		N	111	Y	Y
88	2	2	Y	37	Y	Y
90	2		N	9	Y	Y
91	2	2	Y	7	Y	Y
92	2	2	Y	3	Y	Y

Table A.54 All cyclic codes of length 95 over GF(2), where $\text{ord}_{95}(2) = 36$. BCH
ratio $= 20/31$.

k	d	δ	BCH	Generator polynomial	BKLC	Opt.
1	95	95	Y	39614081257132168796771975167	Y	Y
4	38	20	Y	3833620766819242141623094371	N	N
5	19	19	Y	1277873588939747380541031457	N	N
18	10		N	226674023524131833118723	N	N
19	5	5	Y	75558007841377277706241	N	N

(Continued)

Table A.54 *(Continued)*

k	d	δ	BCH	Generator polynomial	BKLC	Opt.
22	10		N	12186775458268923466917	N	N
23	5	5	Y	7312055977200191114339	N	N
36	16		N	1122326314830562689	N	N
36	16		N	5838001047377778847	N	N
37	16	7	Y	392159847736116095	N	N
37	16	7	Y	573662452757488757	N	N
40	14		N	36490242457326307	N	N
40	14	8	Y	56097043612288385	N	N
41	14	7	Y	18795797962507135	N	N
41	14	7	Y	35854411126976929	N	N
54	10		N	2296642375299	N	N
54	10		N	3323344549281	N	N
55	5	5	Y	1113508228255	N	N
55	5	5	Y	2140664006017	N	N
58	10		N	207711926053	N	N
58	10	6	Y	177092316803	N	N
59	5	5	Y	106996548993	N	N
59	5	5	Y	69599744739	N	N
72	4		N	16252959	N	N
73	4	3	Y	5767157	N	N
76	2		N	524289	N	N
77	2	2	Y	524287	N	N
90	2		N	33	N	N
91	2	2	Y	31	N	N
94	2	2	Y	3	N	N

Table A.55 All cyclic codes of length 97 over GF(2), where $\text{ord}_{97}(2) = 48$. BCH ratio = 6/7.

k	d	δ	BCH	Generator polynomial	BKLC	Opt.
1	97	97	Y	158456325028528675187087900671	Y	Y
48	16		N	587335244108193	N	N
48	16	6	Y	877356107120355	N	N
49	15	5	Y	293918051210657	N	N
49	15	7	Y	547674424979615	N	N
96	2	2	Y	3	Y	Y

Table A.56 Best cyclic codes of length 99 over GF(2), where $\text{ord}_{99}(2) = 30$, Part I. BCH ratio = 61/255.

k	d	δ	BCH	Generator polynomial	BKLC	Opt.
1	99	99	Y	633825300114114700748351602687	Y	Y
2	66	34	Y	271639414334620586035007829723	Y	Y
3	33	33	Y	90546471444873528678335943241	N	N

(Continued)

Table A.56   (*Continued*)

k	d	δ	BCH	Generator polynomial	BKLC	Opt.
6	22		N	11163263602792626823356486153	N	N
7	22	22	Y	8682538357727598640388378119	N	N
8	22		N	3721087867597542274452162051	N	N
9	11	11	Y	1240362622532514091484054017	N	N
10	18		N	9289085981154587700268956363	N	N
11	33		N	4176311130374222309072554549	N	N
11	33	24	Y	5550792788861590692893024399	N	N
12	30		N	2375187271942328140715208033	N	N
12	30	16	Y	1803690907337490434037710490	N	N
13	30	15	Y	1391096271941611297689283559	N	N
16	22		N	1736216808168484794286583311	N	N
17	22		N	7005870231923227461521053	N	N
17	22	20	Y	9420757397839888212146847	N	N
18	22		N	4167032270499769412720955	N	N
18	22	12	Y	2565949898376971724319713	N	N
19	11	11	Y	1426795945702793585002729	N	N
20	18		N	6399544454794632267387361	N	N
21	12		N	4531258214793522142310337	N	N
22	18		N	1963471439434222833352677	N	N
23	9	9	Y	1159193397726856096005477	N	N
26	18		N	1468880658401378332080303	N	N
27	12	9	Y	6207491827289867715093	N	N
27	9		N	8479437589940591373223	N	N
28	18		N	3297599173098761274189	N	N
28	18	10	Y	3618160988718130275363	N	N
29	9	9	Y	1252357035885020570593	N	N
30	12		N	5914692840121049093313	N	N
31	12		N	5247211793195028935111	N	N
32	12		N	2248805054226440972199	N	N
33	11		N	7496016847421469907133	N	N
33	11	8	Y	8416556122714095623311	N	N
36	10		N	10376575094955737097	N	N
37	10		N	8070731946155540487	N	N
38	10		N	3458885119780945923	N	N
39	10		N	1315090674684821577	N	N
39	3	3	Y	1299574925075681801	N	N
40	12		N	1036439936805586023	N	N
41	12		N	389396445184309045	N	N
42	12		N	150324501847720353	N	N
43	11		N	83914393843921065	N	N
43	11	7	Y	129658495642077799	N	N
46	10		N	10630197039991017	N	N
47	10		N	8765433921864863	N	N
47	10	10	Y	8088048165324711	N	N
48	10		N	2385243094624737	N	N
49	10		N	2025914989251175	N	N
49	10	5	Y	1195644390455265	N	N

Table A.57 Best cyclic codes of length 99 over GF(2), where $\text{ord}_{99}(2) = 30$, Part II. BCH ratio $= 61/255$.

k	d	δ	BCH	Generator polynomial	BKLC	Opt.
50	12		N	1125380700473343	N	N
51	12		N	375470531116373	N	N
52	12		N	182528729301093	N	N
53	9		N	73434074106273	N	N
53	9	6	Y	107759788998691	N	N
56	10		N	10408197992169	N	N
57	10		N	5775187388437	N	N
57	10	8	Y	6587212197373	N	N
58	10		N	3372069282339	N	N
58	10	6	Y	3846034586619	N	N
59	9	5	Y	1683752861731	N	N
60	6		N	619700064777	N	N
61	6		N	549740904447	N	N
62	6		N	235603244763	N	N
63	3	3	Y	68856087041	N	N
63	4		N	69946610241	N	N
66	6		N	9682850313	N	N
67	6		N	8455946303	N	N
67	6	6	Y	7531105799	N	N
68	6		N	3623976987	N	N
68	6	4	Y	3227616771	N	N
69	2		N	1227133513	N	N
69	3	3	Y	1075872257	N	N
70	4		N	966003303	N	N
71	4		N	525790559	N	N
72	4		N	234776571	N	N
73	4		N	102477507	N	N
76	2		N	12589059	N	N
77	2		N	4196353	N	N
78	2		N	2444457	N	N
79	2		N	1111521	N	N
80	4		N	1047039	N	N
81	4		N	278497	N	N
82	4		N	149577	N	N
83	4	3	Y	114683	N	N
86	2		N	14139	N	N
87	2		N	4841	N	N
88	2		N	2049	N	N
89	2	2	Y	1193	N	N
90	2		N	513	N	N
91	2		N	511	N	N
92	2		N	219	N	N
93	2	2	Y	73	Y	Y
96	2		N	9	Y	Y
97	2	2	Y	7	Y	Y
98	2	2	Y	3	Y	Y

Table A.58   All cyclic codes of length 101 over GF(2), where $\text{ord}_{101}(2) = 100$. BCH ratio = 3/3.

k	d	δ	BCH	Generator polynomial	BKLC	Opt.
1	101	101	Y	25353012004564588029934064110751	Y	Y
100	2	2	Y	3	Y	Y

Table A.59   All cyclic codes of length 103 over GF(2), where $\text{ord}_{103}(2) = 51$. BCH ratio = 6/7.

k	d	δ	BCH	Generator polynomial	BKLC	Opt.
1	103	103	Y	10141204801825835211973625643007	Y	Y
51	20		N	8202881201280285	Y	?
51	20	4	Y	6497119553438871	Y	?
52	19	8	Y	3127191258723083	Y	?
52	19	8	Y	3673431891067789	Y	?
102	2	2	Y	3	Y	Y

Table A.60   Best cyclic codes of length 105 over GF(2), where $\text{ord}_{105}(2) = 12$, Part I. BCH ratio = 566/32767.

k	d	δ	BCH	Generator polynomial	BKLC	Opt.
1	105	105	Y	40564819207303340847894502572031	Y	Y
2	70		N	17384922517415717506240501102299	Y	Y
3	35	35	Y	5794974172471905835413500367433	N	N
3	60		N	7346384580850211334658059520919	Y	Y
4	56		N	3068947014218725606919476054447	Y	Y
4	56	50	Y	4864014852305515493984115134297	Y	Y
5	49	49	Y	2343492012067788452560939157349	N	N
5	50		N	1393282835771609647857826601311	N	N
6	35	25	Y	856326087690646120702389824751	N	N
6	42		N	887630096488433344155411186381	N	N
7	52		N	617501230119267374244063343899	Y	Y
7	52	46	Y	465738758843024226219478863845	Y	Y
8	45		N	215787245307805821146781012745	N	N
8	45	45	Y	244324886301846249658730289131	N	N
9	42		N	99587780166197082615562558697	N	N
9	42	36	Y	157002495524017834545770525847	N	N
10	35	35	Y	55024205474003630612428414101	N	N
10	40		N	45781269446286853114137010717	N	N
11	30		N	26480323779641176825496864089	N	N
11	40	22	Y	32260770389270433282062970183	N	N
12	21	21	Y	14377401679890147652997818181	N	N
12	32		N	17845113676702934605898262985	N	N
13	15	15	Y	8041491659118891233503829945	N	N
13	40		N	8783096390462903189719451753	N	N
14	25	20	Y	4412481581728224567448342461	N	N
14	32		N	4943102368370935836351279403	N	N

Table A.61 Best cyclic codes of length 105 over GF(2), where $\mathrm{ord}_{105}(2) = 12$, Part II. BCH ratio = 566/32767.

k	d	δ	BCH	Generator polynomial	BKLC	Opt.
15	30	26	Y	1911785352770951521407986497	N	N
15	32		N	2413858716799903772584810433	N	N
16	25	25	Y	6517799634899209108355538239	N	N
16	32		N	9689354420937386476688876007	N	N
17	14	13	Y	3216792758811864885798381113	N	N
17	32		N	5986551319961778752499441645	N	N
18	25	17	Y	2973888954091888077791751163	N	N
18	32		N	2951445705166040133562224949	N	N
19	28	22	Y	9817211237146068442959299 5	N	N
19	32		N	9793320400479003756615077 5	N	N
20	21	21	Y	5914410533208133703572054 5	N	N
20	29		N	6261654344995132960365536 9	N	N
21	15	15	Y	3276420156328856977148129 7	N	N
21	32		N	3609644749291689983060584 1	N	N
22	23	13	Y	1212774276213282889233315 3	N	N
22	32		N	9859437978668053148574643	N	N
23	32		N	9532999202851659150302865	N	N
24	15	15	Y	412924483449441776367086 7	N	N
24	28		N	462512748443196196318464 5	N	N
25	15	13	Y	181799194074016357890224 5	N	N
25	28		N	195565580788291241079448 9	N	N
26	21	12	Y	89332826284842639195355 1	N	N
26	26		N	96814138897474593740829 9	N	N
27	24		N	59372986852072903959013 5	N	N
28	24	11	Y	21857168106730411443242 5	N	N
28	26		N	21000074021605679535811 9	N	N
29	15	11	Y	8443187063492466771708 5	N	N
29	26		N	13284289058906521300111 5	N	N
30	21	15	Y	5971146432464386770024 9	N	N
30	24		N	3947148237305033315062 5	N	N
31	22	18	Y	3258764937370884545919 5	N	N
31	24		N	1907270192444427435199 1	N	N
32	21	17	Y	1117037989087535943620 1	N	N
32	24		N	1023407559006701547041 7	N	N
33	15	13	Y	931623149017899323750 5	N	N
33	24		N	943267194188249781284 3	N	N
34	21	12	Y	460198920897794087840 7	N	N
34	24		N	238183412930448195569 1	N	N
35	15	15	Y	223856598734059697307 7	N	N
35	24		N	232410731657662738959 7	N	N
36	15	14	Y	90552268244596385087 9	N	N
36	24		N	81077470615439060808 7	Y	?
37	15	11	Y	50307776742715838985 1	N	N
37	24		N	55693600594360506807 5	Y	?
38	10	9	Y	25539374535508742584 5	N	N
38	21		N	28642539851296252109 9	N	N
39	22		N	9898675877540866876 1	N	N
39	7	7	Y	10469003029475518514 1	N	N

Table A.62 Best cyclic codes of length 105 over GF(2), where $\mathrm{ord}_{105}(2) = 12$, Part III. BCH ratio = 566/32767.

k	d	δ	BCH	Generator polynomial	BKLC	Opt.
40	15	11	Y	63749435049948324633	N	N
40	21		N	68268338524946617997	N	N
41	15	9	Y	29863557735576931001	N	N
41	20		N	36185370807980506451	N	N
42	20		N	17478734296589857031	N	N
42	20	13	Y	10233260579419809283	N	N
43	12	10	Y	5950713462579842065	N	N
43	20		N	8829717496836187441	N	N
44	16	8	Y	3708832330195975845	N	N
44	20		N	4592318465057059599	Y	?
45	12	12	Y	1637374183535611577	N	N
45	20		N	2094708778676359361	Y	?
46	14	14	Y	1086349306613074157	N	N
46	16		N	992531093394987507	N	N
47	13	13	Y	491170316250340229	N	N
47	18		N	571192242908568039	N	N
48	15	10	Y	178855609818696023	N	N
48	16		N	287657401501596987	N	N
49	11	9	Y	79599817094129895	N	N
49	16		N	101922483073498891	N	N
50	10	9	Y	56474025066240113	N	N
50	16		N	69862735346240355	N	N
51	16		N	21322638559859457	N	N
51	7	6	Y	18518539910222437	N	N
52	13	8	Y	17625958263634183	N	N
52	16		N	9702376919977075	N	N
53	10	7	Y	8342527595107891	N	N
53	16		N	5761710044165671	N	N
54	16		N	2382376269781321	N	N
54	7	7	Y	4124062439857267	N	N
55	10	8	Y	1912699567427227	N	N
55	14		N	2109766186029409	N	N
56	10	8	Y	890780871029501	N	N
56	14		N	969507511986361	N	N
57	14		N	556434802344991	N	N
57	5	5	Y	509630977082343	N	N
58	12	12	Y	261545744328035	N	N
58	14		N	150908516965347	Y	?
59	11	11	Y	72992224094573	N	N
59	14		N	135359195070075	Y	?
60	14		N	59098976673357	Y	?
60	7	7	Y	51478220679245	N	N

(*Continued*)

Table A.62   (*Continued*)

k	d	δ	BCH	Generator polynomial	BKLC	Opt.
61	12		N	34983451878641	N	N
61	8	8	Y	30002877859875	N	N
62	10	7	Y	17301228632605	N	N
62	12		N	10201974999579	N	N
63	12		N	8744844686683	N	N
63	7	6	Y	8682229147687	N	N
64	12		N	4107911133185	N	N
64	8	7	Y	3982153233717	N	N

Table A.63 Best cyclic codes of length 105 over GF(2), where $\mathrm{ord}_{105}(2) = 12$, Part IV. BCH ratio = 566/32767.

k	d	δ	BCH	Generator polynomial	BKLC	Opt.
65	12		N	2074760118571	Y	?
65	5	5	Y	1309632792965	N	N
66	12		N	975842243903	Y	?
66	6	5	Y	683365807535	N	N
67	10		N	305738143607	N	N
67	8	7	Y	375399692751	N	N
68	10		N	265590861827	N	N
68	8	7	Y	197557515159	N	N
69	10		N	102014787709	N	N
69	5	4	Y	105725943569	N	N
70	10		N	56698006731	N	N
70	10	10	Y	55156699325	N	N
71	10		N	21186632633	Y	?
71	9	9	Y	18479566955	N	N
72	9		N	9444751139	N	N
73	8		N	8542629381	N	N
74	8		N	4231667065	N	N
74	8	8	Y	3686025979	N	N
75	7	7	Y	1255021385	N	N
75	8		N	1100633989	N	N
76	8		N	998587237	N	N
77	5	5	Y	490892229	N	N
77	8		N	276702543	N	N
78	6	4	Y	251336381	N	N
78	8		N	256514669	Y	?
79	5	4	Y	96629837	N	N
79	8		N	101487337	Y	?

(*Continued*)

Table A.63 (*Continued*)

k	d	δ	BCH	Generator polynomial	BKLC	Opt.
80	6	6	Y	46121847	N	N
80	8		N	41248975	Y	?
81	5	5	Y	26558765	N	N
81	8		N	31007255	Y	?
82	7		N	12963549	N	N
83	6	4	Y	6032517	N	N
83	8		N	5002381	Y	Y
84	7		N	3918971	Y	?
85	6		N	2070717	Y	?
86	6		N	876821	Y	?
86	6	5	Y	563621	Y	?
87	4		N	521013	N	N
87	4	3	Y	403355	N	N
88	4		N	261081	N	N

Table A.64 Best cyclic codes of length 105 over GF(2), where $\text{ord}_{105}(2) = 12$, Part V. BCH ratio = 566/32767.

k	d	δ	BCH	Generator polynomial	BKLC	Opt.
89	4		N	93347	N	N
89	4	3	Y	113583	N	N
90	4		N	65071	N	N
90	4	3	Y	47327	N	N
91	3	3	Y	32417	N	N
91	4		N	16431	N	N
92	4		N	9953	Y	?
92	4	4	Y	15219	Y	?
93	3	3	Y	7049	N	N
93	4		N	4415	Y	Y
94	4		N	3797	Y	Y
95	4		N	2043	Y	Y
95	4	3	Y	1459	Y	Y
96	4		N	961	Y	Y
97	4		N	503	Y	Y
98	3		N	227	Y	Y
98	3	3	Y	173	Y	Y
99	2		N	93	Y	Y
99	2	2	Y	117	Y	Y
100	2		N	33	Y	Y
101	2		N	29	Y	Y
101	2	2	Y	19	Y	Y
102	2		N	9	Y	Y
102	2	2	Y	11	Y	Y
103	2	2	Y	7	Y	Y
104	2	2	Y	3	Y	Y

Table A.65   All cyclic codes of length 107 over GF(2), where ord$_{107}$(2) = 106. BCH ratio = 3/3.

k	d	δ	BCH	Generator polynomial	BKLC	Opt.
1	107	107	Y	162259276829213363391578010288127	Y	Y
106	2	2	Y	3	Y	Y

Table A.66   All cyclic codes of length 109 over GF(2), where ord$_{109}$(2) = 36. BCH ratio = 11/15.

k	d	δ	BCH	Generator polynomial	BKLC	Opt.
1	109	109	Y	6490371073168534535663312041152511	Y	Y
36	24		N	11828913599551452213765	N	N
36	24		N	16242310892515436001339	N	N
36	24	10	Y	13798768718297540608221	N	N
37	19	11	Y	7100475063678675103235	N	N
37	19	8	Y	5604727470416612382697	N	N
37	19	9	Y	7748199436986147651659	N	N
72	10		N	252864036567	N	N
72	10		N	274412374527	N	N
72	10	4	Y	160599919017	N	N
73	10	4	Y	123958445927	N	N
73	10	4	Y	91741980501	N	N
73	10	4	Y	95742700109	N	N
108	2	2	Y	3	Y	Y

Table A.67   All cyclic codes of length 111 over GF(2), where ord$_{111}$(2) = 36. BCH ratio = 19/31.

k	d	δ	BCH	Generator polynomial	BKLC	Opt.
1	111	111	Y	2596148429267413814265248164610047	Y	Y
2	74	38	Y	1112635041114605920399392070547163	Y	Y
3	37	37	Y	3708783470382019734664464023515721	N	N
36	24		N	4588919719963229659669687	N	N
36	24		N	71769520121566098185945	N	N
36	6		N	56668397794848059424771	N	N
37	24	7	Y	23936819280836201231799	N	N
37	24	7	Y	35074720265228519722053	N	N
37	3		N	18889465931616019808257	N	N
38	22		N	15284203318753090826649	N	N
38	22	12	Y	11330604224919212246259	N	N
38	6		N	13492475665400745876333	N	N
39	22	11	Y	5100117742356395798391	N	N
39	22	11	Y	8815915745508660132945	N	N
39	3	3	Y	8095485399185471944411	N	N
72	4		N	962072674311	N	N
72	6		N	595913335659	N	N
72	6		N	922475101521	N	N

(Continued)

Table A.67 (*Continued*)

k	d	δ	BCH	Generator polynomial	BKLC	Opt.
73	3	3	Y	333887853775	N	N
73	3	3	Y	522191986393	N	N
73	4	3	Y	412316860413	N	N
74	2		N	137438953473	N	N
74	6		N	181136888491	N	N
74	6	4	Y	229015702805	N	N
75	2	2	Y	137438953471	N	N
75	3	3	Y	111207215513	N	N
75	3	3	Y	82440996083	N	N
108	2		N	9	Y	Y
109	2	2	Y	7	Y	Y
110	2	2	Y	3	Y	Y

Table A.68 All cyclic codes of length 113 over GF(2), where $\mathrm{ord}_{113}(2) = 28$. BCH ratio $= 20/31$.

k	d	δ	BCH	Generator polynomial	BKLC	Opt.
1	113	113	Y	10384593717069655257060992658440191	Y	Y
28	28		N	465280374311612745496206333	N	N
28	28		N	509057532459280688959626455	N	N
28	28		N	690691915854505933102041999	N	N
28	28	10	Y	407145179538752552358004177	N	N
29	28	17	Y	360620439929173689708555799	N	N
29	28	7	Y	279533081929462244074338781	N	N
29	28	9	Y	313875437099201126355620851	N	N
29	28	9	Y	379518550794170179192844447	N	N
56	16		N	180073942781267961	N	N
56	16		N	238336713976378827	N	N
56	18		N	205424623425433197	N	N
56	18		N	253277629406588679	N	N
56	18		N	278264821417796847	N	N
56	18	6	Y	197156405528246517	N	N
57	15	5	Y	132082390514795863	N	N
57	15	6	Y	88501161892883641	N	N
57	18	3	Y	107718711090289405	N	N
57	18	5	Y	113768202144835667	N	N
57	18	7	Y	123654965563745755	N	N
57	18	8	Y	93317891902415781	N	N
84	8		N	1021626831	N	N
84	8		N	744827277	N	N
84	8		N	829688739	N	N
84	8	4	Y	899752299	N	N
85	8	3	Y	467462011	N	N
85	8	4	Y	282327393	N	N
85	8	4	Y	325116121	N	N
85	8	5	Y	341591877	N	N
112	2	2	Y	3	Y	Y

Table A.69 Best cyclic codes of length 115 over GF(2), where $\text{ord}_{115}(2) = 44$. BCH ratio = 33/63.

k	d	δ	BCH	Generator polynomial	BKLC	Opt.
1	115	115	Y	415383748682786210282439706337607671	Y	Y
4	46		N	4019842729188253647894577803267171	N	N
5	23	23	Y	13399475763960845492981926010890571	N	N
11	40		N	26140344985722044424502183985679	N	N
11	40	26	Y	3948038843931840476710724174621	N	N
12	35	25	Y	1380055720310803889200149037334	N	N
15	40		N	238722758068628724910543035123	N	N
15	40	24	Y	128442284065638697604193335310	N	N
16	23	23	Y	126205979064438129478973613248	N	N
22	10		N	14855282242312205481164144643	N	N
23	5	5	Y	4951760747437401827054714881	N	N
26	10		N	798671050212749925172810917	N	N
27	5	5	Y	47920265297780400019797104	N	N
44	16		N	3162128876043837867747	N	N
45	16	7	Y	1233575381162991701427	N	N
48	14		N	186628048223390034079	N	N
49	14		N	111496745196563060853	N	N
55	16		N	1825409870870659087	N	N
56	15	6	Y	634136223578079237	N	N
59	14		N	79027798043596117	N	N
59	14	8	Y	96072333809518129	N	N
60	14	7	Y	57648756968095215	N	N
66	8		N	1120934270110501	N	N
67	5	5	Y	376955181172451	N	N
70	8		N	66297256130977	N	N
70	8	6	Y	36727767174287	N	N
71	5	5	Y	22247798630559	N	N
88	4		N	260046879	N	N
89	4		N	92274677	N	N
92	2		N	8388609	N	N
93	2		N	8388607	N	N
99	4		N	87683	N	N
100	4	3	Y	33203	N	N
103	2		N	5279	N	N
104	2	2	Y	2787	N	N
110	2		N	33	Y	Y
111	2	2	Y	31	Y	Y
114	2	2	Y	3	Y	Y

Codes from Difference Sets

Table A.70 All cyclic codes of length 117 over GF(2), where $\text{ord}_{117}(2) = 12$, Part I. BCH ratio $= 380/4095$.

k	d	δ	BCH	Generator polynomial	BKLC	Opt.
1	117	117	Y	16615349947311448411297588253504 3071	Y	Y
2	78	40	Y	71208642631334778905561092515018459	Y	Y
3	39	39	Y	23736214210444926301853697505006153	N	N
6	26		N	29263825738904703659819627060966 49	N	N
7	26	26	Y	22760753352481436179859709936307 27	N	N
8	26		N	97546085796349012199398756869888 3	N	N
9	13		N	32515361932116337399799585623296 1	N	N
12	48		N	73859425907787189832685043350669	N	N
13	45		N	24435900333485249197575603213105	N	N
13	45	20	Y	20369375308733893823972774754655	N	N
14	46		N	19591655358081911977334860255281	N	N
14	46	30	Y	11094901060044659546726452957935	N	N
15	30		N	85432191504123621611400477620 53	N	N
15	39	19	Y	59217922704968465510149646412 43	N	N
18	26		N	82838353024291944252180156676 7	N	N
19	26		N	56880169684004447956241053568 9	N	N
19	26	16	Y	37858908279633620335431672403 5	N	N
20	26		N	29666881873353808927607029090 1	N	N
21	13		N	83911409302845921991483544835	N	N
21	13	13	Y	10980608776916492047068966875 9	N	N
24	36		N	1930726922487680125211971738 5	N	N
25	32	19	Y	7034815485486912388929719409	N	N
25	36		N	5235884355351649912729377505	N	N
26	36		N	4907809999582292840870511229	N	N
26	36	26	Y	3720796449036857107036118505	N	N
27	35		N	2055153688494063668324349077	N	N
27	35	12	Y	1359197785857996093241784773	N	N
30	26		N	303021480918356305189610379	N	N
31	26		N	98645604815169295267339769	N	N
31	26	12	Y	94264234830655779685421017	N	N
32	26		N	76921247274960975848635479	N	N
33	13		N	38567205716359424770632997	N	N
33	13	12	Y	21787517375434285884469151	N	N
36	32		N	3822006177911368113484355	Y	?
37	29		N	2147098566938976248632049	Y	?
37	29	15	Y	2111668510282955197338727	Y	?
38	24	22	Y	113131742144636525658804 3	N	N
38	28		N	9708768221761124184276 31	Y	?
39	26	8	Y	59613591911946162028668 5	N	N
39	27		N	31441329119945199620823 5	N	N

Table A.71   All cyclic codes of length 117 over GF (2), where ord$_{117}$ (2) = 12, Part II. BCH ratio = 380/4095.

k	d	δ	BCH	Generator polynomial	BKLC	Opt.
42	26		N	755477067456441121098653	Y	?
43	23		N	344253050605397707560 85	N	N
43	23	11	Y	234548381084180438064 63	N	N
44	24		N	171202649405092011673 55	Y	?
45	13		N	932325712647128810063 5	N	N
45	13	11	Y	492587577356661094013 7	N	N
48	24		N	960923776809403175565	Y	?
49	24		N	578464958442813336133	Y	?
50	18	18	Y	148171247570924595309	N	N
50	22		N	263606302065404992717	N	N
51	22		N	127870964282251928349	Y	?
51	22	10	Y	138040235714002249753	Y	?
54	20		N	9331917238410928013	N	N
55	19		N	9214897426678019429	N	N
55	19	11	Y	6599068120904152119	N	N
56	18	14	Y	2609579366098051333	N	N
56	20		N	2820840615833246427	Y	?
57	13		N	2260080974718165585	N	N
57	13	10	Y	1254872706832762703	N	N
60	18		N	278683590534977115	N	N
61	16		N	95995335816192337	N	N
62	18		N	57363003469082437	Y	?
63	15		N	31377732940691249	N	N
63	15	6	Y	18019157719233275	N	N
66	14		N	4437043417763447	N	N
67	12	8	Y	1293260341952801	N	N
67	14		N	1325850681292591	N	N
68	14		N	979182289264945	N	N
68	14	10	Y	953082434945807	N	N

Table A.72   All cyclic codes of length 117 over GF (2), where ord$_{117}$ (2) = 12, Part III. BCH ratio = 380/4095.

k	d	δ	BCH	Generator polynomial	BKLC	Opt.
69	13		N	528892790540957	N	N
69	13	5	Y	345778123502243	N	N
72	12		N	35219776537621	N	N
73	12		N	18266670295995	N	N
74	12		N	10032882304761	N	N
75	9		N	8653696321615	N	N
75	9	5	Y	4649878355895	N	N
78	12		N	890697669059	Y	?

(Continued)

Table A.72 (*Continued*)

k	d	δ	BCH	Generator polynomial	BKLC	Opt.
79	10		N	547616211933	N	N
79	10	7	Y	318358384739	N	N
80	10		N	272416988243	N	N
80	10	8	Y	236288812049	N	N
81	9		N	98970047321	N	N
81	9	5	Y	131688770825	N	N
84	8		N	9607908329	N	N
85	8		N	8247772839	N	N
86	8		N	4291601765	N	N
87	8		N	1314811105	N	N
87	8	5	Y	1430545187	N	N
90	8		N	247120611	Y	?
91	6	4	Y	96197443	N	N
91	7		N	105420443	N	N
92	8		N	65334021	N	N
92	8	6	Y	46846449	N	N
93	3		N	28795297	N	N
93	7	5	Y	32119611	N	N
96	4		N	2223871	N	N
97	4		N	1302539	N	N
98	4		N	1029813	N	N
99	4	3	Y	483477	N	N
102	4		N	62103	N	N
103	4		N	25905	N	N
103	4	3	Y	20877	N	N
104	4		N	9777	Y	?
104	4	4	Y	14621	Y	?
105	3	3	Y	7927	N	N
108	2		N	513	N	N
109	2		N	511	N	N
110	2		N	219	N	N
111	2	2	Y	73	Y	Y
114	2		N	9	Y	Y
115	2	2	Y	7	Y	Y
116	2	2	Y	3	Y	Y

Table A.73 Best cyclic codes of length 119, where $\mathrm{ord}_{119}(2) = 24$, Part I. BCH ratio $= 107/511$.

k	d	δ	BCH	Generator polynomial	BKLC	Opt.
1	119	119	Y	6646139978924579364519035301401722 87	Y	Y
3	68		N	120363164972649862507037647190739863	Y	Y
3	68	52	Y	151762251487254174465395294283976349	Y	Y
4	51	51	Y	5756499194344123859032235304266891	N	N

(*Continued*)

Table A.73   (*Continued*)

k	d	δ	BCH	Generator polynomial	BKLC	Opt.
6	34		N	15699543257302155979178823546618243	N	N
7	17	17	Y	5233181085767385326392941182206081	N	N
8	42		N	3209715807966109007896902706004601	N	N
9	35	21	Y	2388271951898953148055989217264087	N	N
11	42		N	6107471537403221505965016856895377	N	N
11	42	22	Y	6462807110525879412001089407296669	N	N
12	35	21	Y	2171583703614069565934381351866273	N	N
14	34		N	5921062486172036769040731050274977	N	N
14	34	18	Y	6615789197486830089108370014285977	N	N
15	17	17	Y	2527297692829375038151434996914577	N	N
16	14		N	1521192326050288629335024979149177	N	N
17	7		N	5070641086834295431116749930497	N	N
19	14		N	1557105289326567846526562325415	N	N
20	7	7	Y	1157857442314731605215013588637	N	N
22	14		N	1996315388517437727638853555653	N	N
23	7	7	Y	1197789233063983147980909940517	N	N
24	24		N	4318803698334367921673094884177	N	N
25	24		N	231028170891042042950485024677	N	N
27	24		N	9900644037890866401359473419	N	N
28	24		N	4304299456695157000699294341	N	N
28	24	11	Y	3083217912716863325240749397	N	N
30	20		N	1005028376230884038696964283	N	N
31	17		N	335113160274713512607150185	N	N
32	24		N	268211336841361901162294067	N	N
33	24		N	91015708100790453644744977	N	N
35	24		N	3826159640082775300149815377	N	N
36	24		N	1750781533614718098679285377	N	N
36	24	8	Y	1351559941428643898691936177	N	N
38	18	14	Y	3901558060656319954300923	N	N
38	20		N	3236231314658424275241703	N	N
39	17	9	Y	208150117697905491825118377	N	N
40	14		N	9741736346102757091771477	N	N
41	7		N	528704280840475034690885	N	N
43	14		N	892621044396730464901757	N	N
44	7		N	71402385206203323229069	N	N
44	7	7	Y	41040623477936222509737	N	N
46	14		N	1026390672642292349429577	N	N
47	7	5	Y	654828666148197812027177	N	N
48	24		N	3171499295753088448515	Y	?
49	24		N	1893447364497478674433	Y	?
51	20		N	529947421865301084155	N	N
52	20		N	270590273150374138775	N	N
54	20		N	73755585864253475461	N	N
55	17		N	27903332860023020885	N	N
56	14		N	16805077392883846621	N	N

Codes from Difference Sets

Table A.74 Best cyclic codes of length 119 over GF(2), where ord$_{119}$(2) = 24, Part II. BCH ratio = 107/511.

k	d	δ	BCH	Generator polynomial	BKLC	Opt.
57	14		N	7964293414777432697	N	N
59	14		N	2197755980793851955	N	N
60	12		N	839760862865084119	N	N
60	12	5	Y	1085359758876918771	N	N
62	12		N	265263430899240221	N	N
62	12	12	Y	207945778362554839	N	N
63	11	5	Y	131111092324377437	N	N
63	5		N	72624942306312321	N	N
64	14		N	45057536232380543	N	N
65	7		N	27037596929275861	N	N
67	12		N	5320853486077965	N	N
68	6	6	Y	3433653274374869	N	N
68	7		N	2765228454139897	N	N
70	12		N	563333412790159	N	N
70	6		N	918125356221195	N	N
71	7	7	Y	562712701340293	N	N
72	8		N	202441285528179	N	N
73	8		N	102671390698821	N	N
75	8		N	32057075296259	N	N
76	8		N	9454794134283	N	N
78	4		N	3948851344405	N	N
79	4		N	1592772301811	N	N
80	8		N	990120951695	N	N
81	8		N	476251694981	N	N
83	8		N	87667644603	N	N
84	8		N	67958680395	N	N
84	8	6	Y	40282377795	N	N
86	4		N	9994487825	N	N
87	4	3	Y	4650627177	N	N
88	6		N	3826368149	N	N
89	6		N	1544514163	N	N
91	6		N	480186325	N	N
92	6		N	244574211	N	N
92	6	5	Y	193657523	N	N
94	4		N	66120937	N	N
94	4	4	Y	54145019	N	N
95	3	3	Y	18225833	N	N
96	4		N	16646271	N	N
97	4		N	5636053	N	N
99	4		N	1441803	N	N
100	4		N	655355	N	N
102	2		N	131073	N	N
103	2		N	131071	N	N
104	4		N	40377	N	N
105	4		N	29847	N	N
107	4		N	5183	Y	Y

Table A.75 Best cyclic codes of length 119 over GF(2), where ord$_{119}$(2) = 24, Part III. BCH ratio = 107/511.

k	d	δ	BCH	Generator polynomial	BKLC	Opt.
108	4		N	3265	Y	Y
108	4	3	Y	2691	Y	Y
110	2		N	633	N	N
111	2	2	Y	313	N	N
112	2		N	129	N	N
113	2		N	127	Y	Y
115	2		N	23	Y	Y
116	2	2	Y	11	Y	Y
118	2	2	Y	3	Y	Y

Table A.76 All cyclic codes of length 121 over GF(2), where ord$_{121}$(2) = 110. BCH ratio = 6/7.

k	d	δ	BCH	Generator polynomial	BKLC	Opt.
1	121	121	Y	265845599156983174580761412056068915	Y	Y
10	22	12	Y	3896125048710061180958887328618499	N	N
11	11	11	Y	12987083495700203936529624428728 33	N	N
110	2		N	2049	N	N
111	2	2	Y	2047	N	N
120	2	2	Y	3	Y	Y

Table A.77 All cyclic codes of length 123 over GF(2), where ord$_{123}$(2) = 20. BCH ratio = 82/255.

k	d	δ	BCH	Generator polynomial	BKLC	Opt.
1	123	123	Y	10633823966279326983230456482242756607	Y	Y
2	82	42	Y	4557353128405425849955909920961181403	Y	Y
3	41	41	Y	1519117709468475283318636640320393801	N	N
20	40		N	171004766011549469187252512172 33	N	N
21	40		N	94486585878316148468107061059 47	N	N
21	40	20	Y	61245662039156074388182945348 63	N	N
22	40		N	48060527983473790267806824399 87	N	N
22	40	24	Y	42235942077602054538520224380 37	N	N
23	40	9	Y	16164602364328534592638129753 77	N	N
40	24		N	19112360006621637847954557	N	N
41	23	13	Y	8017326767629398027027733	N	N
41	24		N	4893153003408120033211435	N	N
42	22	12	Y	4816410452653371068326125	N	N
42	24		N	2731417001355209805227903	N	N
43	23	8	Y	2265403343099293796249307	N	N
43	3		N	2072444262196821302818523	N	N

(Continued)

Table A.77   (*Continued*)

k	d	δ	BCH	Generator polynomial	BKLC	Opt.
60	20		N	17328502283082456591	N	N
61	18	8	Y	7021256595733456521	N	N
61	19		N	5542741280294038425	N	N
62	18		N	4317605462683351173	N	N
62	18	10	Y	3453690590578597379	N	N
63	12		N	1911938570515561399	N	N
63	18	9	Y	1919844913373596161	N	N
80	12		N	15401648137707	N	N
81	10	5	Y	8209485546961	N	N
81	12		N	5044742863491	N	N
82	12		N	4064712672325	Y	?
82	12	8	Y	2785455053111	Y	?
83	11	7	Y	1681580689645	N	N
83	2		N	2199023255551	N	N
100	6		N	8934805	N	N
101	4		N	6777715	N	N
101	4	3	Y	4254593	N	N
102	6		N	3691883	N	N
102	6	4	Y	2497267	N	N
103	3	3	Y	1963601	N	N
120	2		N	9	Y	Y
121	2	2	Y	7	Y	Y
122	2	2	Y	3	Y	Y

Table A.78   All cyclic codes of length 125 over GF(2), where $\mathrm{ord}_{125}(2) = 100$. BCH ratio $= 10/15$.

k	d	δ	BCH	Generator polynomial	BKLC	Opt.
1	125	125	Y	42535295865117307932921825928971026431	Y	Y
4	50	26	Y	41163189546887717354440476705455583203	N	N
5	25	25	Y	13721063182295905784813492235151194401	N	N
20	10		N	41832471054236358881675575296033	N	N
21	10	10	Y	39297169778222034100967964672031	N	N
24	10	6	Y	38029519140214871710614159360 03	N	N
25	5	5	Y	12676506380071623903538053120 01	N	N
100	2		N	33554433	N	N
101	2		N	33554431	N	N
104	2		N	3247203	N	N
105	2	2	Y	1082401	N	N
120	2		N	33	Y	Y
121	2	2	Y	31	Y	Y
124	2	2	Y	3	Y	Y

A.3. Tables of Best Ternary Cyclic Codes of Length up to 79

Table A.79 Best cyclic codes of length 7 over GF(3), where $\text{ord}_7(3) = 6$. BCH ratio = 3/3.

k	d	δ	BCH	Generator polynomial	BKLC	Opt.
1	7	7	Y	1093	Y	Y
6	2	2	Y	5	Y	Y

Table A.80 Best cyclic codes of length 8 over GF(3), where $\text{ord}_8(3) = 2$. BCH ratio = 17/31.

k	d	δ	BCH	Generator polynomial	BKLC	Opt.
1	8		N	4100	Y	Y
1	8	8	Y	3280	Y	Y
2	6		N	1159	Y	Y
2	6	6	Y	1399	Y	Y
3	5		N	310	Y	Y
3	5	5	Y	320	Y	Y
4	4		N	148	Y	Y
4	4	4	Y	113	Y	Y
5	3		N	31	Y	Y
5	3	3	Y	32	Y	Y
6	2		N	11	Y	Y
6	2	2	Y	10	Y	Y
7	2	2	Y	4	Y	Y

Table A.81 Best cyclic codes of length 10 over GF(3), where $\text{ord}_{10}(3) = 4$. BCH ratio = 10/15.

k	d	δ	BCH	Generator polynomial	BKLC	Opt.
1	10	10	Y	29524	Y	Y
2	5	5	Y	7381	N	N
4	4		N	1220	N	N
5	2		N	245	N	N
5	4	3	Y	395	N	N
6	2	2	Y	121	N	N
8	2		N	11	Y	Y
9	2	2	Y	5	Y	Y

Table A.82 Best cyclic codes of length 11 over GF(3), where
$\mathrm{ord}_{11}(3) = 5$. BCH ratio = 6/7.

k	d	δ	BCH	Generator polynomial	BKLC	Opt.
1	11	11	Y	88573	Y	Y
5	6		N	1441	Y	Y
5	6	3	Y	889	Y	Y
6	5	4	Y	314	Y	Y
10	2	2	Y	5	Y	Y

Table A.83 Best cyclic codes of length 13 over GF(3), where
$\mathrm{ord}_{13}(3) = 3$. BCH ratio = 20/31.

k	d	δ	BCH	Generator polynomial	BKLC	Opt.
1	13	13	Y	797161	Y	Y
3	9		N	95074	Y	Y
3	9	8	Y	69190	Y	Y
4	7	7	Y	30935	Y	Y
6	6		N	4145	Y	Y
6	6	5	Y	2987	Y	Y
7	5	5	Y	1156	Y	Y
9	3		N	154	Y	Y
9	3	3	Y	85	Y	Y
10	3	2	Y	41	Y	Y
12	2	2	Y	5	Y	Y

Table A.84 Best cyclic codes of length 14 over GF(3), where
$\mathrm{ord}_{14}(3) = 6$. BCH ratio = 10/15.

k	d	δ	BCH	Generator polynomial	BKLC	Opt.
1	14	14	Y	2391484	Y	Y
2	7	7	Y	597871	N	N
6	4		N	10940	N	N
7	2		N	2189	N	N
7	4	3	Y	3554	N	N
8	2	2	Y	1093	N	N
8	2	2	Y	1366	N	N
12	2		N	11	Y	Y
13	2	2	Y	5	Y	Y

Table A.85 Best cyclic codes of length 16 over GF(3), where $\mathrm{ord}_{16}(3) = 4$. BCH ratio = 39/127.

k	d	δ	BCH	Generator polynomial	BKLC	Opt.
1	16		N	26904200	Y	Y
1	16	16	Y	21523360	Y	Y
2	12		N	7605358	Y	Y
2	12	11	Y	9180238	Y	Y
3	10		N	2034220	Y	Y
3	10	10	Y	2099840	Y	Y
4	8		N	971176	?	N
4	8	8	Y	741506	N	N
5	6		N	345229	N	N
5	6	6	Y	209984	N	N
6	4	4	Y	65620	N	N
6	6		N	91400	N	N
7	6		N	37324	Y	Y
7	6	6	Y	21769	Y	Y
8	5		N	9167	N	N
8	5	5	Y	10069	N	N
9	5		N	4024	Y	Y
9	5	5	Y	2860	Y	Y
10	3		N	811	N	N
10	3	3	Y	740	N	N
11	3		N	457	N	N
11	3	3	Y	368	N	N
12	2		N	83	N	N
12	2	2	Y	101	N	N
13	2		N	31	Y	Y
14	2		N	11	Y	Y
14	2	2	Y	10	Y	Y
15	2	2	Y	4	Y	Y

Table A.86 All cyclic codes of length 17 over GF(3), where $\mathrm{ord}_{17}(3) = 16$. BCH ratio = 3/3.

k	d	δ	BCH	Generator polynomial	BKLC	Opt.
1	17	17	Y	64570081	Y	Y
16	2	2	Y	5	Y	Y

Table A.87 All cyclic codes of length 19 over GF(3), where ord$_{19}(3) = 18$. BCH ratio = 3/3.

k	d	δ	BCH	Generator polynomial	BKLC	Opt.
1	19	19	Y	581130733	?	?
18	2	2	Y	5	?	?

Table A.88 Best cyclic codes of length 20 over GF(3), where ord$_{20}(3) = 4$. BCH ratio = 40/127.

k	d	δ	BCH	Generator polynomial	BKLC	Opt.
1	20		N	2179240250	Y	Y
1	20	20	Y	1743392200	Y	Y
2	10		N	435848050	N	N
2	10		N	479432855	N	N
3	10		N	217924025	N	N
3	10	10	Y	174339220	N	N
4	12		N	77297030	Y	Y
4	12	12	Y	63125510	Y	Y
5	11		N	19345601	N	N
5	11	11	Y	16858213	N	N
6	10	10	Y	9396130	Y	Y
6	8		N	9278942	N	N
7	8		N	2649782	N	N
7	8	6	Y	1712392	N	N
8	5	5	Y	959539	N	N
8	8		N	1035884	N	N
9	6		N	323120	N	N
10	4	4	Y	99829	N	N
10	6		N	109496	N	N
11	5		N	21902	N	N
11	5	5	Y	21901	N	N
12	4		N	9800	N	N
12	4	4	Y	10909	N	N
13	4		N	2320	N	N
14	4		N	1448	Y	Y
14	4	3	Y	1210	Y	Y
15	4		N	362	Y	Y
15	4	3	Y	313	Y	Y
16	2		N	83	N	N
16	2	2	Y	115	N	N
17	2		N	40	Y	Y
18	2		N	11	Y	Y
18	2	2	Y	10	Y	Y
19	2	2	Y	4	Y	Y

Table A.89 Best cyclic codes of length 22 over GF(2), where $\text{ord}_{22}(3) = 5$. BCH ratio = 30/63.

k	d	δ	BCH	Generator polynomial	BKLC	Opt.
1	22	12	Y	19613162255	Y	Y
2	11	11	Y	3922632451	N	N
5	12		N	152170964	Y	Y
6	12		N	82483474	Y	Y
6	12	6	Y	68112727	Y	Y
7	10	7	Y	18423184	N	N
10	9		N	933232	Y	Y
11	7		N	279235	N	N
11	7	5	Y	184297	N	N
12	2		N	110716	N	N
12	7	4	Y	91769	Y	Y
15	4		N	2524	N	N
16	4		N	986	Y	Y
16	4	3	Y	1139	Y	Y
17	2	2	Y	314	N	N
20	2		N	11	Y	Y
21	2	2	Y	4	Y	Y

Table A.90 Best cyclic codes of length 23 over GF(3), where $\text{ord}_{23}(3) = 11$. BCH ratio = 6/7.

k	d	δ	BCH	Generator polynomial	BKLC	Opt.
1	23	23	Y	47071589413	Y	Y
11	9		N	937036	Y	Y
11	9	6	Y	554668	Y	Y
12	8	5	Y	191861	Y	Y
22	2	2	Y	5	Y	Y

Table A.91 Best cyclic codes of length 25 over GF(3), where $\text{ord}_{25}(3) = 20$. BCH ratio = 6/7.

k	d	δ	BCH	Generator polynomial	BKLC	Opt.
1	25	25	Y	423644304721	Y	Y
4	10	6	Y	17505963005	N	N
5	5	5	Y	3501192601	N	N
20	2		N	245	N	N
21	2	2	Y	121	N	N
24	2	2	Y	5	Y	Y

Codes from Difference Sets

Table A.92 Best cyclic codes of length 26 over GF(3), where
$\text{ord}_{26}(3) = 3$. BCH ratio $= 100/1023$.

k	d	δ	BCH	Generator polynomial	BKLC	Opt.
1	26		N	1588666142705	Y	Y
1	26	26	Y	1270932914164	Y	Y
2	13		N	317733228541	N	N
3	18		N	186786248003	Y	Y
3	18	18	Y	138927883907	Y	Y
4	17		N	62479734673	Y	Y
4	17	17	Y	50870032054	Y	Y
5	13		N	13202375855	N	N
6	15		N	6826591462	Y	Y
6	15	15	Y	5807976262	Y	Y
7	14		N	2258845841	Y	Y
7	14	14	Y	1453249039	Y	Y
8	13		N	693874052	Y	Y
8	13	13	Y	530227058	Y	Y
9	11		N	147493546	N	N
10	10		N	83659025	Y	?
10	10	9	Y	58231807	Y	?
11	8		N	27944708	N	N
11	8	8	Y	19357214	N	N
12	9		N	9543695	Y	?
13	8		N	3172784	Y	?
13	8	7	Y	1753363	Y	?
14	7		N	994963	Y	?
14	7	5	Y	764059	Y	?
15	6		N	353060	N	N
16	6		N	98563	Y	?
16	6	6	Y	112043	Y	?
17	5	5	Y	38609	N	N
17	6		N	21914	Y	Y
18	6		N	9952	Y	Y
19	5		N	4268	Y	Y
19	5	5	Y	2792	Y	Y
20	3		N	911	N	N
20	4	4	Y	1274	Y	Y
21	3		N	284	Y	Y
22	3		N	98	Y	Y
22	3	3	Y	131	Y	Y
23	2	2	Y	53	?	?
24	2		N	11	Y	Y
25	2	2	Y	4	Y	Y

Table A.93 Best cyclic codes of length 28 over GF(3), where ord$_{28}$(3) = 6. BCH ratio = 35/127.

k	d	δ	BCH	Generator polynomial	BKLC	Opt.
1	28	15	Y	14297995284350	Y	Y
1	28	28	Y	11438396227480	Y	Y
2	14		N	3145558962557	N	N
2	14	14	Y	2859599056870	N	N
3	14	8	Y	1429799528435	N	N
4	7	7	Y	285959905687	N	N
6	12		N	47973198320	N	N
7	12		N	12285651260	N	N
8	10		N	6103070177	N	N
8	12	6	Y	5539275823	N	N
9	10		N	1286618392	N	N
9	10	10	Y	2204948248	N	N
10	7	5	Y	709075096	N	N
12	8		N	82752044	N	N
13	8		N	25767176	N	N
14	4	4	Y	5978710	N	N
14	8		N	6161411	N	N
15	8		N	2736158	Y	?
15	8	5	Y	1843958	Y	?
16	2		N	657658	N	N
16	4	3	Y	570754	N	N
18	4		N	87560	N	N
19	4		N	20873	N	N
20	4		N	12407	N	N
20	4	3	Y	10930	N	N
21	4		N	2908	Y	?
21	4	3	Y	3374	Y	?
22	2	2	Y	1003	N	N
24	2		N	83	N	N
25	2		N	40	Y	Y
26	2		N	11	Y	Y
26	2	2	Y	10	Y	Y
27	2	2	Y	4	Y	Y

Table A.94 Best cyclic codes of length 31 over GF(3), where ord$_{31}$(3) = 30. BCH ratio = 3/3.

k	d	δ	BCH	Generator polynomial	BKLC	Opt.
1	31	31	Y	308836698141973	Y	Y
30	2	2	Y	5	Y	Y

Table A.95 Best cyclic codes of length 32 over GF(3), where $\mathrm{ord}_{32}(3) = 8$, Part I. BCH ratio = 77/511.

k	d	δ	BCH	Generator polynomial	BKLC	Opt.
1	32		N	1158137618032400	Y	Y
1	32	32	Y	926510094425920	Y	Y
2	24		N	327385731536476	Y	Y
2	24	21	Y	395179153079836	Y	Y
3	20		N	87566502826840	N	N
3	20	17	Y	104514858212680	N	N
4	16		N	41805943285072	N	N
4	16	16	Y	31919402643332	N	N
5	12		N	8756650282684	N	N
5	12	12	Y	13276211718908	N	N
6	12		N	4699367594018	N	N
6	8	8	Y	2824725897640	N	N
7	12		N	984005018198	N	N
7	12	11	Y	1177456986866	N	N
8	10		N	539633706992	N	N
8	10	10	Y	433437443818	N	N
9	10		N	99911441762	N	N
9	10	10	Y	123113624920	N	N
10	6		N	60016307930	N	N
10	8	8	Y	55185897604	N	N
11	6		N	20447192950	N	N
11	6	6	Y	15841193696	N	N
12	6		N	6449579936	N	N
13	6		N	2271948593	N	N
14	4	4	Y	731794274	N	N
14	6		N	709147012	N	N
15	4	4	Y	172186888	N	N
15	6		N	252472838	N	N
16	2		N	43046722	N	N
16	6		N	82722415	N	N
16	6	6	Y	65798663	N	N
17	6		N	27211016	N	N
17	6	6	Y	18921371	N	N
18	5		N	5912920	N	N
18	5	5	Y	4914370	N	N
19	5		N	3076106	N	N
19	5	5	Y	2365276	N	N
20	4		N	982988	N	N
20	4	4	Y	591319	N	N

(*Continued*)

Table A.95   (*Continued*)

k	d	δ	BCH	Generator polynomial	BKLC	Opt.
21	4		N	313615	N	N
22	3		N	73729	N	N
22	3	3	Y	112936	N	N
23	3		N	33217	N	N
23	3	3	Y	26576	N	N
24	2		N	9167	N	N
24	2	2	Y	6644	N	N
25	2		N	4219	N	N
26	2		N	902	N	N
27	2		N	475	N	N

Table A.96 Best cyclic codes of length 32 over GF(3), where $\mathrm{ord}_{32}(3) = 8$, Part II. BCH ratio $= 77/511$.

k	d	δ	BCH	Generator polynomial	BKLC	Opt.
27	2		N	475	N	N
28	2		N	83	N	N
28	2	2	Y	101	N	N
30	2		N	11	Y	Y
30	2	2	Y	10	Y	Y
31	2	2	Y	4	Y	Y

Table A.97 Best cyclic codes of length 34 over GF(3), where $\mathrm{ord}_{34}(3) = 16$. BCH ratio $= 10/15$.

k	d	δ	BCH	Generator polynomial	BKLC	Opt.
1	34	18	Y	10423238562291605	Y	Y
2	17	17	Y	2084647712458321	N	N
16	4		N	516560660	N	N
17	2		N	129140165	N	N
17	4	3	Y	209852765	N	N
18	2	2	Y	64570081	N	N
32	2		N	11	Y	Y
33	2	2	Y	4	Y	Y

Table A.98 Best cyclic codes of length 35 over GF(3), where $\mathrm{ord}_{35}(3) = 12$. BCH ratio $= 31/20$.

k	d	δ	BCH	Generator polynomial	BKLC	Opt.
1	35	35	Y	25015772549499853	Y	Y
4	14	8	Y	1033709609483465	N	N
5	7	7	Y	206741921896693	N	N
6	10		N	114436287966605	N	N
7	5	5	Y	22887257593321	N	N
10	10	6	Y	1229480779481	N	N
11	5	5	Y	472877163601	N	N
12	12		N	118219000112	N	N
13	12	10	Y	42829591849	Y	?
16	8		N	1325276708	N	N
17	7	6	Y	483112588	N	N
18	8		N	146733998	N	N
19	5	5	Y	54072517	N	N
22	6		N	2282144	N	N
22	6	3	Y	3010481	N	N
23	5	4	Y	606256	N	N
24	4		N	264869	N	N
25	4	3	Y	102742	N	N
28	2		N	2189	N	N
29	2	2	Y	1093	N	N
30	2		N	245	N	N
31	2	2	Y	121	Y	Y
34	2	2	Y	5	Y	Y

Table A.99 Best cyclic codes of length 37 over GF(3), where $\mathrm{ord}_{37}(3) = 18$. BCH ratio $= 6/7$.

k	d	δ	BCH	Generator polynomial	BKLC	Opt.
1	37	37	Y	225141952945498681	Y	Y
18	11		N	1325779670	N	N
18	11	3	Y	2269619906	N	N
19	10	5	Y	486448516	Y	?
36	2	2	Y	5	Y	Y

Table A.100 Best cyclic codes of length 38 over GF(3), where $\mathrm{ord}_{38}(3) = 18$. BCH ratio = 10/15.

k	d	δ	BCH	Generator polynomial	BKLC	Opt.
1	38	20	Y	844282323545620055	Y	Y
2	19	19	Y	168856464709124011	N	N
18	4		N	4649045876	N	N
19	2		N	1162261469	N	N
19	4	3	Y	1888674884	N	N
20	2	2	Y	581130733	N	N
36	2		N	11	Y	Y
37	2	2	Y	4	Y	Y

Table A.101 Best cyclic codes of length 40 over GF(3), where $\mathrm{ord}_{40}(3) = 4$, Part I. BCH ratio = 217/8191.

k	d	δ	BCH	Generator polynomial	BKLC	Opt.
1	40		N	7598540911910580500	Y	Y
1	40	40	Y	6078832729528464400	Y	Y
2	30		N	2147977784610820195	Y	Y
2	30	26	Y	2592770423356805395	Y	Y
3	25		N	685721984733393850	N	N
3	25	25	Y	593056851661313600	N	N
4	24		N	269518078524926060	N	N
5	24		N	88020914077284659	Y	Y
5	24	18	Y	69717251627682109	Y	Y
6	24		N	30169111205733439	Y	Y
6	24	23	Y	21544134618881872	Y	Y
7	22		N	9727564704488107	Y	?
7	22	21	Y	9555551491346809	Y	?
8	20		N	3667732925873273	N	N
8	20	20	Y	1986377976354413	N	N
9	15	15	Y	939923230986068	N	N
9	18		N	990318766420418	N	N
10	10	10	Y	410834118805258	N	N
10	18		N	400320902298520	N	N
11	16	15	Y	68971616896750	N	N
11	18		N	114176859380723	Y	?
12	13	13	Y	40585351568087	N	N
12	16		N	44219595416365	N	N
13	15		N	9125278372619	N	N
13	15	13	Y	14658368012507	N	N
14	10	8	Y	4765673191186	N	N
14	14		N	5076208691951	N	N
15	14		N	1583512156688	N	N
15	14	14	Y	1076940698464	N	N
16	11	11	Y	369251122709	N	N
16	13		N	447963372025	N	N

(*Continued*)

Table A.101   (*Continued*)

k	d	δ	BCH	Generator polynomial	BKLC	Opt.
17	12		N	97252565299	N	N
17	12	12	Y	137195022884	N	N
18	10	10	Y	37971685010	N	N
18	12		N	38967116282	N	N
19	10	10	Y	13517694604	N	N
19	11		N	12979514963	N	N
20	11		N	5126578711	N	N
20	8	7	Y	6960317036	N	N
21	10		N	1185364897	Y	?
21	9	9	Y	2229524483	N	N
22	8	8	Y	707638982	N	N
22	9		N	475269983	Y	?
23	8		N	254827805	N	N
23	8	7	Y	172972645	N	N
24	5	5	Y	72731809	N	N
24	8		N	81683981	N	N
25	8		N	28364336	Y	?
25	8	8	Y	15295837	Y	?

Table A.102 Best cyclic codes of length 40 over GF(3), where $\mathrm{ord}_{40}(3) = 4$, Part II. BCH ratio = 217/8191.

k	d	δ	BCH	Generator polynomial	BKLC	Opt.
26	7		N	8866928	Y	?
26	7	7	Y	6763658	Y	?
27	6		N	3188488	N	N
27	6	6	Y	1963912	N	N
28	5	5	Y	996682	N	N
28	6		N	1001243	Y	?
29	6		N	282641	Y	?
30	5		N	91673	Y	?
30	5	4	Y	101647	Y	?
31	5		N	37321	Y	?
31	5	5	Y	26404	Y	?
32	4		N	9953	N	N
32	4	3	Y	10066	N	N
33	4		N	4340	Y	Y
34	4		N	788	Y	Y
34	4	3	Y	1180	Y	Y
35	3		N	449	Y	Y
35	3	3	Y	376	Y	Y
36	2		N	83	N	N
36	2	2	Y	115	N	N
37	2		N	31	Y	Y
38	2		N	11	Y	Y
38	2	2	Y	10	Y	Y

Table A.103 Best cyclic codes of length 41 over GF(3), where $\text{ord}_{41}(3) = 8$. BCH ratio $= 34/63$.

k	d	δ	BCH	Generator polynomial	BKLC	Opt.
1	41	41	Y	18236498188585393201	Y	Y
8	22		N	9507241407762617	Y	?
8	22	9	Y	8970676717431806	Y	?
9	20	9	Y	3017502360555178	N	N
16	14		N	1594359042377	N	N
16	14	8	Y	1170956096486	N	N
17	12		N	500176523995	N	N
17	12	11	Y	348143964778	N	N
24	8		N	250019609	N	N
24	8	5	Y	212995538	N	N
25	7	7	Y	63411511	N	N
25	8		N	58933012	Y	?
32	5		N	39245	Y	?
32	5	3	Y	21737	Y	?
33	5	2	Y	10228	Y	Y
40	2	2	Y	5	Y	Y

Table A.104 Best cyclic codes of length 43 over GF(3), where $\text{ord}_{43}(3) = 42$. BCH ratio $= 3/3$.

k	d	δ	BCH	Generator polynomial	BKLC	Opt.
1	43	43	Y	164128483697268538813	Y	Y
42	2	2	Y	5	Y	Y

Table A.105 Best cyclic codes of length 44 over GF(3), where $\text{ord}_{44}(3) = 10$, Part I. BCH ratio $= 94/511$.

k	d	δ	BCH	Generator polynomial	BKLC	Opt.
1	44	23	Y	615481813864757020550	Y	Y
2	22		N	135405999050246544521	N	N
2	22	22	Y	123096362772951404110	N	N
3	22		N	615481813864757020555	N	N
3	22	22	Y	49238545109180561644	N	N
4	11		N	12309636277295140411	N	N
5	24		N	4775286092195164040	N	N
6	20	16	Y	2162486767691196920	N	N
6	24		N	2004722067942717460	N	N
7	20	14	Y	776185341843542780	N	N
7	22		N	479404510864397099	N	N
8	21		N	275521354347827029	N	N
8	21	12	Y	230424427523104162	N	N
9	11	11	Y	57813908237441347	N	N
10	18		N	29285809021959520	N	N

(Continued)

Codes from Difference Sets

Table A.105 (*Continued*)

k	d	δ	BCH	Generator polynomial	BKLC	Opt.
11	14		N	8762690180198350	N	N
12	14	6	Y	3420472735370780	N	N
12	18		N	2928770365321474	N	N
13	14		N	938069307701542	N	N
13	14	10	Y	1121086785435527	N	N
14	10		N	211296519605566	N	N
14	11	8	Y	365941989093914	N	N
15	12		N	111457285759640	N	N
16	12		N	23772281794496	N	N
17	12		N	8400010349129	N	N
17	12	6	Y	10127416453583	N	N
18	10		N	2996050448168	N	N
18	8	6	Y	4971152105387	N	N
19	10		N	988719236707	N	N
19	4	4	Y	1024591596181	N	N
20	12		N	395507643982	N	N
21	11		N	102353568656	N	N
22	10		N	43329852718	N	N
22	8	5	Y	60053478218	N	N

Table A.106 Best cyclic codes of length 44 over GF(3), where $\mathrm{ord}_{44}(3) = 10$, Part II. BCH ratio = 94/511.

k	d	δ	BCH	Generator polynomial	BKLC	Opt.
23	8	8	Y	19230574691	N	N
23	9		N	13532677312	N	N
24	4		N	6469223686	N	N
24	8	7	Y	6411652262	N	N
25	8		N	2103644477	N	N
26	8		N	680012990	N	N
27	4	4	Y	255182131	N	N
27	7		N	223298174	N	N
28	7		N	70175093	N	N
28	7	5	Y	68184167	N	N
29	2		N	24734198	N	N
29	4	4	Y	16912709	N	N
30	4		N	4848670	N	N
31	4		N	1771480	N	N
32	4		N	885730	N	N
32	4	4	Y	1008710	N	N
33	4		N	346847	N	N
33	4	3	Y	242752	N	N
34	2		N	91769	N	N
34	2	2	Y	60688	N	N
35	4		N	25240	N	N

(*Continued*)

Table A.106 *(Continued)*

k	d	δ	BCH	Generator polynomial	BKLC	Opt.
36	4		N	11251	Y	?
37	2		N	3935	N	N
37	4	3	Y	2303	Y	Y
38	2		N	1139	N	N
39	2	2	Y	314	N	N
40	2		N	83	Y	Y
41	2		N	40	Y	Y
42	2		N	11	Y	Y
42	2	2	Y	10	Y	Y
43	2	2	Y	4	Y	Y

Table A.107 Best cyclic codes of length 46 over GF(3), where $\text{ord}_{45}(3) = 11$. BCH ratio = 28/63.

k	d	δ	BCH	Generator polynomial	BKLC	Opt.
1	46	24	Y	5539336324782813184955	Y	Y
2	23	23	Y	1107867264956562636991	N	N
11	18		N	52218208714169104	N	N
12	18		N	26231761150961344	N	N
12	18	11	Y	29737558987396312	N	N
13	16	10	Y	10229115327400910	N	N
22	15		N	407306454703	Y	?
23	13		N	102214048954	N	N
23	9	6	Y	176611429222	N	N
24	13	6	Y	55979617211	Y	?
24	2		N	47071589413	N	N
33	4		N	2008832	N	N
34	4		N	922850	N	N
34	4	3	Y	1015856	N	N
35	2	2	Y	184642	N	N
44	2		N	11	Y	Y
45	2	2	Y	4	Y	Y

Table A.108 All cyclic codes of length 47 over GF(3), where $\text{ord}_{47}(3) = 23$. BCH ratio = 6/7.

k	d	δ	BCH	Generator polynomial	BKLC	Opt.
1	47	47	Y	13294407179478751643893	Y	Y
23	15		N	304620603343	Y	?
23	15	6	Y	491610829231	Y	?
24	14	5	Y	136552533239	Y	?
24	14	5	Y	98563278383	Y	?
46	2	2	Y	5	Y	Y

Table A.109   All cyclic codes of length 49 over GF(3), where ord$_{49}$(3) = 42.
BCH ratio = 6/7.

k	d	δ	BCH	Generator polynomial	BKLC	Opt.
1	49	49	Y	119649664615308764795041	Y	Y
6	14	8	Y	547345217819344761185	N	N
7	7	7	Y	109469043563868952237	N	N
42	2		N	2189	N	N
43	2	2	Y	1093	N	N
48	2	2	Y	5	Y	Y

Table A.110   Best cyclic codes of length 50 over GF(3), where ord$_{50}$(3) = 20.
BCH ratio = 21/63.

k	d	δ	BCH	Generator polynomial	BKLC	Opt.
1	50	26	Y	448686242307407867981405	Y	Y
2	25	25	Y	897372484614815735962281	N	N
4	20		N	119147139266023495629800	N	N
5	10		N	2978678481650587390745	N	N
5	20	11	Y	4802359184701967425895	N	N
6	10	10	Y	1471102433794779895021	N	N
8	10		N	133736584890434535911	N	N
9	10	10	Y	48631485414703467604	N	N
10	5	5	Y	12157871353675866901	N	N
20	4		N	206738420704580	N	N
21	4		N	102521921742724	N	N
22	4		N	25842302587981	N	N
24	4		N	3389154437780	N	N
25	4		N	854290994644	N	N
26	2		N	529555380901	N	N
26	4	4	Y	425380551871	N	N
28	4		N	38513118611	N	N
29	2		N	17505963005	N	N
29	4	3	Y	17534661305	N	N
30	2	2	Y	3501192601	N	N
40	2		N	59051	N	N
41	2		N	29524	N	N
42	2		N	7381	N	N
44	2		N	1220	N	N
45	2		N	244	N	N
46	2	2	Y	121	Y	Y
48	2		N	11	Y	Y
49	2	2	Y	4	Y	Y

Table A.111 Best cyclic codes of length 52 over GF(3), where $\text{ord}_{52}(3) = 6$, Part I. BCH ratio = 302/32767.

k	d	δ	BCH	Generator polynomial	BKLC	Opt.
1	52		N	40381761807666708118326501	Y	Y
1	52	52	Y	32305409446133336649466120	Y	Y
2	26		N	80763523615333416236653021	N	N
3	36		N	474785581000798403324990	Y	Y
3	36	35	Y	353136040705400631685310	Y	Y
4	34		N	158815102528423766686090	Y	Y
4	34	34	Y	129304796164114361289820	Y	Y
5	26		N	527312993223490429978101	N	N
6	30		N	96974298572512549190201	N	N
6	30	29	Y	147630963921296071024601	N	N
7	28	27	Y	36939640722875114748701	N	N
7	28	28	Y	36954481609863085720101	N	N
8	26		N	19508551059322421656331	N	N
8	26	26	Y	13477660399861489531401	N	N
9	22	21	Y	60626936789339068938501	N	N
9	26		N	61064131137897347540301	N	N
10	18	18	Y	21196468918173415697021	N	N
10	20		N	13396576201707993263021	N	N
11	14	12	Y	6608885977219848121021	N	N
11	20		N	6873715314145926726421	N	N
12	20		N	2377795865373633978421	N	N
12	20	18	Y	2022016065369910139521	N	N
13	18		N	807388568761161143821	N	N
13	18	18	Y	415560165738963211321	N	N
14	17	17	Y	250341169490873631821	N	N
14	18		N	142478115450401614721	N	N
15	16	16	Y	84567100839435142021	N	N
15	18		N	45059423815731716021	N	N
16	18		N	29921859100908178421	N	N
16	18		N	30009361973991196021	N	N
17	16		N	9789377282532112121	N	N
17	16	16	Y	75586351890440867211	N	N
18	12	12	Y	30788476938929432211	N	N
18	18		N	31186635329647880211	Y	?
19	15	15	Y	9527346815834791211	N	N
19	16		N	10085377175726954211	N	N
20	14		N	3690919170345434211	N	N
20	14	14	Y	2781161692955269211	N	N
21	14		N	991911145308842211	N	N
21	14	12	Y	1035228064895827211	N	N
22	13		N	409424316295003211	N	N
22	13	13	Y	278399761152532211	N	N
23	12		N	98655195409021211	N	N
23	12	10	Y	112427656106837211	N	N
24	12		N	45690201268673211	N	N
24	12	11	Y	23345146651469211	N	N
25	10	10	Y	14379791099957211	N	N
25	12		N	9941412040801211	N	N
26	12		N	5054627437486211	N	N
26	9	9	Y	3523089117100211	N	N

Table A.112   Best cyclic codes of length 52 over GF(3), where
$\text{ord}_{52}(3) = 6$, Part II. BCH ratio $= 302/32767$.

k	d	δ	BCH	Generator polynomial	BKLC	Opt.
27	10	10	Y	1577444953444	N	N
27	11		N	1024465873708	N	N
28	11		N	564012305582	N	N
28	8	8	Y	541336258139	N	N
29	10		N	99047066483	N	N
29	8	7	Y	130479282700	N	N
30	10		N	32013867407	Y	?
30	9	9	Y	47386937548	N	N
31	8	7	Y	12541922417	N	N
31	9		N	10469539138	N	N
32	8		N	6955534445	N	N
32	8	6	Y	5641110743	N	N
33	8		N	2321206715	N	N
33	8	8	Y	1367154002	N	N
34	7		N	772893185	N	N
34	7	5	Y	523227929	N	N
35	6		N	257593256	N	N
35	6	6	Y	200515579	N	N
36	6		N	86071247	N	N
36	6	6	Y	63580712	N	N
37	4	4	Y	28514735	N	N
37	6		N	28510570	N	N
38	5	5	Y	8669951	N	N
38	6		N	4879844	N	N
39	6		N	3176533	Y	?
39	6	6	Y	1908344	Y	?
40	5		N	998999	N	N
40	5	4	Y	727592	N	N
41	4	4	Y	348362	N	N
41	5		N	205034	N	N
42	5		N	99908	Y	?
42	5	5	Y	117053	Y	?
43	4		N	39280	N	N
43	4	3	Y	20815	N	N
44	4		N	9989	Y	?
44	4	4	Y	10490	Y	?
45	3		N	4360	N	N
45	3	3	Y	2992	N	N
46	2	2	Y	901	N	N
46	3		N	1000	N	N
47	3		N	433	Y	Y
47	3	3	Y	349	Y	Y
48	2		N	103	Y	Y
49	2		N	50	Y	Y
49	2	2	Y	34	Y	Y
50	2		N	11	Y	Y
50	2	2	Y	10	Y	Y
51	2	2	Y	4	Y	Y

Table A.113 Best cyclic codes of length 53 over GF(3), where $\mathrm{ord}_{53}(3) = 52$. BCH ratio = 3/3.

k	d	δ	BCH	Generator polynomial	BKLC	Opt.
1	53	53	Y	969162283384000948398361	Y	Y
52	2	2	Y	5	Y	Y

Table A.114 Best cyclic codes of length 55 over GF(3), where $\mathrm{ord}_{55}(3) = 20$. BCH ratio = 29/63.

k	d	δ	BCH	Generator polynomial	BKLC	Opt.
1	55	55	Y	8722460550456008953558525 3	Y	Y
4	22	12	Y	36043225415107475014704 65	N	N
5	11	11	Y	72086450830214950029409 3	N	N
5	30		N	14190628806980805552570 01	N	N
6	25	20	Y	30921980884052553390055 4	N	N
9	22		N	12069415303739011301920	N	N
9	22	11	Y	16134870532600173004888	N	N
10	10		N	4923882306377795125805	N	N
10	11	11	Y	3580997870723682936245	N	N
11	5	5	Y	984776461275559025161	N	N
14	10	6	Y	52901214858408983201	N	N
15	5	5	Y	20346414357645723121	N	N
20	12		N	64650907532831408	N	N
21	12		N	31348720985770012	N	N
24	10		N	1223451762896495	N	N
25	12		N	374002133526238	N	N
26	10	7	Y	83629896889325	N	N
26	6		N	102525401382074	N	N
29	10		N	3288896752189	N	N
30	10	6	Y	1168429671428	N	N
30	6		N	1674835305365	N	N
31	5	3	Y	364957557871	N	N
34	6		N	19022362793	N	N
34	6	3	Y	14244983753	N	N
35	5	4	Y	4252215121	N	N
40	4		N	21435029	N	N
41	4		N	8325862	N	N
44	2		N	177149	N	N
45	4		N	94492	N	N
46	4	3	Y	27104	N	N
49	2		N	1441	N	N
50	2		N	245	N	N
50	2	2	Y	314	N	N
51	2	2	Y	121	Y	Y
54	2	2	Y	5	Y	Y

Codes from Difference Sets

Table A.115 Best cyclic codes of length 56 over GF(3), where ord$_{56}(3) = 6$, Part I. BCH ratio = 249/8191.

k	d	δ	BCH	Generator polynomial	BKLC	Opt.
1	56		N	3270922706421003357584444700	Y	Y
1	56	56	Y	26167381651368026860675576 0	Y	Y
2	42		N	92463400408340070522935953	Y	Y
2	42	36	Y	1116102650312922852990400 33	Y	Y
3	35		N	29518082960384664446493790	N	N
3	35	35	Y	25529152830602953034805440	N	N
4	28		N	98925467218586443009871 08	N	N
5	21		N	29518082960384664446493 79	N	N
5	21	21	Y	25529152830602953034805 44	N	N
6	30		N	8556362228074943316731 50	N	N
7	30		N	4737976071183895653195 10	N	N
8	30		N	14492234234305540928449 7	N	N
8	30	30	Y	12132426310643882409217 7	N	N
9	29		N	5122269548416426421728 0	N	N
9	29	29	Y	31328500845012244739252	N	N
10	28		N	17441604238556150742563	N	N
10	28	28	Y	14016108068253509727679	N	N
11	21		N	5855925614727222334841	N	N
11	21	15	Y	3144197794835749249664	N	N
12	14	14	Y	1924312504067646314543	N	N
12	22		N	1966825950325069474411	N	N
13	22		N	647942348400087100063	N	N
14	22		N	214848913950060402395	N	N
15	16	16	Y	43384372776118315432	N	N
15	22		N	72225103938466305805	N	N
16	17	12	Y	22650049189140394816	N	N
16	21		N	22653650740746984730	N	N
17	18		N	7176036295097328607	N	N
17	18	12	Y	4767520402054721372	N	N
18	14	8	Y	2267413859274620491	N	N
18	16		N	2467160589635769961	N	N
19	14	14	Y	548266644749778832	N	N
19	16		N	870768242350076824	N	N
20	16		N	296433358183818043	N	N
20	7	7	Y	266245228233080723	N	N
21	16		N	98992421970818098	N	N
22	16		N	33038810140065833	N	N
23	15		N	9248149384631953	N	N
23	15	11	Y	11039739619440745	N	N
24	14		N	3683686397878402	N	N
24	8	8	Y	3689777563932482	N	N
25	14		N	800797271049559	N	N
25	14	12	Y	756311781380159	N	N
26	13		N	404866865532007	N	N
26	7	6	Y	300450808979051	N	N
27	12		N	99058610574635	N	N
28	12		N	45610252033351	N	N
29	12		N	9969344330521	N	N
29	12	9	Y	10417830388360	N	N

Table A.116 Best cyclic codes of length 56 over GF(3), where $\mathrm{ord}_{56}(3) = 6$, Part II. BCH ratio = 249/8191.

k	d	δ	BCH	Generator polynomial	BKLC	Opt.
30	11		N	4938492360286	N	N
30	8	8	Y	5046191117966	N	N
31	10		N	992672212115	N	N
31	10	10	Y	941089566484	N	N
32	10		N	561418326958	N	N
32	7	5	Y	332216140429	N	N
33	10		N	103258132723	N	N
34	10		N	32014024295	Y	?
35	9		N	20414933518	N	N
36	8		N	6966821716	N	N
36	8	5	Y	3980226364	N	N
37	8		N	2235933340	N	N
37	8	8	Y	2023808308	N	N
38	5	5	Y	675292943	N	N
38	8		N	761038526	Y	?
39	8		N	256854262	Y	?
40	6		N	85255012	N	N
41	6		N	24419344	N	N
42	5	5	Y	8801726	N	N
42	6		N	5547265	Y	?
43	5		N	3158293	N	N
43	5	5	Y	1695922	N	N
44	4		N	993383	N	N
44	4	3	Y	1029713	N	N
45	4		N	350917	N	N
46	4		N	98909	N	N
47	4		N	39070	N	N
48	4		N	9770	Y	?
48	4	3	Y	11036	Y	?
49	3		N	4318	N	N
49	3	3	Y	2696	N	N
50	2		N	902	N	N
50	2	2	Y	1003	N	N
51	2		N	310	N	N
52	2		N	113	Y	Y
53	2		N	31	Y	Y
54	2		N	11	Y	Y
54	2	2	Y	10	Y	Y
55	2	2	Y	4	Y	Y

Codes from Difference Sets

Table A.117   Best cyclic codes of length 58 over GF(3), where ord$_{58}$(3) = 28. BCH ratio = 10/15.

k	d	δ	BCH	Generator polynomial	BKLC	Opt.
1	58	30	Y	29438304357789030021826002305	Y	Y
2	29	29	Y	588766087155780604365200461	N	N
28	4		N	274521509459540	N	N
29	2		N	68630377364885	N	N
29	4	3	Y	111524363217935	N	N
30	2	2	Y	34315188682441	N	N
56	2		N	11	Y	Y
57	2	2	Y	4	Y	Y

Table A.118   All cyclic codes of length 59 over GF(3), where ord$_{59}$(3) = 28. BCH ratio = 6/7.

k	d	δ	BCH	Generator polynomial	BKLC	Opt.
1	59	59	Y	7065193045869367252382405533	Y	Y
29	18		N	390640891274893	Y	?
29	18	3	Y	242576052016381	Y	?
30	17	6	Y	108531558226613	Y	?
30	17	6	Y	91096951570550	Y	?
58	2	2	Y	5	Y	Y

Table A.119   Best cyclic codes of length 61 over GF(3), where ord$_{61}$(3) = 10. BCH ratio = 55/127.

k	d	δ	BCH	Generator polynomial	BKLC	Opt.
1	61	61	Y	63586737412824305271441649801	?	?
10	31		N	4293469233431552037242480	N	N
10	31	11	Y	2759131622077582440418247	N	N
11	31	8	Y	893120896158112798766164	Y	?
20	20	9	Y	37530442054008887096	N	N
20	21		N	42580289861133028166	Y	?
21	21		N	19187786343716404996	Y	?
21	21	6	Y	13383211024142974414	Y	?
30	15		N	694543308588668	N	N
30	15	6	Y	1190597227011581	N	N
31	14		N	335780949034516	N	N
31	14	5	Y	252092463798547	N	N
40	10		N	19966021061	Y	?
40	10	5	Y	19839255818	Y	?
41	10	4	Y	6810631684	Y	?
41	8		N	4337495182	N	N
50	6		N	351824	Y	Y
50	6	3	Y	285758	Y	Y
51	5	2	Y	109663	Y	?
60	2	2	Y	5	Y	Y

Table A.120   Best cyclic codes of length 62 over GF(3), where $\text{ord}_{62}(3) = 30$. BCH ratio $= 10/15$.

k	d	δ	BCH	Generator polynomial	BKLC	Opt.
1	62	32	Y	2384502652980911447679061867555	Y	Y
2	31	31	Y	4769005305961822895358123735 1	N	N
30	4		N	2470693585135796	N	N
31	2		N	617673396283949	N	N
31	4	3	Y	1003719268961414	N	N
32	2	2	Y	308836698141973	N	N
60	2		N	11	Y	Y
61	2	2	Y	4	Y	Y

Table A.121   Best cyclic codes of length 64 over GF(3), where $\text{ord}_{64}(3) = 16$, Part I. BCH ratio $= 131/2047$.

k	d	δ	BCH	Generator polynomial	BKLC	Opt.
1	64		N	214605238768282030291115568 0800	Y	Y
1	64	64	Y	171684191014625624232892454 4640	Y	Y
2	48		N	606652370079119202700982788 792	Y	Y
2	48	41	Y	732274948870308683847001657 912	Y	Y
3	40		N	162262497605286413146941039 280	N	N
3	40	33	Y	193668142303083783433445756 560	N	N
4	32		N	774672569212335133733783026 24	N	N
4	32	32	Y	591472975141850473729172175 44	N	N
5	24		N	275376900167019413457272605 96	N	N
5	24	24	Y	167496771721585974861358492 16	N	N
6	16	16	Y	523427411629956171441745288 0	N	N
6	24		N	870802302655146072540748115 6	N	N
7	24		N	297720279056331669352205137 6	N	N
7	24	21	Y	218185156816735616541390717 2	N	N
8	20		N	999952153641135410727479264	N	N
8	20	20	Y	100035098586578487443628888 4	N	N
9	20		N	320980174397888392849982176	N	N
9	20	17	Y	276550264571938135688590508	N	N
10	12		N	915859674124590203727844 82	N	N
10	16	16	Y	888598196519005143227833 36	N	N
11	12		N	378890613416990523369139 00	N	N
11	12	12	Y	293540517342005289683880 32	N	N
12	12		N	988713572312921734119479 0	N	N
13	12		N	420996661086253671735830 6	N	N
14	12		N	999695341734964406603672	N	N
15	12		N	467837265950720511267596	N	N
15	8	8	Y	319065779719570967047696	N	N
16	12		N	153286305065580447438430	N	N
16	12	12	Y	121926250938458708687246	N	N

(*Continued*)

Table A.121 (*Continued*)

k	d	δ	BCH	Generator polynomial	BKLC	Opt.
17	12		N	50422562007170494291472	N	N
17	12	11	Y	41751459655589064207466	N	N
18	10		N	99841487513614676621522	N	N
18	10	10	Y	11054321648008883079940	N	N
19	10		N	5700086521048284287252	N	N
19	10	10	Y	43829041802067294383392	N	N
20	8		N	18214966093990944463896	N	N
20	8	8	Y	1095726045051682359598	N	N
21	8		N	581134926526770428830	N	N
22	6		N	206721084050156444575	N	N
22	6	6	Y	123114661347316382480	N	N
23	6		N	70371338269300351943	N	N
23	6	6	Y	49245864538926552992	N	N
24	6		N	23242268137290224273	N	N
25	6		N	7819449210836713238	N	N
26	6		N	2583613433652339424	N	N
27	6		N	880358848732682006	N	N
28	4	4	Y	187155039074036042	N	N
28	6		N	277674451086812413	N	N

Table A.122 Best cyclic codes of length 64 over GF(3), where ord$_{64}$(3) = 16, Part II. BCH ratio = 131/2047.

k	d	δ	BCH	Generator polynomial	BKLC	Opt.
29	6		N	97818078210848872	N	N
30	6		N	32010846837891911	N	N
31	4	4	Y	7412080755407368	N	N
31	6		N	9859872727730800	N	N
32	6		N	3560774670918107	N	N
33	6		N	970076876160359	N	N
34	6		N	380664365768527	N	N
34	6	6	Y	211578475515203	N	N
35	6		N	133763839356058	N	N
35	6	6	Y	101828243234528	N	N
36	5		N	42321971542216	N	N
36	5	5	Y	25450479922241	N	N
37	5		N	9251833519661	N	N
38	5		N	4878752207147	N	N
38	5	5	Y	2859168590470	N	N
39	5		N	1436558926514	N	N
39	5	5	Y	1143667436188	N	N
40	4		N	539931206132	N	N
40	4	4	Y	285916859047	N	N

(*Continued*)

Table A.122 *(Continued)*

k	d	δ	BCH	Generator polynomial	BKLC	Opt.
41	4		N	99941644327	N	N
42	4		N	57135630617	N	N
43	4		N	20451241229	N	N
44	3		N	6372325538	N	N
44	3	3	Y	3960902125	N	N
45	3		N	2152795390	N	N
46	3		N	473638609	N	N
46	3	3	Y	430532840	N	N
47	3		N	215266417	N	N
47	3	3	Y	172213136	N	N
48	2		N	82722415	N	N
48	2	2	Y	43053284	N	N
49	2		N	28053950	N	N
50	2		N	9180238	N	N
51	2		N	2018330	N	N
52	2		N	531523	N	N
53	2		N	345229	N	N
54	2		N	93904	N	N
55	2		N	37978	N	N
56	2		N	9167	N	N
56	2	2	Y	6644	N	N
57	2		N	4219	N	N
58	2		N	902	N	N
59	2		N	310	N	N
60	2		N	83	Y	Y
60	2	2	Y	101	Y	Y
61	2		N	31	Y	Y
62	2		N	11	Y	Y
62	2	2	Y	10	Y	Y

Table A.123 Best cyclic codes of length 65 over GF(3), where $\mathrm{ord}_{65}(3) = 12$, Part I. BCH ratio $= 146/1023$.

k	d	δ	BCH	Generator polynomial	BKLC	Opt.
1	65	65	Y	515052573043876872698677363392 1	Y	Y
3	45		N	691581717036715201821476811118	Y	Y
3	45	36	Y	447042536312060434743063029590	Y	Y
4	26		N	212831641753668129214329489005	N	N
4	35	30	Y	140056959960059749153423576997	N	N

(Continued)

Table A.123 *(Continued)*

k	d	δ	BCH	Generator polynomial	BKLC	Opt.
5	13		N	4256632835073362584286589780l	N	N
6	30		N	2678120122869620612819766234 5	N	N
6	30	21	Y	19299263708109907769584177907	N	N
7	25	25	Y	74690153487027296222428221l6	N	N
7	26		N	56151906182882739445625462 38	N	N
8	13		N	27470017777164072619850127 80	N	N
8	13	13	Y	1827576501414684835674936434	N	N
9	15		N	54919230505167129575314868 5	N	N
10	26		N	3398740039439221327807118 48	N	N
10	26	14	Y	2179682005557094552004078l8	N	N
11	13	13	Y	928847729310368710750693 51	N	N
12	18		N	3804924036657863942458399 7	N	N
13	15	11	Y	110265635724706045397066 05	N	N
13	18		N	11439293319136150989222 322	N	N
14	13	8	Y	411159828595241492368530 5	N	N
15	18		N	1035884001855814893471409	N	N
16	18		N	4631210357665812589275 68	N	N
17	13		N	10340095672010808593499 1	N	N
18	18		N	2715268324439199043267 1	N	N
19	18		N	16747443117781026541738	N	N
19	18	10	Y	11864402567721806000344	N	N
20	13		N	35494038518018764408 07	N	N
21	15		N	1142390105896006176808	N	N
22	14		N	6352931287758048378 68	N	N
22	14	12	Y	383184405407910348977	N	N
23	13		N	20833677818967249772 3	N	N
23	13	10	Y	1108440465850731646 21	N	N
24	12		N	40893273115389615707	N	N
25	12		N	2391570133188360639 7	N	N
25	9	8	Y	16786682410179568549	N	N
26	11	9	Y	7236048249036361484	N	N
26	7		N	4773128180412693503	N	N
27	12		N	1371764950263949963	N	N
28	12		N	896063208216913838	N	N
29	10		N	238986567004515412	N	N
29	5	5	Y	157509493213699255	N	N
30	12		N	59253988277579042	N	N
31	12		N	30087304873256953	N	N
32	10		N	11115853717819853	N	N
33	9		N	3528125585891704	N	N
34	10		N	1019354685586424	N	N
34	9	5	Y	1094537963503460	N	N
35	10		N	368598487370164	N	N
35	10	7	Y	294329641364644	N	N
36	6		N	107863682966102	N	N

Table A.124 Best cyclic codes of length 65 over GF(3), where $\text{ord}_{65}(3) = 12$, Part II. BCH ratio = 146/1023.

k	d	δ	BCH	Generator polynomial	BKLC	Opt.
37	5	5	Y	30943678059757	N	N
37	9		N	44274858155569	N	N
38	7		N	11414932472069	N	N
38	7	5	Y	12397947813650	N	N
39	6		N	2838242701030	N	N
40	6		N	986270168786	N	N
40	6	5	Y	1361112893834	N	N
41	5	5	Y	331321682215	N	N
41	6		N	312149115736	N	N
42	6		N	106472819372	N	N
43	6		N	31720288321	N	N
44	6		N	11728077812	N	N
45	6		N	6926129755	N	N
46	6		N	2299744607	N	N
46	6	4	Y	1200001370	N	N
47	6		N	771744034	N	N
48	4		N	192913325	N	N
49	3	3	Y	67654882	N	N
49	6		N	86012713	N	N
50	3		N	14467493	N	N
50	5	4	Y	25604054	N	N
51	4		N	5307190	N	N
52	3	3	Y	2853674	N	N
52	4		N	1777166	N	N
53	2		N	797161	N	N
53	3	2	Y	618613	N	N
54	4		N	197243	N	N
55	4		N	98251	N	N
56	2		N	21677	N	N
57	3		N	10042	N	N
58	3		N	3872	N	N
58	3	3	Y	3146	N	N
59	2		N	1156	N	N
60	2		N	245	N	N
61	2		N	85	Y	Y
61	2	2	Y	121	Y	Y
62	2	2	Y	53	Y	Y
64	2	2	Y	5	Y	Y

Table A.125 Best cyclic codes of length 67 over GF(3), where $\text{ord}_{67}(3) = 22$. BCH ratio $= 11/15$.

k	d	δ	BCH	Generator polynomial	BKLC	Opt.
1	67	67	Y	4635473157394891854288096270529 3	Y	Y
22	22		N	5556587028842867436725	N	N
22	22	5	Y	5093342083005353705525	N	N
23	22	8	Y	1049018088623514962527	Y	?
44	11		N	153530269388	Y	?
44	11	3	Y	117414954065	Y	?
45	10	3	Y	42992737015	Y	?
66	2	2	Y	5	Y	Y

Table A.126 Best cyclic codes of length 68 over GF(3), where $\text{ord}_{68}(3) = 16$. BCH ratio $= 42/127$.

k	d	δ	BCH	Generator polynomial	BKLC	Opt.
1	68	35	Y	17383024340230844453580361014485 0	Y	Y
2	34		N	38242653548507857797876794231867	N	N
2	34	34	Y	34766048680461688907160722028970	N	N
3	34	18	Y	17383024340230844453580361014485	N	N
4	17	17	Y	34766048680461688907160722202897	N	N
16	24		N	1219469297172561451293515 0	N	N
17	24		N	2462007739537910525437168	N	N
18	20		N	979882286557624687913990	N	N
18	24	8	Y	1181216782968604744084360	N	N
19	20		N	247186762707168282400937	N	N
19	20	6	Y	246202139814972255767923	N	N
20	17	10	Y	119155079860880313907726	N	N
32	8		N	295200480118499408	N	N
33	8		N	93896055573030767	N	N
34	4	4	Y	28534461258802276	N	N
34	8		N	20902350440652926	N	N
35	8		N	9410236645925252	N	N
35	8	5	Y	6390135062491054	N	N
36	2		N	2293112483704153	N	N
36	4	4	Y	1906472256638236	N	N
48	4		N	4702597646	N	N
49	4		N	1181733865	N	N
50	4		N	759149864	N	N
50	4	3	Y	645700810	N	N
51	4		N	162330182	N	N
51	4	3	Y	161857861	N	N
52	2	2	Y	58755868	N	N
64	2		N	83	Y	Y
65	2		N	40	Y	Y
66	2		N	11	Y	Y
66	2	2	Y	10	Y	Y
67	2	2	Y	4	Y	Y

Table A.127 Best cyclic codes of length 70 over GF(3), where ord$_{70}$(3) = 12, Part I.
BCH ratio = 164/1023.

k	d	δ	BCH	Generator polynomial	BKLC	Opt.
1	70	36	Y	156447219062077600082232491303655	Y	Y
2	35	35	Y	312894438124155200164446498260731	N	N
4	28		N	415440386616545313861478889943980	N	N
5	28		N	167447910932178978546208327888645	N	N
5	28	28	Y	205176680737150950927505900499884	N	N
6	14	14	Y	640117330398962677480441962300	N	N
6	20		N	458243283286043800860033943571	N	N
7	20		N	228807632997554076902703106589	N	N
8	14		N	466310638038979433926149773861	N	N
9	14		N	169567504741447066882236281404	N	N
10	20		N	758006913866008212711739548	N	N
10	7	7	Y	423918761853617667205590703	N	N
11	20		N	169482486698926401699440358	N	N
11	20	20	Y	164321597941609446018504188	N	N
12	10	10	Y	475114806286076497086514480	N	N
12	24		N	589718112963294725080419853	N	N
13	24		N	265674147357178083355984106	N	N
13	24	20	Y	214283065616499379181018009	N	N
14	14		N	85355101371512916430650842	N	N
14	24	19	Y	54005360483764590999697506	N	N
15	14		N	25210173394771407472863918	N	N
15	14	14	Y	31034109874686646464644939576	N	N
16	24		N	99351437904052494158918006	N	N
16	7	6	Y	7760521663519876399751904	N	N
17	24		N	2119436431347395223359231	N	N
17	24	15	Y	2055635316141106807261494	N	N
18	14	14	Y	948028571547912227727344	N	N
18	20		N	913806302727096206714687	N	N
19	20		N	389073762166526464371807	N	N
20	10	10	Y	97537545331369750943987	N	N
20	14		N	10879636946450515220048	N	N
21	10	10	Y	38997596307478860386352	N	N
21	14		N	27201238325415158602609	N	N
22	20		N	13775794807709146980629	N	N
22	5	5	Y	9749399076869715096588	N	N
23	20		N	4700528070081976454124	N	N
23	20	11	Y	284281442764740021290	N	N
24	10	9	Y	172483549176571787908	N	N
24	18		N	165933449408695124865	N	N
25	18		N	51208141795488624554	N	N
26	18		N	19079880993910653783	N	N
27	14		N	4315415889065553445	N	N
27	14	8	Y	4209036033477963499	N	N
28	16		N	2187490240057460715	N	N
28	7	7	Y	2146402336340189304	N	N
29	16		N	682460800165639056	N	N
30	14	8	Y	210386915396967315	N	N
30	16		N	122083189966565632	N	N
31	16		N	75551478588590663	N	N
32	12		N	22656248843193467	N	N

Table A.128 Best cyclic codes of length 70 over GF(3), where
$\text{ord}_{70}(3) = 12$, Pat II. BCH ratio = 164/1023.

k	d	δ	BCH	Generator polynomial	BKLC	Opt.
33	10	6	Y	887418969864884203	N	N
33	12		N	452581030498037957	N	N
34	16		N	251364125448442916	N	N
34	5	5	Y	282369501927605269	N	N
35	14		N	99690108497592229	N	N
36	10		N	27970209874852316	N	N
36	10	6	Y	21175008248859106	N	N
37	10		N	11058945476472928	N	N
38	10		N	2274276533109911	N	N
39	10		N	1033730676894764	N	N
40	10		N	252511501424333	N	N
40	7	6	Y	387967589823643	N	N
41	10		N	114444416739181	N	N
42	5	5	Y	45734975068879	N	N
42	8		N	31002697626532	N	N
43	8		N	9488088783611	N	N
44	10		N	4158056619041	N	N
45	7		N	1233289939331	N	N
45	7	5	Y	1039560788717	N	N
46	5	4	Y	473685919843	N	N
46	8		N	303043292393	N	N
47	8		N	99444702442	N	N
48	4		N	61365838403	N	N
48	8	5	Y	56290507777	N	N
49	4		N	20350118198	N	N
50	4		N	3879251918	N	N
51	4		N	2273325679	N	N
52	4		N	728765039	N	N
52	4	4	Y	573114418	N	N
53	4		N	253437371	N	N
54	4		N	83104108	N	N
54	4	4	Y	63394216	N	N
55	4		N	16149628	N	N
56	4		N	9218986	N	N
57	4		N	2995481	N	N
57	4	3	Y	1774588	N	N
58	2	2	Y	884206	N	N
58	4		N	1059116	N	N
59	4		N	331664	N	N
60	2		N	85027	N	N
60	4	3	Y	66673	N	N
61	2		N	29524	N	N
62	2		N	10940	N	N
63	2		N	2188	N	N
64	2		N	980	N	N
64	2	2	Y	1093	N	N
65	2		N	244	N	N
66	2	2	Y	121	Y	Y
68	2		N	11	Y	Y
69	2	2	Y	4	Y	Y

Table A.129   All cyclic codes of length 71 over GF(3), where $\mathrm{ord}_{71}(3) = 35$. BCH ratio $= 6/7$.

k	d	δ	BCH	Generator polynomial	BKLC	Opt.
1	71	71	Y	3754733257489862401973357979128773	Y	Y
35	18		N	250292001780904501	Y	?
35	18	8	Y	150240565477396285	Y	?
36	17	7	Y	50060115723561974	Y	?
36	17	7	Y	75017190660445883	Y	?
70	2	· 2	Y	5	Y	Y

Table A.130   Best cyclic codes of length 73 over GF(3), where $\mathrm{ord}_{73}(3) = 12$. BCH ratio $= 60/127$.

k	d	δ	BCH	Generator polynomial	BKLC	Opt.
1	73	73	Y	33792599317408761617760221812158961	Y	Y
12	34		N	201001020643024185952190227706	N	N
12	34	14	Y	244087285136168310072406844270	N	N
13	33	9	Y	609259113863407591546460010808	N	N
24	24		N	4766767861468358607222 11	Y	?
24	24	8	Y	2686392163912783472401 52	Y	?
25	24		N	146441825064828547405783	Y	?
25	24	11	Y	151171290842454484528669	Y	?
36	18		N	755184048973189421	Y	?
36	18	6	Y	452607100023170297	Y	?
37	16		N	259036294150620430	N	N
37	17	5	Y	225958687268352196	Y	?
48	11		N	1644978270902	Y	?
48	11	5	Y	1100785337354	Y	?
49	10		N	478611359422	Y	?
49	10	3	Y	365284250875	Y	?
60	6		N	2963192	Y	?
60	6	3	Y	2408534	Y	?
61	5	2	Y	594388	N	N
72	2	2	Y	5	Y	Y

Table A.131   Best cyclic codes of length 74 over GF(3), where $\mathrm{ord}_{74}(3) = 18$. BCH ratio $= 24/63$.

k	d	δ	BCH	Generator polynomial	BKLC	Opt.
1	74	38	Y	126722247440282856066600831795596105	Y	Y
2	37	37	Y	253444494988056571213320166359119221	N	N
18	22		N	1021973316161638283527927784	N	N
19	20	6	Y	339395223861949731212088347	N	N
19	22		N	257135188495931661015627956	N	N
20	20	10	Y	79048087375590318905265253	N	N
36	16		N	2260786221629073059	N	N

(Continued)

Table A.131   (*Continued*)

k	d	δ	BCH	Generator polynomial	BKLC	Opt.
37	14		N	454051336673941220	N	N
37	14	5	Y	731970282581053640	N	N
38	14	5	Y	279818288423836072	N	N
38	2		N	225141952945498681	N	N
54	4		N	3895153940	N	N
55	4		N	2281138349	N	N
55	4	3	Y	1638405665	N	N
56	2	2	Y	485033284	N	N
72	2		N	11	Y	Y
73	2	2	Y	4	Y	Y

Table A.132   Best cyclic codes of length 76 over GF(3), where ord$_{76}(3) = 18$. BCH ratio $= 35/127$.

k	d	δ	BCH	Generator polynomial	BKLC	Opt.
1	76	39	Y	11405002269625457045994074861603649550	Y	Y
2	38		N	25091004993176005501186964695528028	N	N
2	38	38	Y	22810004539250914091988149723207299	N	N
3	38	20	Y	11405002269625457045994074861603649	N	N
4	19	19	Y	22810004539250914091988149723207299	N	N
18	20		N	71473775385116312054301548333	N	N
19	20		N	17884685185449035124359386100	N	N
20	18		N	88794878363182625864916431333	N	N
20	20	8	Y	10192599784595883500099645299	N	N
21	18		N	17893659182154091462923666644	N	N
21	18	14	Y	34472967065302230385962719222	N	N
22	18	7	Y	107751935784886273210152724	N	N
36	8		N	23945015316957904904	N	N
37	8		N	7638062869137498323	N	N
38	4	4	Y	2315536232797740760	N	N
38	8		N	1686872928452976659	N	N
39	8		N	763829530883953232	N	N
39	8	5	Y	537819087966449132	N	N
40	2		N	185742111180036412	N	N
40	4	4	Y	153878427462190888	N	N
54	4		N	42385757768	N	N
55	4		N	10596459125	N	N
56	4		N	6822659147	N	N
56	4	3	Y	5811307330	N	N
57	4		N	1451902096	N	N
57	4	3	Y	1721799935	N	N
58	2	2	Y	529580515	N	N
72	2		N	83	Y	Y
73	2		N	40	Y	Y
74	2		N	11	Y	Y
74	2	2	Y	10	Y	Y
75	2	2	Y	4	Y	Y

Table A.133 Best cyclic codes of length 77 over GF(3), where ord$_{77}$(3) = 30. BCH ratio = 29/63.

k	d	δ	BCH	Generator polynomial	BKLC	Opt.
1	77	77	Y	273720054471010969103857796678487588 1	Y	Y
5	42		N	4453169684810572143640376695084287 7	N	N
5	42	15	Y	2747305933238444577166061680728613 3	N	N
6	22		N	1252150294926857132222588273918058 5	N	N
6	35	28	Y	1202139491597024679997298080768763 3	N	N
7	11	11	Y	250430058985371426444517654783611 7	N	N
10	14		N	154516644164141989716876359995985	N	N
11	22		N	42268716157017485783120999007634	N	N
11	22	12	Y	56180100653944010229317075833522	N	N
12	11	11	Y	12567548758460292404079160047143	N	N
16	14	8	Y	1837800891247800426637837 33061	N	N
17	7	7	Y	7068464965932250757794891 5022	N	N
30	12		N	3438434170860459985519 1	N	N
31	12		N	1224932833860768425362 3	N	N
35	12		N	150891699340177348597	N	N
36	10		N	72095475626540846255	N	N
36	11	7	Y	44476085523610637744	N	N
37	10		N	15709910758877894449	N	N
40	6		N	549114539696319794	N	N
41	10		N	194285832996670960	N	N
41	6	3	Y	240276347321672620	N	N
42	10	6	Y	69003189077615555	N	N
46	6		N	845316587591897	N	N
46	6	3	Y	1122893682818759	N	N
47	5	4	Y	251192549110813	N	N
60	4		N	193623857	N	N
61	4		N	74490145	N	N
65	4		N	687253	N	N
66	2		N	177149	N	N
66	4	3	Y	244832	N	N
67	2		N	88573	N	N
70	2		N	2189	N	N
71	2		N	889	N	N
71	2	2	Y	1093	N	N
72	2	2	Y	314	N	N
76	2	2	Y	5	Y	Y

Table A.134 Best cyclic codes of length 79 over GF(3), where ord$_{79}$(3) = 78. BCH ratio = 3/3.

k	d	δ	BCH	Generator polynomial	BKLC	Opt.
1	79	79	Y	246348049023909872193472017010638829 33	Y	Y
78	2	2	Y	5	Y	Y

References

Antweiler, M. and Bomer, L. (1992). Complex sequences over GF(p^M) with a two-level autocorrelation function and a large linear span, *IEEE Trans. Inform. Theory* **8**, pp. 120–130.

Arasu, K. T. (2011). Sequences and arrays with desirable correlation properties, in *Information Security, Coding Theory and Related Combinatorics*, Eds., Crnkovic, D. and Tonchev, V. D. (IOS Press), pp. 136–171.

Arasu, K. T., Dillon, J. F. and Player, K. J. (2004). New p-ary sequences with ideal autocorrelation, in *Sequences and their Applications — SETA 2004*, Eds., Helleseth, T., Sarwate, D., Song, H.-Y. and Yang, K. (Springer-Verlag, Berlin), pp. 1–5.

Arasu, K. T., Ding, C., Helleseth, T., Kumar, P. V. and Martinsen, H. (2001). Almost difference sets and their sequences with optimal autocorrelation, *IEEE Trans. Inform. Theory* **47**, pp. 2834–2943.

Arasu, K. T. and Player, K. J. (2003). A new family of cyclic difference sets with Singer parameters in characteristic three, *Designs, Codes and Cryptography* **28**, pp. 75–91.

Aschbacher, M. (2000). *Finite Group Theory* (Cambridge University Press, Cambridge).

Assmus, E. F. and Key, J. D. (1992a). *Designs and their Codes, Cambridge Tracts in Mathematics*, Vol. 103 (Cambridge University Press, Cambridge).

Assmus, E. F. and Key, J. D. (1992b). Hadamard matrices and their designs, *Trans. Amer. Math. Soc.* **330**, 1, pp. 269–293.

Assmus, E. F. and Mattson, H. F. (1969). New 5-designs, *J. Comb. Theory* **6**, pp. 122–151.

Ball, S. and Zieve, M. (2004). Symplectic spreads and permutation polynomials, in *Finite Fields and Applications, Lecture Notes in Comput. Sci.*, Vol. 2948 (Springer, Berlin), pp. 79–88.

Baumert, L. D. (1971). *Cyclic Difference Sets, Lecture Notes in Mathematics*, Vol. 182 (Springer, Berlin).

Baumert, L. D., Mills, W. H. and Ward, R. L. (1982). Uniform cyclotomy, *J. Number Theory* **14**, pp. 67–82.

Beth, T., Jungnickel, D. and Lenz, H. (1999). *Design Theory,* Vol. I, 2nd Edition, *Encyclopedia of Mathematics and its Applications*, Vol. 69 (Cambridge University Press, Cambridge).

Bierbrauer, J. (2010). New semifields, PN and APN functions, *Designs, Codes and Cryptography* **54**, pp. 189–200.

Blokhuis, A., Coulter, R. S., Henderson, M. and O'Keefe, C. M. (2001). Permutations amongst the Dembowski–Ostrom polynomials, in *Finite Fields and Applications: Proceedings of the Fifth International Conference on Finite Fields and Applications* (Springer-Verlag, Berlin), pp. 37–42.

330 *References*

Bose, R. C. (1942). An affine analogue of Singer's theorem, *J. Indian Math. Soc.* **6**, pp. 1–15.

Bose R. C. and Chowla, S. (1962). Theorems in the additive theory of numbers, *Comment. Math. Helv.* **37**, pp. 141–147.

Bose, R. C. and Ray-Chaudhuri, D. K. (1960). On a class of error correcting binary group codes, *Inform. and Control* **3**, pp. 68–79.

Brandstatter, N. and Winterhof, A. (2005). Some notes on the two-prime generator of order 2, *IEEE Trans. Inform. Theory* **51**, 10, pp. 3654–3657.

Budaghyan, L. and Helleseth, T. (2008). New perfect nonlinear multinomials over $F_{p^{2k}}$ for any odd prime p, in *Proc. of SETA 2008,* LNCS 5203 (Springer-Verlag), pp. 403–414.

Cai, Y. and Ding, C. (2009). Binary sequences with optimal autocorrelation, *Theoretical Computer Science* **410**, pp. 2316–2322.

Calderbank, A. R., Cameron, P. J., Kantor, W. M. and Seidel, J. J. (1997). Z_4-Kerdock codes, orthogonal spreads, and extremal Euclidean line-sets, *Proc. London Math. Soc.* **75**, 3, pp. 436–480.

Canteaut, A., Charpin, P. and Kyureghyan, G. (2008). A new class of monomial bent functions, *Finite Fields and their Applications* **14**, pp. 221–241.

Cao, X. (2007). A note on cyclic difference sets with Singer parameters, *Southeast Asian Bulletin of Mathematics* **31**, pp. 667–681.

Carlet, C. (2004). On the confusion and diffusion properties of Maiorana–McFarlands and extended Maiorana–McFarlands functions, *J. Complexity* **20**, pp. 182–204.

Carlet, C. (2010). Boolean functions for cryptography and error-correcting codes, in *Boolean Models and Methods in Mathematics, Computer Science, and Engineering*, Eds., Crama, Y. and Hammer, P. L. (Cambridge University Press, Cambridge), pp. 257–397.

Carlet, C. and Ding, C. (2004). Highly nonlinear mappings, *J. Complexity* **20**, 2, pp. 205–244.

Carlet, C., Ding, C. and Yuan, J. (2005). Linear codes from highly nonlinear functions and their secret sharing schemes, *IEEE Trans. Inform. Theory* **51**, 6, pp. 2089–2102.

Carlet, C. and Mesnager, S. (2011). On Dillon's class H of bent functions, Niho bent functions and o-polynomials, *J. Combin. Theory Ser A* **118**, pp. 2392–2410.

Chang, A., Gaal, P., Golomb, S. W., Gong, G. and Kumar, P. V. (1998). On a sequence conjectured to have ideal 2-level autocorrelation function, in *Proc. of ISIT 1998* (IEEE, Cambridge), p. 468.

Chang, A., Golomb, S. W., Gong, G. and Kumar, P. V. (1998). On ideal autocorrelation sequences arising from hyperovals, In *Sequences and their Applications — Proc. of SETA'98*, Eds., Ding, C., Helleseth, T. and Niederreiter, H. (Springer, Singapore), pp. 17–38.

Chen, Y. Q. and Polhill, J. (2011). Paley type group schemes and planar Dembowski–Ostrom polynomials, *Discrete Mathematics* **311**, 14, pp. 1349–1364.

Cherowitzo, W. (1998). α-flocks and hyperovals, *Geom. Dedicata* **72**, pp. 221–246.

Cherowitzo, W., Penttila, T., Pinneri, I. and Royle, G. F. (1996). Flocks and ovals, *Geometriae Dedicata* **60**, 1, pp. 17–37.

Cherowitzo, W. E., OKeefe, C. M. and Penttila, T. (2003). A unified construction of finite geometries associated with q-clans in characteristic two, *Advances in Geometry* **3**, pp. 1–21.

Conway, J. H., Harding, R. H. and Sloane, N. J. A. (1996). Packing lines, planes, etc.: packings in Grassmannian spaces, *Exper. Math.* **5**, 2, pp. 139–159.

Coulter, R. and Matthews, R. W. (1997). Planar functions and planes of Lenz-Barlotti Class II, *Designs, Codes and Cryptography* **10**, pp. 167–184.

Davis, A. (1992). Almost difference sets and reversible difference sets, *Arch. Math.* **59**, pp. 595–602.

Delsarte, P., Goethals, J. M. and Seidel, J. J. (1977). Spherical codes and designs, *Geometriae Dedicate* **67**, 3, pp. 363–388.

Dembowski, P. and Ostrom, T. G. (1968). Planes of order n with collineation groups of order n^2, *Math. Z.* **193**, pp. 239–258.

Dickson, L. E. (1896). The analytic representation of substitutions on a power of a prime number of letters with a discussion of the linear group, *Ann. of Math.* **11**, pp. 65–120, pp. 161–183.

Dillon, J. F. (1974). Elementary Hadamard difference sets, Ph.D. thesis, University of Maryland, USA.

Dillon, J. F. (1975). Elementary Hadamard difference sets, in *Proc. Sixth Southeastern Conf. on Combinatorics, Graph Theory and Computing*, pp. 237–249.

Dillon, J. F. (1985). Variations on a scheme of McFarland for noncyclic difference sets, *J. Combin. Theory Ser A* **40**, pp. 9–20.

Dillon, J. F. (1999). Multiplicative difference sets via additive characters, *Designs, Codes and Cryptography* **17**, pp. 225–236.

Dillon, J. F. and Dobbertin, H. (2004). New cyclic difference sets with Singer parameters, *Finite Fields and their Applications* **10**, pp. 342–389.

Dillon, J. F. and Shatz, J. R. (1987). Block designs with the symmetric difference property, in *Proc. NSA Mathematical Sciences Meeting*, Ed., Ward, R. L. (Ft. George Meade, MD), pp. 159–164.

Ding, C. (1994). The differential cryptanalysis and design of the natural stream ciphers, in *Proc. of FSE'93*, LNCS 809 (Springer-Verlag), pp. 101–115.

Ding, C. (1995). Binary cyclotomic generators, in *Proc. of FSE'94*, LNCS 1008 (Springer-Verlag), pp. 29–60, 1995.

Ding, C. (1997). *Cryptographic Counter Generators*, TUCS Series in Dissertation 4, Turku Centre for Computer Science, Turku, Finland.

Ding, C. (1998). Autocorrelation values of the generalized cyclotomic sequences of order 2, *IEEE Trans. Inform. Theory* **44**, pp. 1698–1702.

Ding, C. (2006). Complex codebooks from combinatorial designs, *IEEE Trans. Information Theory* **52**, 9, pp. 4229–4235.

Ding, C. (2012). Cyclic codes from the two-prime sequences, *IEEE Trans. Inform. Theory* **58**, 6, pp. 3881–3890.

Ding, C. (2013). Cyclic codes from cyclotomic sequences of order four, *Finite Fields and their Applications* **23**, pp. 8–34.

Ding, C. and Feng, T. (2007). A generic construction of complex codebooks meeting the Welch bound, *IEEE Trans. Information Theory* **53**, 11, pp. 4245–4250.

Ding, C. and Feng, T. (2008). Codebooks from cyclic almost difference sets, *Designs, Codes and Cryptography* **46**, pp. 113–126.

Ding, C., Golin, M. and Kløve, T. (2003). Meeting the Welch and Karystinos-Pados bounds on DS-CDMA binary signature sets, *Designs, Codes and Cryptography* **30**, pp. 73–84.

Ding, C., Helleseth, T. and Lam, K. Y. (1999). Several classes of sequences with three-level autocorrelation, *IEEE Trans. Inform. Theory* **45**, 7, pp. 2606–2612.

Ding, C., Helleseth, T. and Martinsen, H. M. (2001). New families of binary sequences with optimal three-level autocorrelation, *IEEE Trans. Inform. Theory* **47**, pp. 428–433.

Ding, C., Helleseth, T. and Shan, W. (1998). On the linear complexity of Legendre sequences, *IEEE Trans. Inform. Theory* **44**, pp. 1276–1278.

Ding, C., Lam, K. Y. and Xing, C. (1999). Enumeration and construction of all duadic codes of length p^m, *Fundamenta Informaticae* **38**, 1, pp. 149–161.

Ding, C. and Pless, V. (1999). Cyclotomy and duadic codes of prime lengths, *IEEE Trans. Inform. Theory* **45**, 2, pp. 453–466.

Ding,C., Pott, A. and Wang, Q. (2013). Skew Hadamard difference sets from Dickson polynomials of order 7, arXiv:1305.1831.

Ding, C., Pott, A. and Wang, Q. (2014). Constructions of almost difference sets from finite fields, *Designs, Codes and Cryptography*, DOI: 10.1007/s10623-012-9789-9.

Ding, C., Qu, L., Wang, Q. and Yuan, P. (2014). Permutation trinomials over finite fields with even characteristic, arXiv:1402.5734.

Ding, C., Wang, Z. and Xiang, Q. (2007). Skew Hadamard difference sets from the Ree-Tits slice symplectic spreads in $PG(3, 3^{2h+1})$, *J. Combin. Theory Ser A* **114**, 5, pp. 867–887.

Ding, C., Xiao, G. and Shan, W. (1991). *The Stability Theory of Stream Ciphers, Lecture Notes in Computer Science*, Vol. 561 (Springer-Verlag, Heidelberg).

Ding, C. and Yang, J. (2013). Hamming weights in irreducible cyclic codes, *Discrete Mathematics* **313**, pp. 434–446.

Ding, C. and Yin, J. (2008). Sets of optimal frequency hopping sequences, *IEEE Trans. Inform. Theory* **54**, 8, pp. 3741–3745.

Ding, C. and Yuan, J. (2006). A family of skew Hadamard difference sets, *J. Comb. Theory Ser A* **113**, pp. 1526–1535.

Dobbertin, H. (1999a). Almost perfect nonlinear power functions on $GF(2^n)$: The Welch case, *IEEE Trans. Inform. Theory* **45**, 4, pp. 1271–1275.

Dobbertin, H. (1999b). Kasami power functions, permutation polynomials and cyclic difference sets, in *Difference Sets, Sequences and their Correlation Properties*, Eds., Pott, A. *et al.* (Kluwer), pp. 133–158.

Dobbertin, H. (2002). Uniformly representable permutation polynomials, in *Sequences and their Applications — SETA 01, Proc. of SETA 01*, Eds., Helleseth, T., Kumar, P. V. and Yang, K. (Springer, London), pp. 1–22.

Drakakis K. (2009). A review of the available construction methods for Golomb rulers, *Adv. Math. Commun.* **3**, 3, pp. 235–250.

Evans, R., Hollmann, H. D. L., Krattenthaler, C. and Xiang, Q. (1999). Gauss sums, Jacobi sums, and p-ranks of cyclic difference sets, *J. Combin. Theory Ser A* **87**, pp. 74–119.

Feng, T. (2011). Non-abelian skew Hadamard difference sets fixed by a prescribed automorphism, *J. Combin. Theory Ser A* **118**, pp. 27–36.

Feng, T. and Xiang, Q. (2012). Cyclotomic constructions of skew Hadamard difference sets, *J. Combin. Theory Ser A* **119**, pp. 245–256.

Gauss, C. F. (1801). *Disquisitiones Arithmeticae* (Leipzig) (English translation, Yale, New Haven, 1966; Reprint by Springer Verlag, Berlin, Heidelberg, and New York, 1986).

Gilbert, E. N. (1952). A comparison of signaling alphabets, *Bell System Tech. J.* **31**, pp. 504–522.

Glynn, D. (1983). Two new sequences of ovals in finite desarguesian planes of even order, *Lecture Notes in Mathematics*, **1036**, pp. 217–229.

Goethals, J. M. and Delsarte, P. (1968). On a class of majority logic decodable cyclic codes, *IEEE Trans. Inform. Theory* **14**, pp. 182–188.

Gong, G. and Youssef, A. M. (2002). Cryptographic properties of the Welch-Gong transformation sequence generators, *IEEE Trans. Inform. Theory* **48**, 11, pp. 2837–2846.

Gordon, B., Mills, W. H. and Welch, L. R. (1962). Some new difference sets, *Cand. J. Math.* **14**, pp. 614–625.

Gordon, D. M. (1994). The prime power conjecture is true for $n < 2,000,000$, *Electronic J. Combinatorics* **1**, R6.

Griesmer, J. H. (1960). A bound for error-correcting codes, *IBM J. Research Develop.* **4**, pp. 532–542.

Gurak, S. J. (2004). Period polynomials for $\mathbb{F}_q$ of fixed small degree, *Centre de Recherches Mathématiques CRM Proceedings and Lecture Notes* **36**, pp. 127–145.

Hall Jr., J. (1956). A survey of difference sets, *Proc. Amer. Math. Soc.* **7**, pp. 975–986.

Hamada, N. (1973). On the p-rank of the incidence matrix of a balanced or partially balanced incomplete block design and its applications to error-correcting codes, *Hiroshima Math. J.* **3**, 154–226.

Hamada, N. and Ohmori, H. (1975). On the BIB-design having the minimum p-rank, *J. Combin. Theory Ser A* **18**, pp. 131–140.

Hartmann, C. R. P. and Tzeng, K. K. (1972). Generalizations of the BCH bound, *Inform. and Control* **20**, pp. 489–498.

Hayashi, H. S. (1965). Computer investigation of difference sets, *Math. Comp.* **19**, pp. 73–78.

Helleseth, T. and Gong, G. (2002). New nonbinary sequences with ideal two-level autocorrelation, *IEEE Trans. Inform. Theory* **48**, 11, pp. 2868–2872.

Helleseth, T., Kumar, P. V. and Martinsen, H. M. (2001). A new family of ternary sequences with ideal two-level autocorrelation, *Designs, Codes and Cryptography* **23**, pp. 157–166.

Hocquenghem, A. (1959). Codes correcteurs d'erreurs, *Chiffres (Paris)* **2**, pp. 147–156.

Hoshi, A. (2006). Explicit lifts of quintic Jacobi sums and period polynomials for F_q, *Proc. Japan Acad. Ser A* **82**, pp. 87–92.

Hou, X. D. (1998). Cubic bent functions, *Discrete Mathematics* **189**, pp. 149–161.

Hu, H., Shao, S., Gong, G. and Helleseth, T. (2013). The proof of Lin's conjecture via the decimation-Hadamard transform, arXiv:1307.0885v1.

Huffman, W. C. and Pless, V. (2003). *Fundamentals of Error-Correcting Codes* (Cambridge University Press, Cambridge).

Ipatov, V. P. (2004). On the Karystinos-Pados bounds and optimal binary DS-CDMA signature ensembles, *IEEE Comm. Letters* **8**, 2, pp. 81–83.

Johnson, E. C. (1966). Skew-Hadamard abelian group difference sets, *J. Algebra* **4**, pp. 388–402.

Jungnickel, D. (1982). On automorphism groups of divisible designs, *Can. J. Math.* **34**, pp. 257–297.

Jungnickel, D. (1992). Difference sets, in *Contemporary Design Theory: A Collection of Surveys*, Eds., Dinitz, J. H. and Stinson, D. R. (John Willey & Sons, Inc.)

Jungnickel, D. and Tonchev, V. D. (1992). On symmetric and quasi-symmetric designs with the symmetric difference property and their codes, *J. Combin. Theory Ser A* **59**, pp. 40–50.

Jungnickel, D. and Tonchev, V. D. (1999). Decompositions of difference sets, *J. Algebra* **27**, 1, pp. 21–39.

Jungnickel, D. and Vedder, K. (1984). On the geometry of planar difference sets, *Europ. J. Combin.* **5**, pp. 143–148.

Kantor, W. M. (1975). Symplectic groups, symmetric designs and line ovals, *J. Algebra* **33**, pp. 43–58.

Kantor, W. M. (2003). Commutative semifields and symplectic spreads, *J. Algebra* **270**, pp. 96–114.

Karystinos, G. N. and Pados, D. A. (2003). New bounds on the total-squared-correlation and perfect design of DS-CDMA binary signature sets, *IEEE Trans. Communications* **3**, pp. 260–265.

Kelly, J. B. (1954). A characteristic property of quadratic residues, *Proc. Amer. Math. Soc.* **5**, pp. 38–46.

Kim, J.-H. and Song, H.-Y. (2001). On the linear complexity of Halls sextic residue sequences, *IEEE Trans. Inform. Theory* **47**, 5, pp. 2094–2096.

Klapper, A. (1995). d-form sequences: Families of sequences with low correlation values and large linear spans, *IEEE Trans. Inform. Theory* **41**, pp. 423–431.

Klapper, A., Chan, A. H. and Goresky, M. (1993). Cascaded GMW sequences, *IEEE Trans. Inform. Theory* **39**, pp. 177–183.

Klemm, M. (1986). Über den p-Ranken von Inzidenzmatrizen, *J. Combin. Theory Ser A* **43**, pp. 138–139.

König, H. (1979). Cubature formulas on spheres, in *Advances in Multivarate Approximation*, New York, Wiley, pp. 201–211.

Lander, E. S. (1981). Symmetric designs and self-dual codes, *J. London Math. Soc.* **2**, 24, pp. 193–204.

Lander, E. S. (1983). *Symmetric Designs: An Algebraic Approach* (Cambridge University Press, Cambridge).

Lehmer, E. (1953). On residue difference sets, *Canad. J. Math.* **5**, pp. 425–432.

Lempel, A., Cohn, M. and Eastman, W. L. (1977). A class of binary sequences with optimal autocorrelation properties, *IEEE Trans. Inform. Theory* **23**, pp. 38–42.

Leon, J. S., Masley, J. M. and Pless, V. (1984). Duadic codes, *IEEE Trans. Inform. Theory* **30**, pp. 709–714.

Lidl, R. and Niederreiter, H. (1997). *Finite Fields*, 2nd Edition (Cambridge University Press, Cambridge).

Lidl, R., Mullen, G. L. and Turnwald, G. (1993). *Dickson Polynomials* (Longman, England).

Lin, H. A. (1998). *From cyclic Hadamard difference sets to perfectly balanced sequences*, Ph.D. Dissertation University of Southern California, California, USA.

Lint van, J. H. and Wilson, R. W. (1986). On the minimum distance of cyclic codes, *IEEE Trans. Inform. Theory* **32**, pp. 23–40.

Love, D. J., Heath, R. W. and Strohmer, T. (2003). Grassmannian beam forming for multiple-input multiple-output wireless systems, *IEEE Trans. Inform. Theory* **49**, 10, pp. 2735–2747.

Ma, S. L. (1994). A survey of partial difference sets, *Designs, Codes and Cryptography* **4**, pp. 221–261.

MacWilliams, J. and Mann, H. B. (1968). On the p-rank of the design matrix of a difference set, *Inform. Control* **12**, pp. 474–488.

MacWilliams, F. J. and Sloane, N. J. A. (1977). *The Theory of Error-Correcting Codes* (North-Holland, Amsterdam).

Maschietti, A. (1998). Difference sets and hyperovals, *Designs, Codes and Cryptography* **14**, pp. 89–98.

Massey, J. L. and Mittelholzer, T. (1993). Welch's bound and sequence sets for code-division multiple-access systems, in *Sequences II: Methods in Communication, Security and Computer Science* (Springer, Heidelberg and New York), pp. 63–78.

McEliece, R. J. (1974). Irreducible cyclic codes and Gauss sums, in *Proc. NATO Advanced Study Inst., Breukelen, 1974*, Part 1: Theory of designs, Finite geometry and coding theory, Math. Centre Tracts, No. 55 (Math. Centrum, Amsterdam), pp. 179–196.

McFarland, R. L. (1973). A family of difference sets in noncyclic groups, *J. Combin. Theory Ser A* **15**, pp. 1–10.

McFarland, R. L. and Rice, B. F. (1978). Translates and multipliers of abelian difference sets, *Proc. Amer. Math. Soc.* **68**, pp. 375–379.

Menon, P. K. (1960). Difference sets in abelian groups, *Proc. Amer. Math. Soc.* **11**, pp. 368–376.

Menon, P. K. (1962). On difference sets whose parameters satisfy a certain relation, *Proc. Amer. Math. Soc.* **13**, pp. 739–745.

Mertens, S. and Bessenrodt, C. (1998). On the ground states of the Bernasconi model, *J. Phys. A: Math. Gen.* **31**, pp. 3731–3749.

Michael, T. S. and Wallis, W. D. (1998). Skew-Hadamard matrices and the Smith normal form, *Designs, Codes and Cryptography* **13**, pp. 173–176.

Myerson, G. (1981). Period polynomials and Gauss sums for finite fields, *Acta Arith.* **39**, pp. 251–264.

Momihara, K. (2013). Inequivalence of skew Hadamard difference sets and triple intersection numbers modulo a prime, *Electronic J. of Combinatorics* **20**, 4, P35.

Muzychuk, M. E. (2010). On skew Hadamard difference sets, arXiv:1012.2089.

Newhart, D. W. (1988). On minimum weight codewords in QR codes, *J. Combin. Theory Ser A* **48**, pp. 104–119.

No, J. S. (2004). New cyclic difference sets with Singer parameters constructed from d-homogeneous functions, *Designs, Codes and Cryptography* **33**, pp. 199–213.

No, J. S., Chung, H. and Yun, M. S. (1998). Binary pseudorandom sequences of period $2^m - 1$ with ideal autocorrelation generated by the polynomial $z^d + (z + 1)^d$, *IEEE Trans. Inform. Theory* **44**, pp. 1278–1282.

No, J. S., Chung, H., Song, H. Y., Yang, K., Lee, J. D. and Helleseth, T. (2001). New construction for binary sequences of period $p^m - 1$ with optimal autocorrelation using $(z + 1)^d + az^d + b$, *IEEE Trans. Inform. Theory* **47**, pp. 1638–1644.

No, J.-S., Golomb, S. W., Gong, G., Lee, H.-K. and Gaal, P. (1998). Binary pseudorandom sequences of period $2^m - 1$ with ideal autocorrelation, *IEEE Trans. Inform. Theory* **44**, pp. 814–817.

No, J. S., Shin, D. J. and Helleseth, T. (2004). On the p-rank and characteristic polynomials of cyclic difference sets, *Designs, Codes and Cryptography* **33**, pp. 23–37.

Nowak, K., Olmez, O. and Song, S. Y. (2013). Almost difference sets, normally regular digraphs and cyclotomic schemes from cyclotomy of order twelve, arXiv:1310.1164v1.

Paley, R. A. E. C. (1933). On orthogonal matrices, *J. Math. Phys. MIT* **12**, pp. 311–320.

Parker, C., Spence, E. and Tonchev, V. D. (1994). Designs with the symmetric difference property on 64 points and their groups, *J. Combin. Theory* **67**, pp. 23–43.

Payne, S. E. (1985). A new infinite family of generalized quadrangles, *Congr. Numer.* **49**, pp. 115–128.

Pless, V. (1986). Q-codes, *J. Comb. Theory Ser A* **43**, pp. 258–276.

Pless, V. (1993). Duadic codes and generalizations, in *Proc. of Eurocode 1992*, CISM Courses and Lectures No. 339, Eds., Camion, P., Charpin, P. and Harari, S. (Vienna: Springer, Berlin), pp. 3–16.

Plotkin, M. (1960). Binary codes with specified minimum distances, *IRE Trans. Inform. Theory* **6**, pp. 445–450.

Pollack, P. (2009). *Not Always Buried Deep: A Second Course in Elementary Number Theory* (AMS Press, Providence).

Pott, A. (1992). On abelian difference set codes, *Designs, Codes and Cryptography* **2**, pp. 263–271.

Pott, A. (1995). *Finite Geometry and Character Theory* (Springer-Verlag, New York).

Reed, I. S. and Solomon, G. (1960). Polynomial codes over certain finite fields, *J. SIAM* **8**, pp. 300–304.

Roos, C. (1982a). A generalization of the BCH bound for cyclic codes, including the Hartmann-Tzeng bound, *J. Comb. Theory Ser A* **33**, pp. 229–232.

Roos, C. (1982b). A new lower bound on the minimum distance of a cyclic code, *IEEE Trans. Inform. Theory* **29**, pp. 330–332.

Rothaus, O. S. (1976). On bent functions, *J. Comb. Theory Ser A* **20**, pp. 300–305.

Rushanan, J. J. (1986). Generalized Q-codes, Ph.D. Thesis, California Institute of Technology, USA.

Sarwate, D. (1999). Meeting the Welch bound with equality, in *Sequences and their Applications, Proc. of SETA'98* (Springer, London), pp. 79–102.

Segre, B. (1962). Ovali e curve nei piani di galois di caratteristica due, *Atti dell Accad. Naz. Lincei Rend* **32**, 8, pp. 785–790.

Sidelnikov, V. M. (1969). Some k-valued pseudo-random sequences and nearly equidistant codes, *Probl. Inform. Trans.* **5**, pp. 12–16.

Singer, J. (1938). A theorem in finite projective geometry and some applications to number theory, *Trans. Amer. Math. Soc.* **43**, pp. 377–385.

Singleton, R. C. (1964). Maximum distance q-ary codes, *IEEE Trans. Inform. Theory* **10**, pp. 116–118.

Smith, K. J. C. (1969). On the p-rank of the incidence matrix of points and hyperplanes in a finite projective geometry, *J. Comb. Theory* **7**, pp. 122–129.

Spence, E. (1977). A family of difference sets, *J. Comb. Theory Ser A* **22**, pp. 103–106.

Stanton, R. G. and Sprott, D. A. (1958). A family of difference sets, *Canad. J. Math.* **10**, pp. 73–77.

Storer, T. (1967). *Cyclotomy and Difference Sets* (Markham, Chicago).

Strohmer, T. and Heath Jr., R. W. (2003). Grassmannian frames with applications to coding and communication, *Applied Computational Harmonic Analysis* **14**, 3, pp. 257–275.

Swanson C. N. (2000). Planar cyclic difference packings, *J. Combin. Des.* **8**, 6, pp. 426–434.

Tang, X. and Ding, C. (2010). New classes of balanced quaternary and almost balanced binary sequences with optimal autocorrelation value, *IEEE Trans. Inform. Theory* **56**, 12, pp. 6398–6405.

Tilborg van, H. C. A. (1980). On the uniqueness resp. nonexistence of certain codes meeting the Griesmer bound, *Inform. and Control* **44**, pp. 16–35.

Turyn, R. (1964). The linear generation of the Legendre sequences, *J. Soc. Ind. Appl. Math.* **12**, 1, pp. 115–117.

Tze, T. W., Chanson, S., Ding, C., Helleseth, T. and Parker, M. (2003). Logarithm authentication codes, *Information and Computation* **184**, 1, pp. 93–108.

Varshamov, R. R. (1957). Estimate of the number of signals in error correcting codes, *Dokl. Akad. Nauk SSSR* **117**, pp. 739–741.

Wang, X. and Wang, J. (2011). Partitioned difference families and almost difference sets, *J. Statistical Planning and Inference* **141**, pp. 1899–1909.

Welch, L. (1974). Lower bounds on the maximum cross correlation of signals, *IEEE Trans. Inform. Theory* **20**, 3, pp. 397–399.

Weldon Jr., E. J. (1966). Difference set cyclic codes, *The Bell System Technical Journal*, pp. 1045–1055.

Weng, G., Feng, R. and Qiu, W. (2007). On the ranks of bent functions, *Finite Fields and their Applications* **13**, pp. 1096–1116.

Weng, G. and Hu, L. (2009). Some results on skew Hadamard difference sets, *Designs, Codes and Cryptography* **50**, pp. 93–105.

Weng, G., Qiu, W., Wang, Z. and Xiang, Q. (2007). Pseudo-Paley graphs and skew Hadamard difference sets from presemifields, *Designs, Codes and Cryptography* **44**, pp. 49–62.

Whiteman, A. L. (1962). A family of difference sets, *Illinois J. Math.* **6**, pp. 107–121.

Wilbrink, H. A. (1985). A note on planar difference sets, *J. Combin. Theory Ser A* **38**, pp. 94–95.

Wootters, W. and Fields, B. (2005). Optimal state-determination by mutually unbiased measurements, *Ann. Physics* **191**, 2, pp. 363–381.

Xia, P., Zhou, S. and Giannakis, G. B. (2005). Achieving the Welch bound with difference sets, *IEEE Trans. Inform. Theory* **51**, 5, pp. 1900–1907.

Xiang, Q. (1999). Recent results on difference sets with classical parameters, in *Difference Sets, Sequences and their Correlation Properties*, Eds., Pott, A, *et al.*, Kluwer, Amsterdam, pp. 419–437.

Xiang, Q. (2005). Recent results on p-ranks and Smith normal forms of some 2-(v, k, λ) designs, in *Coding Theory and Quantum Computing*, Eds., Evans, *et al.*, Contemporary mathematics 381, American Mathematical Society, Providence, RI, pp. 53–67.

Yang, J. and Xia, L. (2010). Complete solving of explicit evaluation of Gauss sums in the index 2 case, *Sci. China Ser A* **53**, pp. 2525–2542.

Yu, N. Y., Feng, K. Q. and Zhang, A. (2014). A new class of near-optimal partial Fourier codebooks from an almost difference set, *Designs, Codes and Cryptography* June 2014, 71(3), pp. 493–501, DOI: 10.1007/s10623-012-9753-8.

Zha, Z., Kyureghyan, G. and Wang, X. (2009). Perfect nonlinear binomials and their semifields, *Finite Fields Appl.* **15**, pp. 125–133.

Zha, Z. and Wang, X. (2009). New families of perfect nonlinear polynomial functions, *J. Algebra* **322**, 1, pp. 3912–3918.

Zhang, A. and Feng, K. (2012). Two classes of codebooks nearly meeting the Welch bound, *IEEE Trans. Inform. Theory* **58**, 4, pp. 2507–2511.

Zhang, Y., Lei, J. G. and Zhang, S. P. (2006). A new family of almost difference sets and some necessary conditions, *IEEE Trans. Inform. Theory* **52**, 5, pp. 2052–2061.

Index

Printed in the United States
By Bookmasters